AF774562

Energy Conversion, Modeling and Storage

Energy Conversion, Modeling and Storage

Edited by **Nora Ayling**

SYRAWOOD
PUBLISHING HOUSE
New York

Published by Syrawood Publishing House,
750 Third Avenue, 9th Floor,
New York, NY 10017, USA
www.syrawoodpublishinghouse.com

Energy Conversion, Modeling and Storage
Edited by Nora Ayling

© 2016 Syrawood Publishing House

International Standard Book Number: 978-1-68286-206-3 (Hardback)

This book contains information obtained from authentic and highly regarded sources. Copyright for all individual chapters remain with the respective authors as indicated. All chapters are published with permission under the Creative Commons Attribution License or equivalent. A wide variety of references are listed. Permission and sources are indicated; for detailed attributions, please refer to the permissions page and list of contributors. Reasonable efforts have been made to publish reliable data and information, but the authors, editors and publisher cannot assume any responsibility for the validity of all materials or the consequences of their use.

The publisher's policy is to use permanent paper from mills that operate a sustainable forestry policy. Furthermore, the publisher ensures that the text paper and cover boards used have met acceptable environmental accreditation standards.

Trademark Notice: Registered trademark of products or corporate names are used only for explanation and identification without intent to infringe.

Printed in China.

Contents

Preface

Every book is a source of knowledge and this one is no exception. The idea that led to the conceptualization of this book was the fact that the world is advancing rapidly; which makes it crucial to document the progress in every field. I am aware that a lot of data is already available, yet, there is a lot more to learn. Hence, I accepted the responsibility of editing this book and contributing my knowledge to the community.

Energy conversion and storage technology are crucial to sustainable energy harvesting & production, and meeting energy requirements for future generations. Chapters included in this book provide a detailed explanation of the various concepts and applications of alternative energy extraction and production, energy modeling and storage, energy conservation, assessing alternative energy potential, etc. The aim of this book is to present researches that have transformed this discipline and aided its advancement. This book will serve as a reference to a broad spectrum of readers.

While editing this book, I had multiple visions for it. Then I finally narrowed down to make every chapter a sole standing text explaining a particular topic, so that they can be used independently. However, the umbrella subject sinews them into a common theme. This makes the book a unique platform of knowledge.

I would like to give the major credit of this book to the experts from every corner of the world, who took the time to share their expertise with us. Also, I owe the completion of this book to the never-ending support of my family, who supported me throughout the project.

Editor

1

A numerical investigation of flow profile and performance of a low cost Crossflow turbine

Chiyembekezo S. Kaunda[1], Cuthbert Z. Kimambo[2], Torbjorn K. Nielsen[1]

[1] Waterpower Laboratory, Department of Energy and Process Engineering, Norwegian University of Science and Technology, Trondheim 7491, Norway.
[2] Department of Mechanical and Industrial Engineering, University of Dar es Salaam, P.O. Box 35091, Dar es Salaam, Tanzania.

Abstract
Low efficiency is the main drawback of a Crossflow turbine, despite the turbine being an important low cost technology for micro hydropower generation. Poor flow profile has been mentioned by other Crossflow turbine performance investigators as one of the reasons for the underperformance. This paper has investigated, using numerical method, the flow profile in the turbine at best efficiency point and at operating conditions that are away from best efficiency point. Numerical method has also been used to calculate and predict the efficiency of the turbine.
The flow physics in a Crossflow turbine runner is a two-phase with a movable free surface. Such flow physics is difficult to analyse even numerically. A procedure for numerical analysis was followed and ANSYS CFX® was used to solve the governing equations and to process the simulation results. Actual pictures of the flow were taken so as to compare the actual flow with the numerically determined flow profile. Turbine efficiency results from the previous performance evaluation experiment conducted on the model Crossflow turbine were compared with the numerically obtained efficiency results. It has been observed that the numerically obtained flow profile compare favorably with the actual flow pictures. The numerical analysis over-predicts the efficiency, especially for runner speeds that were more than the best efficiency point speed.
Pockets of negative pressures and flow circulation have been observed in the flow profile. At constant head and valve opening, the velocity profile was found to vary more with runner speed than the pressure profile. The numerically obtained flow pattern showed positions where the flow gives maximum efficiency. Therefore, the study has shown that numerical method is a superior design tool for Crossflow turbines.

Keywords: Micro hydropower; Crossflow turbine; Turbine performance, Numerical analysis; Flow analysis.

1. Introduction

1.1 General background

Microhydropower is an important renewable energy source especially in countries that have perennial rivers and topographies that are mountainous. Currently, most of the installed microhydropower (MHP) plants are used to generate electricity. If the turbine shaft is extended from both sides, mechanical power can also be generated together with electrical power. The sizes of MHP plants defined in terms of installed electrical capacity are numerous but with no international consensus. However, the definition of 'up to 100 kW installed capacity' seems to be common amongst countries and international development organisations. On the basis of off-grid rural power supply, such a capacity is robust enough to provide power to support domestic and some small-scale business applications. The MHP systems, therefore, are useful in supporting social and economic development activities in rural communities.

Crossflow turbine (CFT) is arguably the simplest turbine to design and manufacture. For this reason, they are among the low cost MHP technologies. They are widely used for rural power supply in some countries of Asia such as Nepal. International charitable organisations such as Practical Action (formerly called Intermediate Technology Group), Swiss Centre for Appropriate Technology (SKAT) and German Technical Operation Agency (GTZ) have been very pivotal in the development of the CFT technology in developing countries. These organisations, like SKAT, were interested in the development of a very low cost CFT, thereby modifying some basic parts of the original Crossflow turbine as designed by Prof. Banki. Therefore, it can be said that the CFT technology has been promoted basically as 'an appropriate technology' and not necessarily as a special 'industry-made' turbine.

The CFT can be applied in a wide range of flow and head. However, it has a general drawback of inferior performance compared to other turbines used in MHP applications. This has contributed to the turbine not being promoted by turbine manufacturers in the same way as Francis, Pelton and Kaplan turbines. The original CFT turbine, as proposed by Prof. Banki, had a maximum theoretical efficiency of 87.8% [1]. However, upon building and testing the turbine, Mockmore and Merryfield in 1949 obtained an efficiency of 68% [1]. This indicated the need for CFT performance improvement. Consequently, the CFT has been subjected to several performance improvement studies (such as those in Table 1).

Table 1. Some Crossflow experimental studies and maximum efficiency levels attained

Investigators	Maximum efficiency (%)
Mockmore and Merryfield [1]	68
Nakase *et. al.* [2]	82
Johnson *et. al.* [3]	80
Durgin and Fay [4]	66
Khosrowpanah [5]	80
Hothersall [6]	75
Ott and Chappel [7]	79
Desai and Aziz [8]	88
Olgun [9]	72
Andrade *et. al.* [10]	78

Most of the CFT performance improvement studies have concentrated on optimising its geometric design parameters (such as number of blades, runner diameter, width, nozzle and nozzle entry arc) using laboratory based experiments. As it can be seen from Table 1, different investigators have reported diverse values of maximum efficiencies. Despite the fact that these results may have depended on the laboratory settings and other controls of design parameters, the results still indicate room for further performance improvement.

The other room for performance improvement is in the area of improving turbine flow profile [10]. Poor flow profile is associated with efficiency reduction. Flow profile can be known from flow characterisation. Flow in the turbine can be characterised by solving the Navier-Stokes equations in the flow domain using analytical and numerical methods. Analytical solution, which gives a continuous solution to the Navier-Stokes equations, is nearly impossible due to non-linearity of the equations and the complexity of the turbine flow. An alternative is the numerical analysis, which gives discrete solution to the equations.

Experimental methods are also used in analysing flow profile in the rotating runner. These methods include visualising the flow by photographing using high speed camera. Sometimes a stroboscope is used to 'bring the flow to a halt or very low speed' and the flow pattern can then be visualised and photographed. The visibility in the flow visualisation experiments can be enhanced by injecting a dye in the flow. Pitot tubes can be used to characterise the flow velocity profile, from which, using the Bernoulli's equation, pressure profile can be determined. Experimental methods have the advantage that

they can capture the exact flow profile and are therefore, also used to validate analytical and numerical flow characterisation results.
Experimental, as with analytical methods, have limitations in charactering flow with complex flow geometries as in CFT. Currently, numerical analysis is becoming an important industrial tool in characterising the flow during the turbine design process. With advancements in computational fluid dynamic (CFD) solvers and tools for post-processing of the numerical results, the capacity of numerical method in the design of complex turbines has been enhanced.
The determination of flow field in terms of pressure and velocity fields is important in the design of a high performance turbine because regions of poor flow (for example, regions of flow recirculation and negative pressure) can be identified. Once regions of poor flow profile are identified, the turbine performance can be improved through optimization of geometric and dynamic design parameters.

1.2 Numerical method overview
In general, a numerical analysis procedure involves the following steps: i) development of governing equations for the flow physics; ii) discretising the equations in both time and space, iii) discretising the flow domain (meshing); iv) solving the discretised equations using a computerised solver (or program) and v) post-processing of the results using a computerised program. It is necessary to validate the numerical results by comparing them with experimental results.
As stated already, numerical analysis is currently the standard way of designing turbines in the industry. Numerical analysis for predicting the performance of impulse turbine runners poses a special challenge because of complexity of the flow physics. In impulse turbines like CFT, the jet flow in the runner is intermittent and there is a movable free surface boundary between the jet and surrounding air. This means that the actual flow is unsteady and two phase with a movable free surface. Two approaches are being used to numerically analyse such a flow. These approaches are particle tracking method and a classical CFD method where a free surface is tracked.
The particle tracking method does not require discretisation of the flow domain; it uses Lagrange principle of following the path of a fluid particle. This method has been used by various turbine designers to design Pelton turbine runners, for example Zoppe *et. al.* in 2006 [11]. The classical CFD method where a free surface is tracked is generally referred to as volume-of-fluid method. In terms of computational costs, the classical CFD is superior to particle tracking method. This study employs a classical CFD method.
Some previous investigators have attempted to numerically study the flow through the nozzle and the runner of CFTs. Fukutomi *et. al.* in 1991 and 1995 [12, 13] numerically characterized the flow in the CFT runner and calculated unsteady forces acting on the blades, assuming only water in the turbine. Pereira and Borges in 1996) [14] used two-dimensional CFD to simulate the flow inside the CFT nozzle. Arzola *et. al.* in 2008 [15] conducted three-dimensional free surface CFD simulations inside the nozzle where influence of angles of attack on performance was investigated. Choi *et. al.* in 2008 [16] employed water-air free surface CFD simulations to investigate effect of varying nozzle shape, blade angle and number of blades on performance.

2. Crossflow turbine technology
2.1 Crossflow turbine history
Crossflow turbine is first reported to be invented by an Australian Engineer, Anthony Mitchell, who patented the original design in 1903 [17]. It is also reported that a Hungarian mechanical engineer, Professor Donat Banki, refined it through series of experiments between 1916 and 1918 in West Germany [17]. JLA & Co. Limited in 2010 states that by 1920, the turbine was well known in Europe [18]; may be as a result of Banki's research publications. It is further reported that a German engineer, Fritz Ossberger, collaborated with Mitchell and worked on the development of the turbine. In 1933, Ossberger obtained the German Imperial patent for the turbine [19]. Therefore, it is convenient to state that Mitchell, Banki and Ossberger played important roles in developing the Crossflow turbine technology. This explains why the Crossflow turbine is also known as Banki or Mitchell or Ossberger turbine or their combinations.
The original CFT design was developed based on impulse principle [1]. A gap was left between the nozzle and the runner, as shown in Figure 1(a), so that the jet from the nozzle entered the runner at an atmospheric pressure. Recent modifications in the design have removed the gap between the nozzle and

the runner resulting in the nozzle following the runner closely, as shown in Figure 1(b). This causes the jet entering the turbine to have a positive pressure [20, 21].

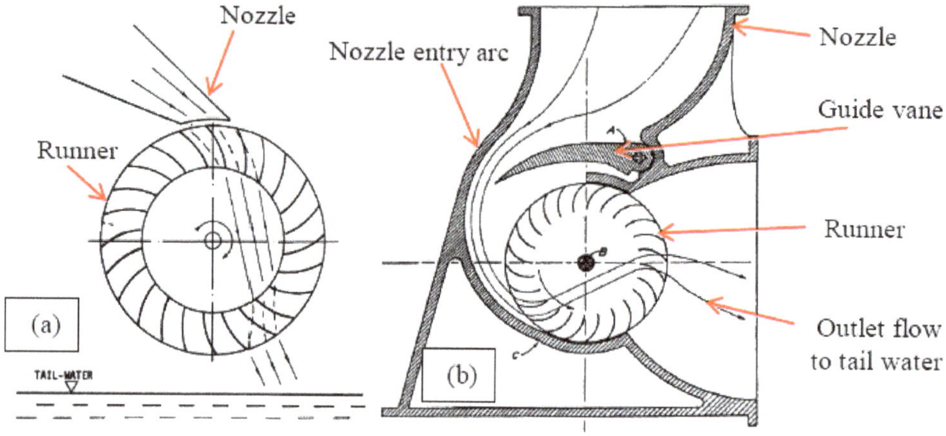

Figure 1. (a) Pure impulse Crossflow turbine where a gap is left between the runner and nozzle; (b) Modified Crossflow turbine where the nozzle wall follow the runner closely [1]

2.2 Crossflow turbine physical description

Crossflow turbine is basically composed of runner and nozzle (Figure 1). For low cost projects, the runner is simply a fabricated squirrel-cage-shaped device that is made of two or more circular discs joined together by curved horizontal blades. The runner blades can be cut from a standard sheet metal or steel pipe and then be bent into the required blade profile. In some cases, to improve on the structural integrity of the runner, more than two equally spaced discs are employed. The discs are then keyed into the shaft which is supported by two bearing housing.

The CFT nozzle is traditionally a convergent pipe with rectangular cross-section at inlet and outlet. It has four walls: two are parallel fixed plane walls and the other two walls can be made of cylindrical or spiral logarithmic shapes. The nozzle, apart from converting pressure energy into kinetic energy in form of water jet, also directs and regulates flow (jet) into the runner using a guide vane. The nozzle can be fabricated from the sheet metal and other materials. For example, Durgin and Fay in 1984 [4] used relatively cheap Plexiglas for nozzle walls.

To simplify the design, some versions of the CFT nozzles do not have guide vanes. Instead, the flow is controlled by one of the curved wall of the nozzle that is made movable. To simplify the design further, the movable wall can be a simple 'flap' which is operated by a screw (Figure 2). In this project, the last version of the nozzle is used.

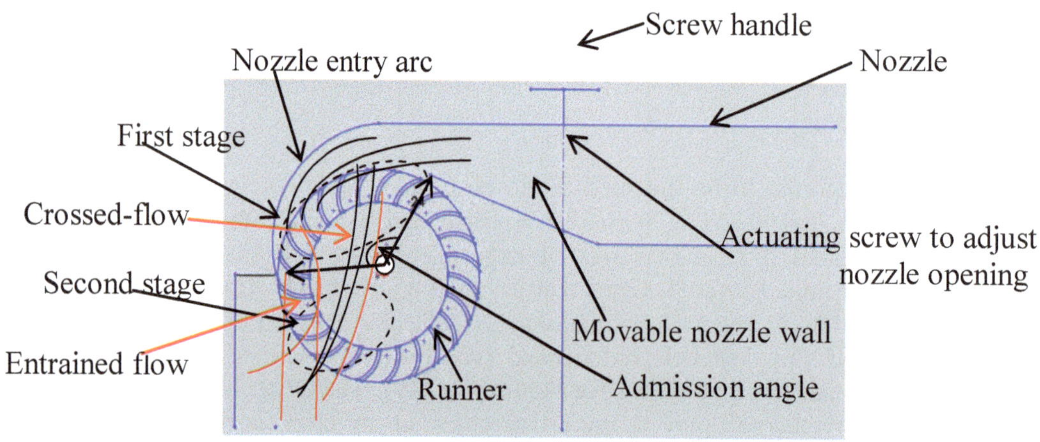

Figure 2. Schematic diagram of a low-cost Crossflow turbine used in this project and some basic flow terminology

The nozzle can be oriented in vertical, horizontal or slanting positions. For relatively high cost company-manufactured CFT, the runner can be partitioned so that the jet, through the nozzle, is channelled to the specific portion depending on flow availability (and sometimes on power requirement). Technically, these are the most versatile and efficient Crossflow turbines, but the requirement of the nozzle with more than one guidevanes to guide the portion of water is an extra cost. This extra cost cannot be justified for low cost MHP systems.

2.3 Flow description in the Crossflow turbine runner

The jet flow in the CFT runner can be described in the following manner. The jet issued from the nozzle is directed to the runner. The jet then exchanges energy with the runner in two distinct stages (Figure 2). Firstly, the water jet flows through the runner blades from the periphery towards the centre. This is the first stage. Then, after crossing the open centre space of the runner, the jet flows from the inside of the runner towards its outer periphery, from where it is discharged to the tailrace water. This is the second stage. This flow phenomenon has made CFT to be regarded as a two-stage partial 'impulse' turbine. Due to partial flow in the runner blades and flow in the open centre space, the jet flow inside of the runner is essentially two phase, namely water and air.

The first stage may operate with a degree of reaction. The second stage operates on pure impulse principle because the jet exits first stage and enters second stage at an atmospheric pressure due to presence of open space between stages. The flow within the runner blade can be modelled as radial flow. Because of this, there is a possibility for the jet to strike the shaft in the open space, thereby increasing losses and spoiling flow incident angles to the second stage.

Not all of the flow from the nozzle crosses the runner. The flow that crosses the runner blades is known as the crossed-flow. Some of the flow from the nozzle is entrained within the blades. The remaining flow gets leaked through the clearance between the runner periphery and the turbine housing. It is only the crossed-flow that takes part in energy transfer to the runner blades. Therefore entrained and leakage flows both reduce the power output of the turbine.

The flow can be entrained if it is decelerated in the radial direction by back pressure or centrifugal forces of the flow. The entrained flow is flushed out tangentially from the runner. The entrained flow analysis is an area that has not been studied extensively. The CFT theory from Prof. Banki CFT design did not even include it. The published experimental results of Durgin and Fay in 1984 [4] provide some information that the entrained flow depends on flow angle, flow velocity, runner velocity, runner geometry and nozzle geometry.

The leakage flow can be limited to smallest practically possible amount. Complete elimination is not possible because clearance is required between rotating runner and housing. The leakage flow also depends on the flow pressure, especially if the turbine operates more on reaction mode than on impulse mode.

3. Research objectives

The main objective of this study is to numerically characterize the two-phase flow in the Crossflow turbine. Specifically, the study will:

a) At constant head, determine using computational fluid dynamic approach the flow pressure and velocity fields in the turbine both at best efficiency and outside the best efficiency conditions.
b) Analyze the effect of flow pattern on efficiency and compare with experimentally obtained efficiency values.

The goal of this study is to achieve a better use of microhydropower resources with efficient low-cost Crossflow turbine designs.

4. Model Crossflow turbine description

The dimensions of the Crossflow turbine used in this study are those of the actual simplified Crossflow turbine that was tested at Waterpower Laboratory at the Norwegian University of Science and Technology in 2013. The performance test established the best efficiency point and the general performance of the turbine with varied head, flow and runner speed from the best efficiency point. The experimental results are used to validate the numerical results on performance assessment. The best operating point of 79% was identified at 5m head, 350 rpm and a nozzle opening of 80%.

As already stated, the simplified CFT used in this study does not have a guide vane. The flow in the nozzle is controlled by a flap that is operated via a screw handle (Figures 3 and 4). Figure 3 shows the schematic drawing of the simplified CFT. The normalized design parameters are given in Table 2.

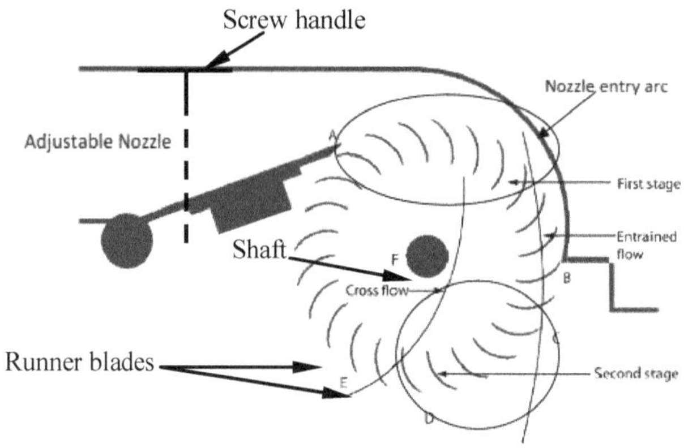

Figure 3. Schematic sketch of the model Crossflow turbine

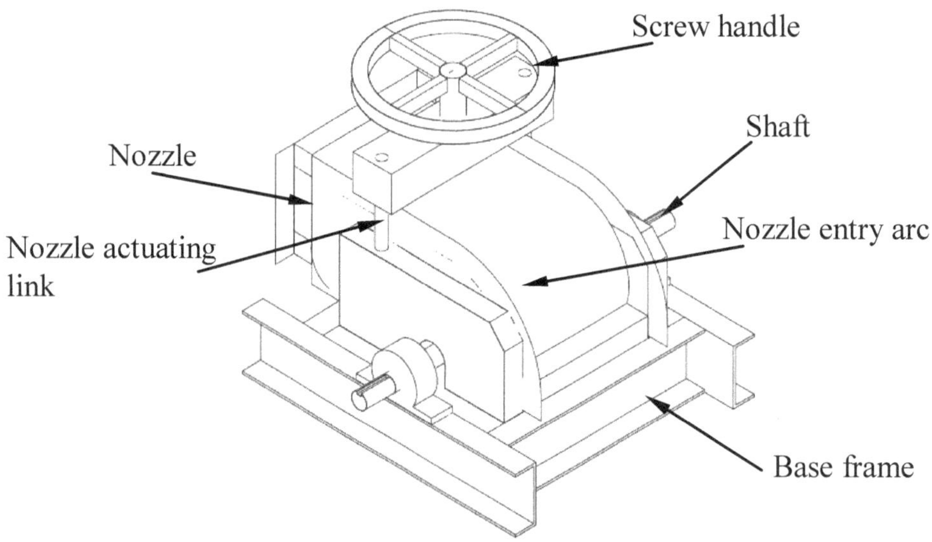

Figure 4. Pictorial sketch of the model Crossflow turbine housing

Table 2. Details of the Crossflow turbine design parameters used in the numerical study

Design parameter	Normalized expression	Value
Diameter ratio D_1 is outside diameter of runner D_2 is the inside diameter of the runner	$\frac{D_2}{D_1}$	0.693
Angle of attack	α	16 degrees
Blade angle (outside runner periphery)	β_1	30 degrees
Blade angle (inside runner periphery)	β_2	90 degrees
Radius of the blade (r_b)	$\frac{r_b}{D_1}$	0.157
Radius from centre of the runner to the centres of where blades are drawn	$\frac{R}{D_1}$	0.378
Admission arc	φ	90 degrees
Number of blades	n_b	24
Nozzle opening	$\frac{l}{D_1}$	0.254
Nozzle width where W is the runner width	$\frac{B}{W}$	1
Radius of the nozzle entry arc	$\frac{r_{(\emptyset)}}{D_1}$	0.567

5. Methodology

The numerical analysis for this study was performed in ANSYS Workbench® 2012 where ANSYS CFX Project with its subprograms, namely: Geometry, Mesh, ANSYS CFX-Pre, ANSYS CFX solver and ANSYS Post-processing, was used. As stated already, numerical analysis require coming up with a flow domain which has to be discretised (or meshed) in space and time (for unsteady state analysis). The flow domain (Figure 5) was designed in 'Geometry' sub-program using the ANSYS Design Modeler.

The flow domain consists of the nozzle, turbine casing and the runner. The nozzle and turbine casing were designed in ANSYS Design Modeler while the runner was designed in Solidworks® 2012. In the ANSYS Design Modeler, the runner was imported and positioned at the required central position.

This numerical analysis involves the runner, which has a rotating frame of reference while the nozzle and turbine casing have non-rotating frame of reference (stationary frame of reference). The numerical problem can be solved by using the 'multiple frame of reference' as detailed in ANSYS Theory Guide of 2010 [22]. The 'multiple frame of reference' requires dividing the flow domain into flow zones with the aim of isolating the zone that has a rotating frame of reference. To ensure continuity of flow, the zones (or domains) are interfaced. In this case, three flow zones were created. The zone with rotating frame of reference is isolated, as shown in Figure 6. The other two zones are assigned names: 'turbine casing zone' and 'non-rotating zone' and are given non-rotating frame of reference.

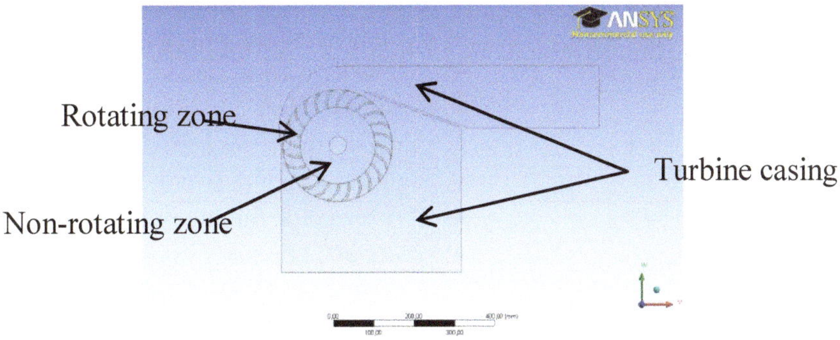

Figure 5. Crossflow turbine runner showing flow zones

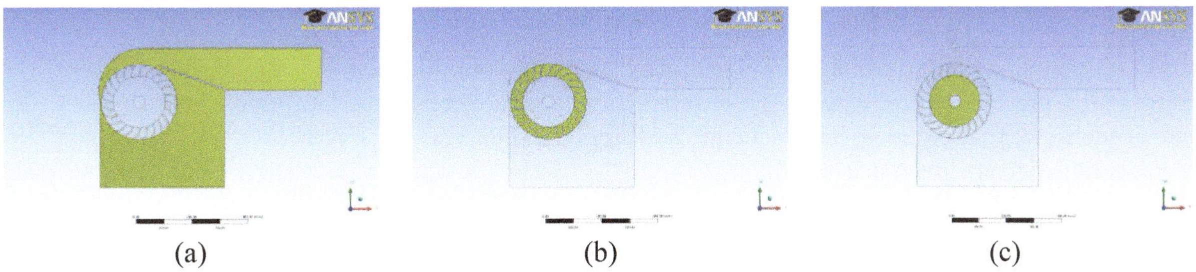

Figure 6. Specified flow zones by shading: (a) Turbine casing zone; (b) Rotating zone; (c) Non-rotating zone

The flow domain was meshed using the ANSYS Mesh sub-program of the ANSYS CFX Project. The mesh properties are given in Section 5.2. The mesh sensitivity analysis was performed (Section 5.3) to make sure that the solution obtained is independent of the mesh.

Modeling of the physics of the flow problem was done in the ANSYS CFX-Pre, a subprogram after meshing in the ANSYS CFX. Modeling of the flow physics included creating interfaces for flow zones (or domains), assigning boundary conditions, assigning initial conditions, applying governing equations with appropriate turbulence models and choosing the flow materials (water and air) and their phase properties. The solution control and declaration of variables were also done in the ANSYS CFX-Pre.

The domains were interfaced using the general connection model in the ANSYS CFX-Pre. The meshes in the domains were connected using the general grid interface model. Frozen rotor principle was used to model frame mixing process for the rotating and non-rotating frames of flow domains at the interfaces. In the frozen rotor model, the frame of reference and/or pitch is changed but relative orientations of the components across the interface remain fixed. In this way, the relative positions of components during calculations remain fixed.

The frozen rotor model produces a steady state solution to the multiple frame of reference problems, with some account of the interaction between the two frames but falls short of modelling the transient effects in the vicinity of the interface. The ANSYS CFX performs appropriate equation transformations when the frame of reference changes. The frozen rotor model requires least amount of computational cost compared to other frame mixing models that are available in ANSYS CFX. Frozen rotor model is recommended in frame mixing situations where the circumferential variation of the flow is large relative to the component pitch and where the flow domain has a rotating element [22].

The boundary conditions were prescribed at inlet, outlet, wall and interface locations of the turbine, as shown in Table 3.

Table 3. Details of the boundary conditions for the ANSYS CFX simulations

Location (s)	Assigned boundary condition type	Boundary condition details
Inlet	Inlet	• static pressure according the operational head for the particular simulation • volume fraction of 1 for water and 0 for air
Outlet	Opening	• atmospheric pressure • volume fraction of 0 for water and 1 for air
runner, nozzle and casing walls	Wall	• all walls were assigned 'non-slip' condition • wall boundary condition default details
Non-rotating zone and rotating zone interface	Interface	• Frozen rotor frame mixing model • Interface boundary condition default details
Turbine casing zone and rotating zone interface	Interface	• Frozen rotor frame mixing model • Interface boundary condition default details

A standard homogenous free surface model was used when modeling two-phase flow in the turbine runner. In the standard homogenous free surface model, both fluids (air and water) are assumed to share the same velocity, pressure and turbulence fields. Flow properties for air and water necessary for simulation were obtained from the ANSYS Library. Salient parameters for two-phase simulation are given in Table 4 in 'Fluid pairs'.

A second order upwind discretisation scheme was used for the governing equations. Viscous fluxes were computed with a high resolution scheme. To ensure computational robustness, locations where gradient changed abruptly, a first-order upwind scheme was applied. The rest of solver controls and salient simulation parameters are given in Table 4.

5.1 Governing equations

The equations used in this numerical flow analysis are the Reynolds Averaged Navier Stokes (RANS) equation with turbulence models based on eddy viscosity concept. As already stated, this is a two-phase flow problem with a movable free surface and that one of the flow domain has a rotating frame of reference. In this case, the effective density of each phase is given in terms of its volume fraction. For the rotating zone, the RANS for the momentum equations are modified to take into account centrifugal and Coriolis forces, as detailed in ANSYS Theory Guide [22]. These forces cannot be ignored in a rotating frame of reference. The modified RANS equations are given in equations (1) and (2). The Table 5 lists the meaning of the symbols used in the modified RANS equations and the modified pressure equation (3).

$$\frac{\partial(\alpha_p\rho_p)}{\partial t} + \nabla\cdot\left(\alpha_p\rho_p\vec{V_r}\right) = S_p \qquad (1)$$

$$\frac{\partial\left(\alpha_p\rho_p\vec{V_r}\right)}{\partial t} + \nabla\cdot\left(\alpha_p\rho_p\vec{V_r}\otimes\vec{V_r}\right) - \nabla\cdot\left(\mu_{eff}\,\nabla\vec{V_r}\right) + \alpha_p\rho_p(2\vec{\omega}\times\vec{V_r}$$
$$+\vec{\omega}\times\vec{\omega}\times\vec{r} + \vec{\alpha}\times\vec{r} + \vec{a} = -\nabla P' + \nabla\cdot\left(\mu_{eff}\,\nabla\vec{V_r}\right)^T + S_{Mp} \qquad (2)$$

$$P' = P + \frac{2}{3}\rho_p k + \frac{2}{3}\mu_{eff}\nabla \cdot \vec{V} \tag{3}$$

Table 4. Solver control and some parameters used in the ANSYS CFX numerical analysis

Basic settings	• Advection scheme set to high resolution option • Turbulence numeric set to high resolution option • Convergence control set to 500 as maximum iterations. After execution of the maximum number of iterations, the domain flux imbalances are checked from the simulation report. If the domain flux imbalances are less than 0.2%, then the solution is accepted as converged. • Convergence criteria type of root-mean square of and residual target is $1e^{-4}$ for all the fluxes (mass, momentum and turbulence)
Advanced settings	• Global dynamic model control with turbulence control • Body forces control by averaging type of volume-weighted • Velocity-pressure coupling according to the Rhie-Chow option of 4th power • Choose volume fraction coupling option for multiphase control • Choose pressure-based solver
Fluid pairs	• The air/water fluid is used as the fluid pair with air and water being the primary and secondary fluid respectively. • The air/water surface tension coefficient is taken as equal to $0.07199 Nm^{-1}$ at 25^0C [23] • Surface tension model with continuum surface force with air as the primary fluid • Volume fraction smoothing type of the volume-weighted • No inter-phase mass transfer
Domain initialization	• The initial Cartesian velocity in all the three flow domains is: $u = 0; v = 0; w = 0$. • The initial static pressure in all the flow domains is zero relative pressure (zero gauge pressure). • The initial turbulence in all the three flow domains is taken to be of low intensity (1%). • In all the flow domains, the initial volume fraction for air is 1 and for water is 0.

Table 5. Meaning of the symbols used in the modified RANS equations

Symbol	Meaning (or representation)
t (s)	time
α_p (-)	Volume fraction for the particular phase p
ρ_p (kg/m^3)	Density for the particular phase p
S_p (kg/s)	Continuity equation source term for the particular phase
S_{Mp} (N/m^3)	Momentum equation source term for the particular phase
$\vec{V}_r$ (m/s)	Linear equivalent of rotational velocity
∇ (-)	Del operator
$\otimes$ (-)	Dyadic symbol
μ_{eff} (kg/m.s)	Effective viscosity
$\vec{\alpha}$ (m/s^2)	Angular acceleration
$\vec{a}$ (m/s^2)	Linear acceleration
$\vec{\omega}$ (rad/s)	Angular velocity
P (N/m^2)	Pressure
P' (N/m^2)	Modified pressure
k (J/kg)	Turbulent kinetic energy

Equation (1) is the continuity equation for the two-phase flow. Equation (2) is the modified momentum equation. In both equations, the subscript 'p' refers to property of each phase. In the modified momentum equation, the expression $2\vec{\omega} \times \vec{V_r}$ is the Coriolis acceleration, $\vec{\omega} \times \vec{\omega} \times \vec{r}$ is the centripetal acceleration, $\vec{\alpha} \times \vec{r}$ is the acceleration due to unsteady change in rotational speed of the runner and $\vec{a}$ is the acceleration due to unsteady change of linear relative speed, V_r. The effective viscosity used in the momentum equation is given by equation (4). The effective viscosity is obtained by applying the eddy viscosity concept.

$$\mu_{eff} = \mu + \mu_t \tag{4}$$

where μ (kg/m.s) is the mean viscosity and μ_t (kg/m.s) is the turbulent viscosity.
The turbulent viscosity is evaluated depending on the type of the turbulent model employed in the simulation. This numerical scheme has employed the $k - \varepsilon$ turbulence model for generating initial solution that is used as initial condition for the subsequent numerical analysis using the $k - \omega$ model. These two turbulence models are widely used in industrial applications despite having some technical differences between them. The $k - \varepsilon$ model, the turbulent viscosity, given by equation (5), is linked to the turbulent kinetic energy (k) and turbulent dissipation (ε). On the other hand, for the $k - \omega$ model, the turbulent viscosity is calculated from turbulent kinetic energy and turbulent frequency (equation 6).

$$\mu_t = C_\mu \rho \frac{k^2}{\varepsilon} \tag{5}$$

$$\mu_t = \rho \frac{k}{\omega} \tag{6}$$

where C_μ (-) is a constant, ρ (kg/m^3) is the density of the fluid, k (J/kg) is the turbulent kinetic energy, ε (m^2/s^3) is the turbulent dissipation rate and ω (s^{-1}) is the mean turbulent frequency. The values of k, ε and ω are calculated from the differential transport equations for the turbulence kinetic energy and turbulence dissipation rate. The $k - \varepsilon$ model is popular for industrial applications because of its good convergence rate and relatively low memory requirements [24, 25]. It performs well for external flow problems around complex geometries. However, the model does not accurately compute flow fields that exhibit adverse pressure gradients, flow with strong curvature and jet flow.
The $k - \omega$ has relatively large memory requirements compared to $k - \varepsilon$, however, it has difficulties in converging and is sensitive to the solution initial guess [24]. The model is useful in many cases where the $k - \varepsilon$ model is not accurate, such as flows that exhibit strong curvature, separated flows and jets. In such flow phenomena, the $k - \varepsilon$ model is often recommended to be used for the initial condition for solving the $k - \omega$ model. This is the procedure that was used in the numerical analysis in this numerical study, as stated earlier.
For the volume fraction treatment of multiphase flow, due to the coexistence assumption of the free surface homogeneous multiphase flow, the effective fluid properties in each of the control volume is given in equation (7) for density and equation (8) for viscosity.

$$\rho = \textstyle\sum_{p=w,a} \alpha_p \rho_p \tag{7}$$

$$\mu = \textstyle\sum_{p=w,a} \alpha_p \mu_p \tag{8}$$

where α (-) is the volume fraction, subscript p is the phase (water w, or air a), ρ (kg/m^3) and μ (kg/m.s) are the density and viscosity of the averaged mixture that are used in the RANS modified equations. The air density ρ_a (kg/m$^{3)}$ is assumed to be a function of the pressure P (N/m^2) in the equation of state, equation (9).

$$\rho_a = \rho_{a,0} e^{\gamma (P - P_0)} \tag{9}$$

where the sub-index *0* marks the reference state values and γ (-) is the air compressibility coefficient.

For continuity, in each control volume, the sum of volume fraction must be equal to 1, as given by equation (10).

$$\sum_{p=w,a} \alpha_p = 1 \tag{10}$$

The volume fraction gives the proportion of the control volume that is occupied by the secondary fluid, which is water in this project. In the two phase flow problem, the secondary fluid is chosen as that which has the largest density.

5.2 Mesh properties

For numerical solutions, it is important to make sure that the mesh is of good quality so as to produce results that make sense with respect to the nature of the physical process. Also, a good quality mesh reduces computational costs. Table 6 lists mesh properties used in this study. Table 7 shows the geometric qualities of the mesh with reference to the acceptable values that are suggested in the ANSYS CFX Best Practices Guide for Turbomachinery of 2009 [24].

Table 6. Mesh properties used in the ANSYS CFX simulation study

Mesh property	Value/Action
Sizing	
Use advanced size function	On Curvature
Relevance Centre	Medium
Initial size seed	Active assembly
Smoothing	Medium
Transition	Slow
Span angle centre	Fine
-curvature normal angle	Default (18^0)
-min size	$3.5e^{-5}$m
-max face size	$3.5e^{-3}$ m
-maximum size	$0.75e^{-2}$ m
-growth rate	Default (1.2%)
Minimum edge length	$4.314e^{-3}$ m
Inflation	
Use Automatic Inflation	Program controlled
Inflation Option	Smooth transition
-Transition ratio	0.77
-maximum layers	3
-growth rate	Default (1.2%)
Inflation Algorithm	Pre
Patch conforming options	
Triangle surface mesher	Program controlled
Advanced	
Shape checking	CFD
Element mid-side nodes	Dropped
Mesh morphing	Enabled
Rigid body behaviour	Dimensionally reduced
Extra retries for assembly	Yes
Defeaturing	
Pinch tolerance	Default ($6.75e^{-5}$)m
Generate pinch on refresh	No
Automatic mesh defeaturing	On
Defeaturing tolerance	Default ($3.75e^{-5}$m)

Table 7. Mesh geometrical quality parameters at best efficiency point

Domain name	Orthogonal angle (acceptable > 20^0)	Expansion factor (acceptable < 20)	Aspect ratio (acceptable < 100)
Non-rotating zone	37.5	11	6
Rotating zone	20.3	19	22
Turbine casing	21.6	9	24
Global	21.9	19	24

As can be seen from Table 7, the mesh used in this project has favourable characteristics. The average values on orthogonality for the rotating zone and turbine casing is almost on the lowest acceptable values, possibly due to the presence of runner blades (walls) in the rotating zone. The CFX solver calculates the stated important measures of mesh quality (orthogonal angle, expansion factor and aspect ratio) at the start of a simulation run and updates them each time the mesh is deformed.

5.3 Mesh sensitivity analysis and simulations

After obtaining a good quality mesh, it is important to perform a mesh sensitivity analysis to make sure that the solution is mesh independent, as it has already been stated. The sensitivity analysis involves identifying a physical output variable to be monitored and recording its value with respect to changes in the number of mesh elements. Then the rate of change in value of the monitored variable with respect to the number of mesh elements can be evaluated from the plotted graph.

Outlet velocity from the turbine was the monitored variable for the sensitivity analysis. After convergence of the solution, the value of the outlet velocity from the monitored graph (Figure 7) was recorded together with the number of mesh elements (which were changed by varying the size of the mesh in ANSYS Meshing sub-program). Ten simulations with different number of mesh elements were performed. Then, outlet velocity was plotted against number of mesh elements (Figure 8). It can be seen from Figure 8 that after 600 $x10^3$ number of mesh elements, the outlet velocity was observed to be constant (at 3 m/s) with respect to increase in number of elements.

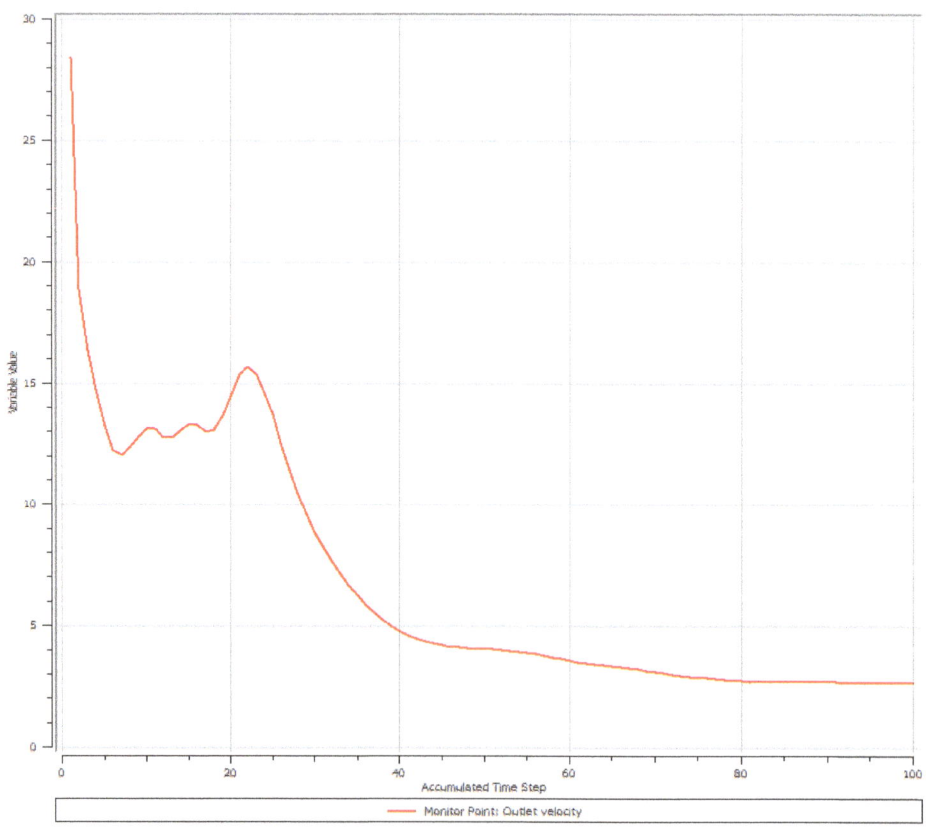

Figure 7. Graph of monitored outlet velocity during simulation against number of iterations

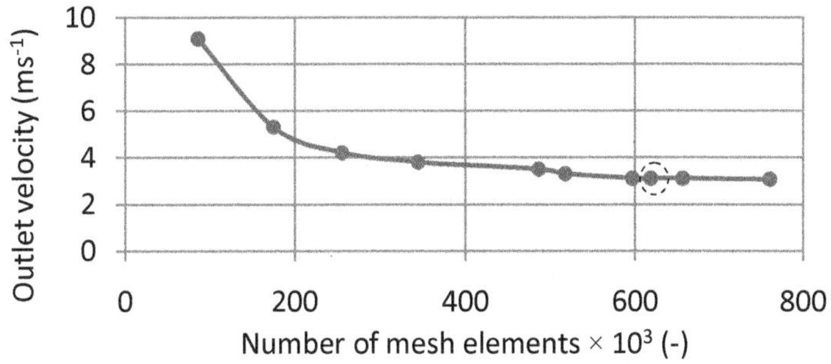

Figure 8. Graph of outlet velocity against number of mesh elements

The number of mesh elements for the final numerical analysis reported in this study was 618695 which according to the sensitivity analysis (Figure 8), is a good number of mesh for the simulation. Large number of mesh elements would cause a delay in the convergence of the solution. Table 8 shows mesh statistics for the three computational zones.

Table 8. Mesh statistics for the three flow computational zones

Computational zone	Mesh statistics	
	Nodes	Elements
Turbine Casing	133 544	492 668
Rotating	22 758	71 320
Non-rotating	14 696	54 687
Total	170 998	618 695

With respect to the first specific objective, the pressure and velocity distribution simulations were performed at best efficiency point and results processed in the ANSYS CFX-Post. The simulations at out of best efficiency points were achieved by changing the rotational speeds of the runner at constant head and valve opening. In the ANSYS CFX-Pre, the runner rotational speed was entered as the rotational speed for the 'rotating zone' (Figure 6b), while the velocity of the runner walls (representing blades) was given a relative velocity of zero.

During each simulation session, the pressure and velocity distributions were processed in ANSYS CFX – Post by creating a mid-plane across the flow domain (turbine geometry) representing the average flow plane. Apart from determining the flow pattern, the efficiency of the turbine at each simulation session was numerically determined. This was to analyze the effect of flow pattern on efficiency and to compare with the experimentally obtained values. The efficiency in the numerical analysis was calculated according to the Euler's turbine equation for the Crossflow turbine; equation (11).

$$\eta_h = \frac{1}{gH}\{(u_1 V_{u1} - u_2 V_{u2}) - (u_3 V_{u3} - u_4 V_{u4})\} \quad (11)$$

where u (m/s) and V_u (m/s) is the runner and flow peripheral speeds respectively. Subscript 1 and 2 refers to the inlet and outlet of first stage respectively; subscript 3 and 4 refers to the inlet and outlet of second stage respectively. g (m/s^2) is the acceleration due to gravity and H (m) is the effective head at the entry to the turbine. In ANSYS CFX-Post, the velocities were obtained through creating circular locations at outer diameter and inner diameter of the runner where the resultant efficiency was computed through integration.

6. Results and discussions

Flow profile at best efficiency point, in terms of velocity contours reveal that the flow in the runner does not strike the shaft, as shown in Figure 9a. Actual turbine flow picture (Figure 9b) taken using a high speed camera (Photron FASTCAM SA5) shows flow profile that is consistent with the numerically generated flow profile. Therefore the flow profile depicted in the picture validates the numerically generated flow profile. It was not possible to take a picture for the full part of the flow in the second stage due to mechanical challenges associated with positioning the camera for this purpose. The flow past the second stage was observed using stroboscope and the flow pattern in the second stage also mimics the numerically generated pattern.

Turbine losses are at minimum level at best efficiency point. In this numerically generated flow profile, validated with actual pictures, it has been observed that at best efficiency point, the jet does not strike the shaft. Therefore, avoidance of jet from striking the shaft is important in the design high efficient CFT. The flow pattern at best efficiency point supports the theory proposed by Prof. Banki [1] that at the best efficiency point, a clear identifiable jet exists. However, this observation is in disagreement with findings from other investigators. For example, using photographic method, Durgin and Fay in 1984 [4] found out that inside the CFT runner, a free standing jet does not exist.

From Figure 9(a), it can be seen that the flow in the nozzle outlet has relatively high velocity compared to the first stage. This is due to transfer of kinetic energy from the jet to the runner. It can also be seen from the contours of the velocity field that velocity is relatively high in the second stage than in the first stage. This implies acceleration of jet and that the first stage acts as a nozzle. This is a required flow pattern for CFT design because back pressure is prevented.

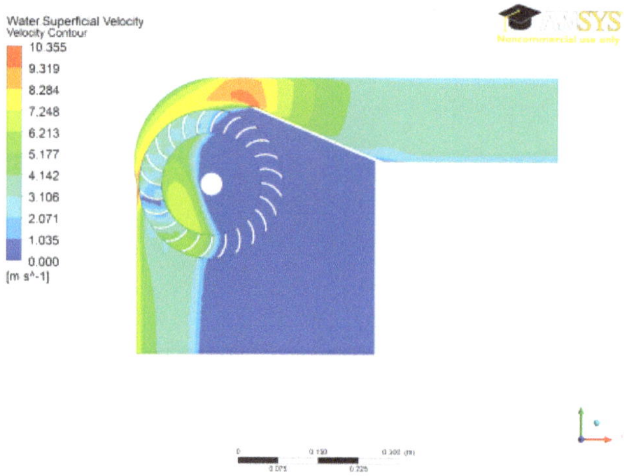

(a)

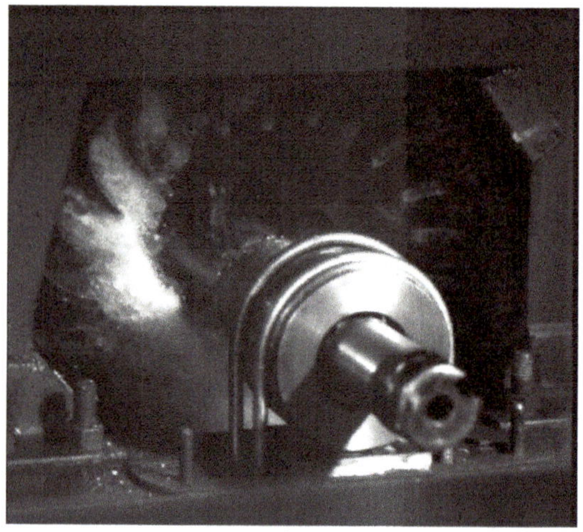

(b)

Figure 9. (a) Velocity flow profile at best efficiency point using an ANSYS CFX; (b) Picture of flow profile using a high speed camera

The vector plot in Figure 10 gives a further qualitative description of the flow pattern at best efficiency point. It can be seen that the demarcation between first and second stage is short and the jet in the first stage cuts several runner blades. In the first stage, the jet covers more blades than the second stage. Therefore a considerable portion of the runner is utilized for energy transfer. This flow profile is unlike that observed in typical impulse turbine runners such as Pelton and Turgo turbines where relatively small portion of the runner is used for energy transfer.

It can also be seen from Figure 10, that at the outlet of both stages, the velocity vectors cross each other. This flow pattern supports the hypothesis that the jet leaves the stages (within blade spacing) with different flow directions resulting in flow streamlines crossing and colliding against each other. Flow pattern where the streamlines cross each other is poor for energy transfer process. It creates localized vorticities and foam in the runner that reduce efficiency.

The pressure contour diagram in Figure 11 shows that at best efficiency point, the first stage has positive gauge pressure indicating that it is operating with a degree of reaction. This is in agreement with experimental observation that at large nozzle openings, the CFT operates with a degree of reaction. It can also be seen that locations of negative pressure are identified in the runner, especially at the demarcation between stages and in the second stage. Flow pattern with pockets of negative pressure is associated with cavitation.

The flow velocity profiles for runner speeds away from the best efficiency speed (350 rpm) display some flow patterns worth noting.

It can be seen from the velocity contours (Figure 12) that as the runner speed increases, the jet in the runner drifts way from the shaft. The change in absolute velocity vector of the jet at first stage outlet as a result of change in runner periphery velocity may explain the drifting away of the jet from the shaft. The

absolute velocity is the vector sum of relative velocity and runner periphery velocity. It can also be seen that the demarcation between the two stages progressively becomes small as the runner speed increases.
The velocity contour diagrams also reveal that as the runner speed increases beyond the best efficiency point, at high speeds, there is back flow. This implies that at these speeds, the CFT is operating with a pumping effect. This can be seen with speeds of 500 rpm and 600 rpm velocity contour diagrams and the 600 rpm velocity vector diagram. The flow with back pressure is due to the increased contribution of the centrifugal forces on the flow as the runner speed increases.
Referring to velocity contour and vector at 100 rpm in Figures 13(a) and 14(a) respectively, it can be seen that at lower runner speeds, the flow in the runner strikes the shaft, resulting in increased incidence losses. Figure 14(b) shows the actual picture of the flow in the runner taken at 100 rpm. It can be seen that the jet strikes the shaft and the actual flow pattern mimics the numerically generated flow profile.
Numerical flow analysis results when the CFT operates outside the best efficiency point are presented in Figure 12.
The flow pattern with back pressure reduces the resultant flow that passes through the runner to generate power. Also, back pressure reduces the amount of pressure energy that could be imparted to the runner due to reaction effect. This numerical observation is in agreement with the results obtained by Varga in 1959 [26]. He experimentally found out that the increased speed of the modified CFT runner affects the flow through the runner. However, this is a deviation from the operation of a typical impulse turbine as originally proposed by Prof. Banki.
In Figure 15, the pressure contours when the CFT operates away from the best efficiency point do not show significant differences. At all operating speeds, the jet enters the first stage with positive gauge pressure. It can be seen that the static pressure in the nozzle entry arc increases with runner rotational speed. The increase in static pressure at constant head indicates presence of back pressure. Again, this pressure profile does not support the operating principle of a pure impulse turbine.

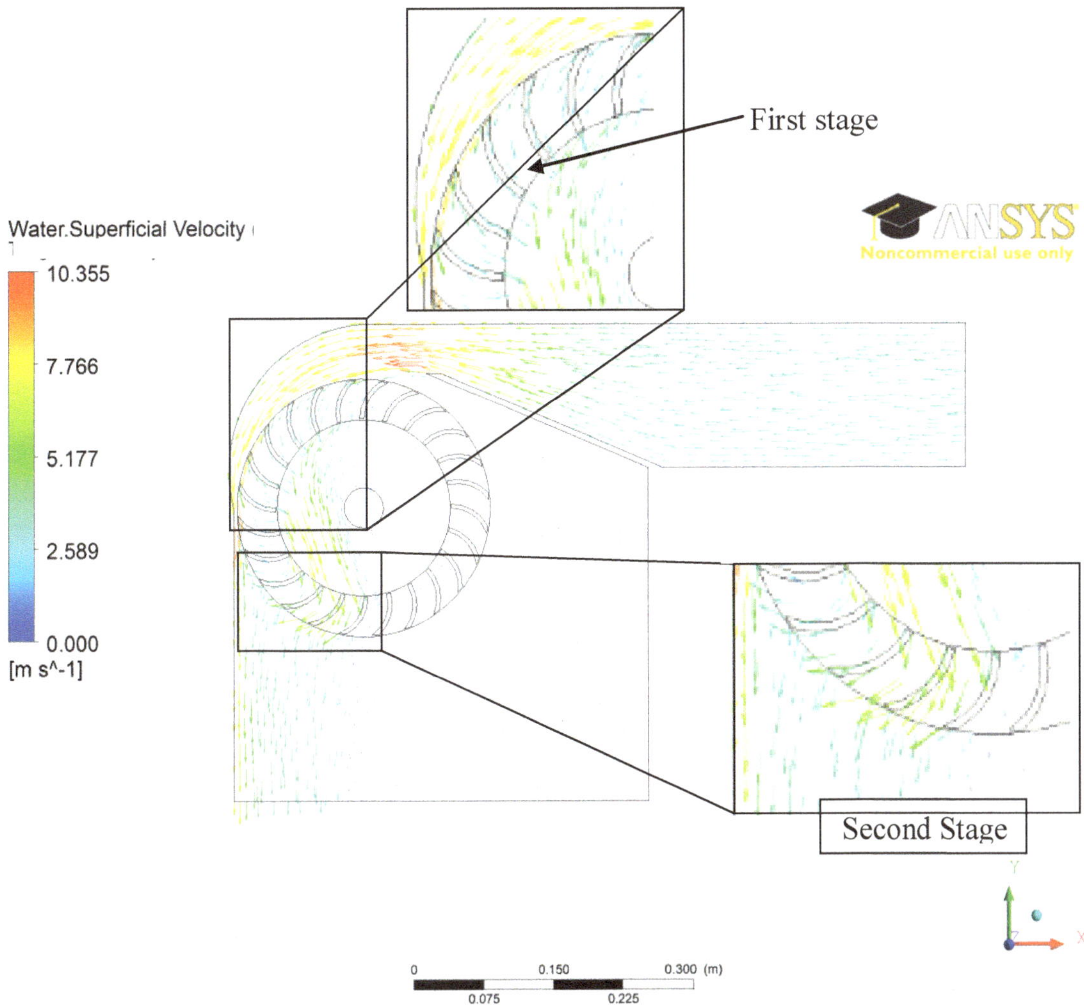

Figure 10. Vector plot of flow in the Crossflow turbine at best efficiency point

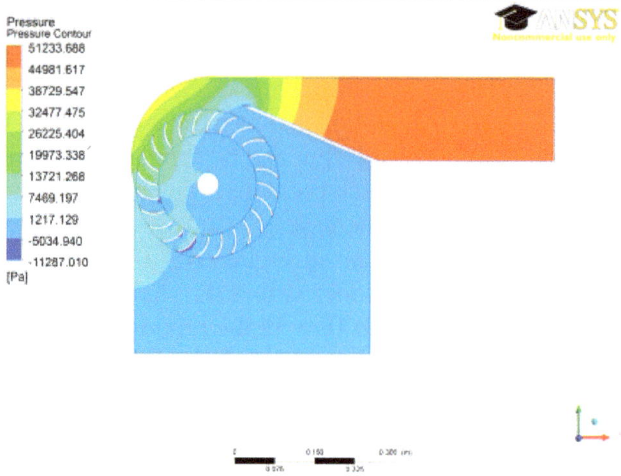

Figure 11. Pressure contour at best efficiency point

100 rpm

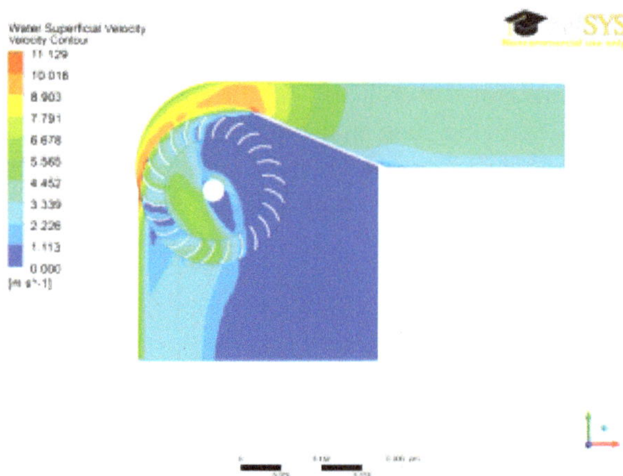

200rpm

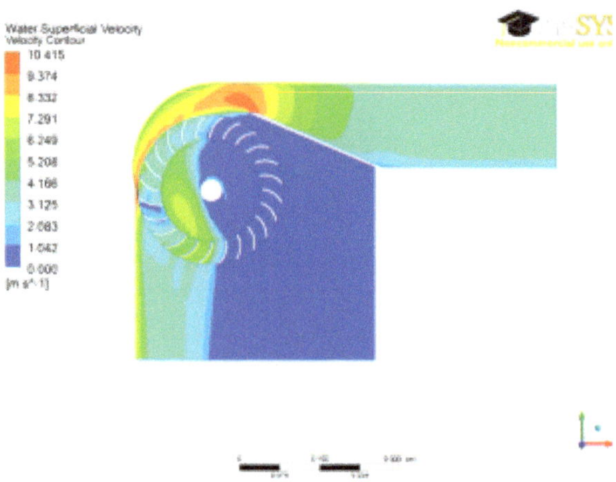

300rpm

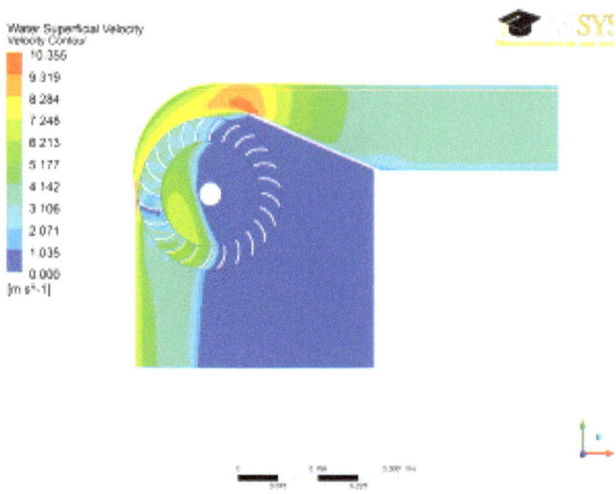

350rpm

Figure 12. (Continued)

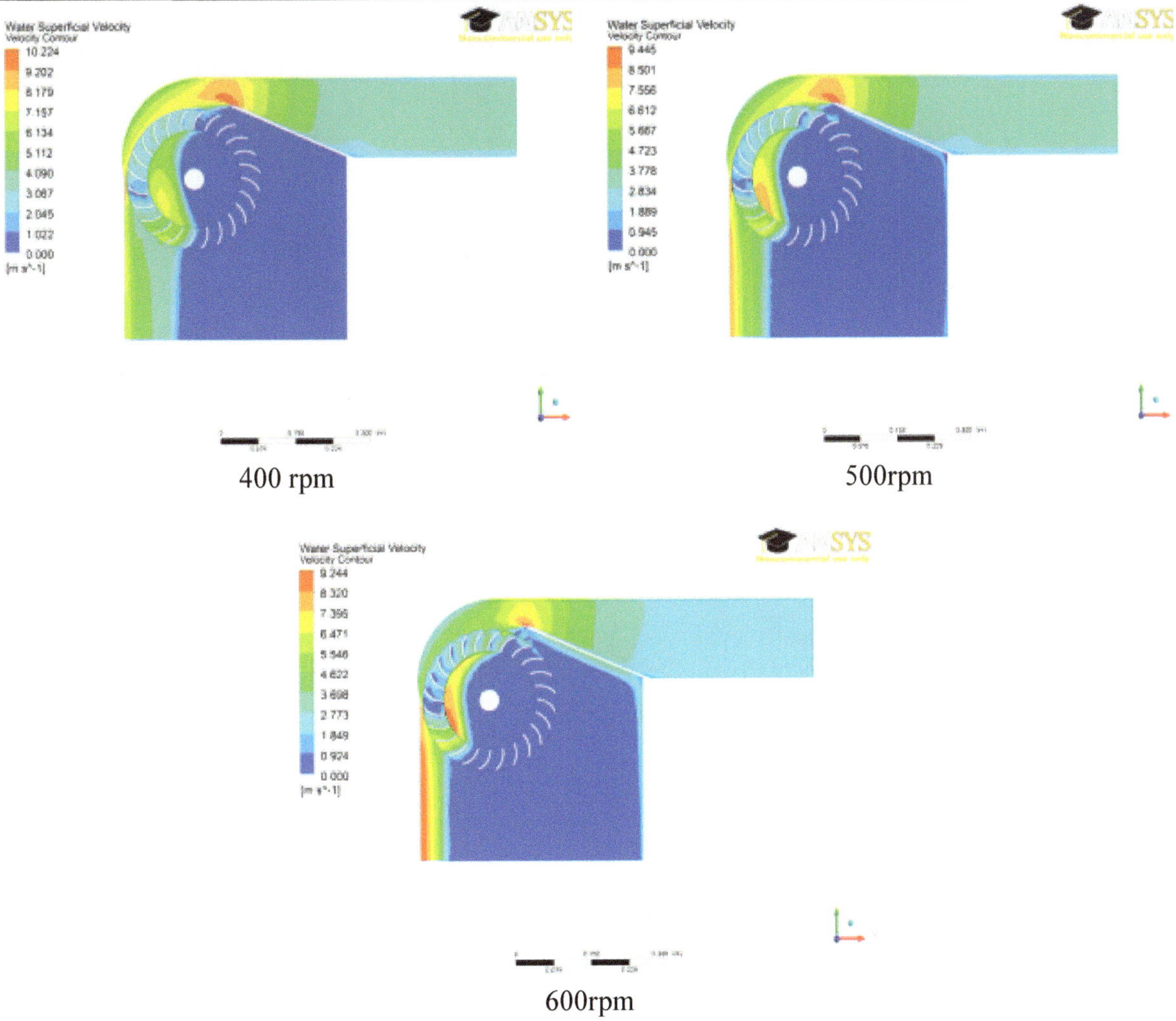

Figure 12. velocity contours for various runner speeds away from best efficiency point

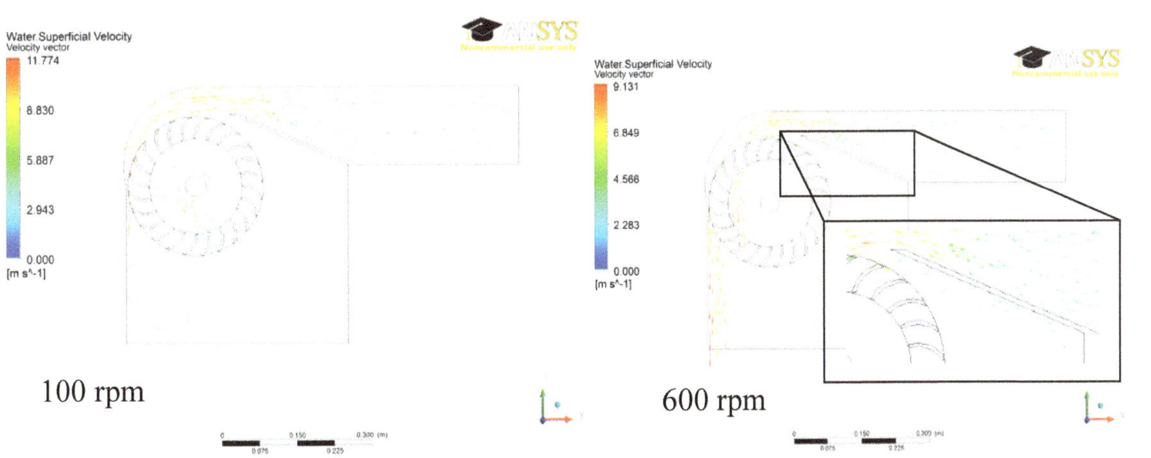

(a) (b)

Figure 13. (a) Velocity vector diagram at 100 rpm runner speed; (b) Velocity vector diagram at 600 rpm runner speed

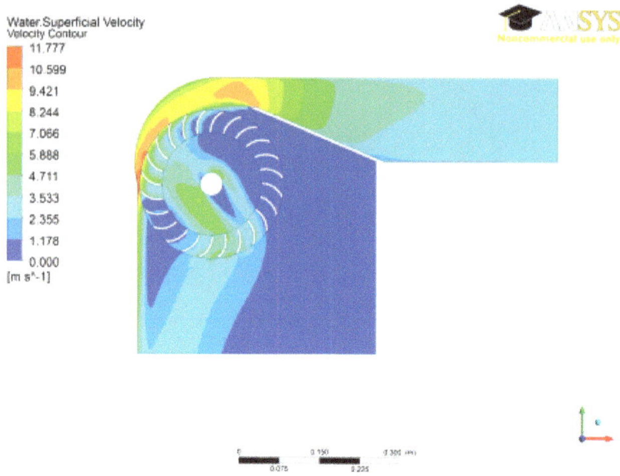

(a)

(b)

Figure 14. (a) The velocity contour at 100rpm; (b) picture of flow in the runner at 100 rpm

The hydraulic efficiency was numerically evaluated from equation (11) and compared with the turbine efficiency that was measured experimentally on the same CFT model in 2013 at Waterpower Laboratory of the Norwegian University of Science and Technology. This was to evaluate the extent to which the numerical analysis predicts the actual turbine performance. The tests in Figure 16 were performed at 5m head and valve opening of 80%.

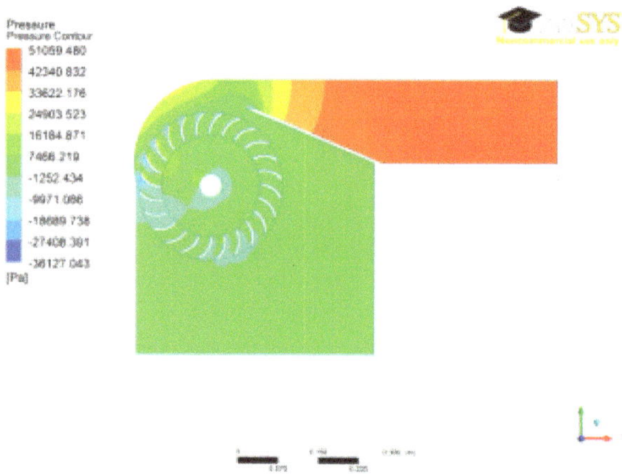

100 rpm

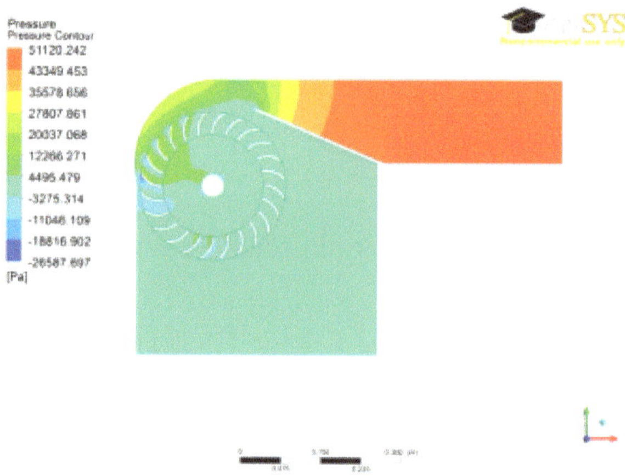

200rpm

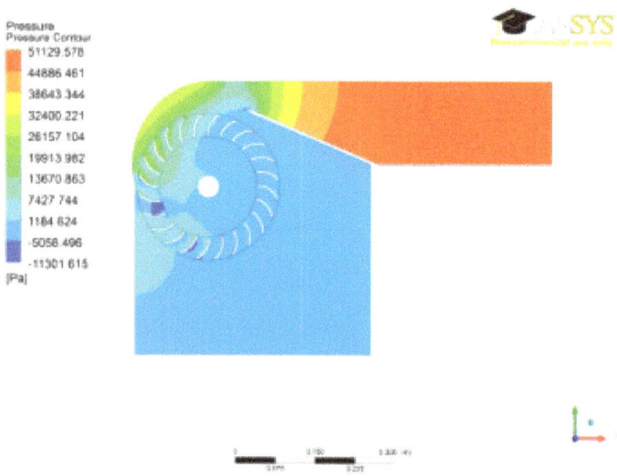

300rpm

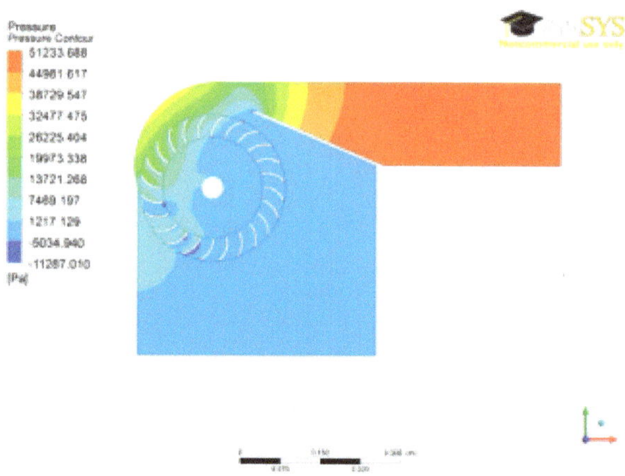

350rpm

Figure 15. (Continued)

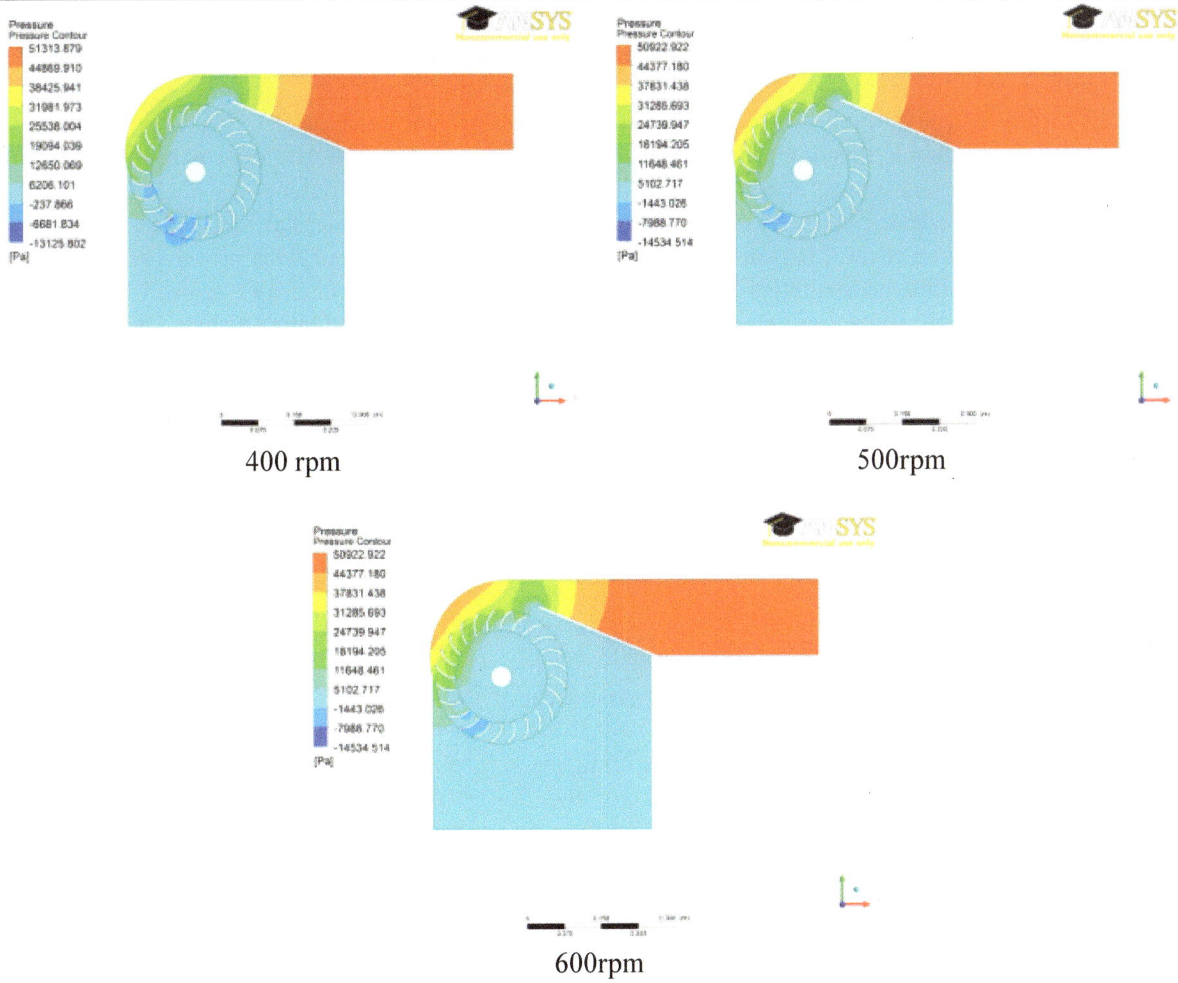

Figure 15. Pressure contours for various runner speeds away from the best efficiency point

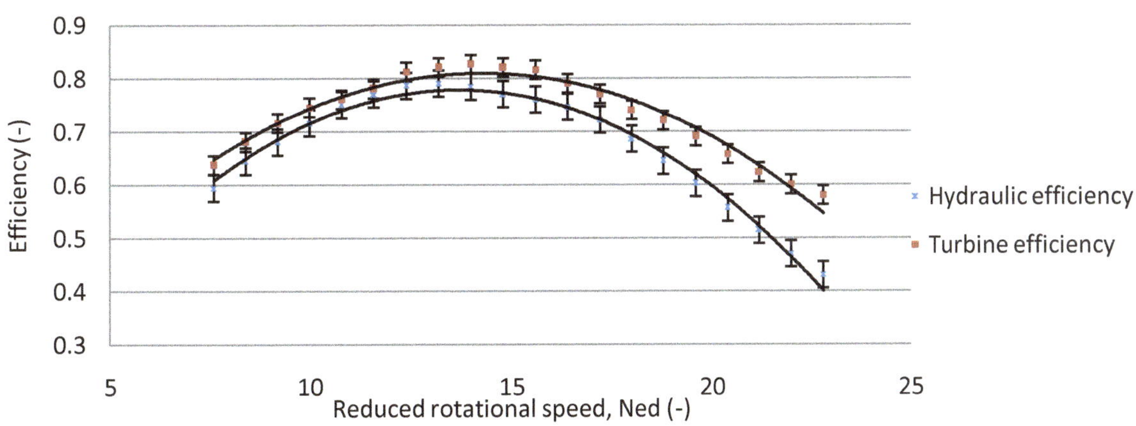

Figure 16. Variation of hydraulic efficiency and turbine efficiency with runner rotational speed.

It can be seen from Figure 16 that the graphs for experimental turbine efficiency and numerically determined hydraulic efficiency show similar trends. As expected from Euler's turbine equation, the relationship between the efficiency and rotational speed is quadratic for both graphs. The quadratic relationship with maximum values corresponding to the maximum efficiency speed can be seen from both graphs. The maximum efficiency values corresponding to reduced speed of around 13 are observed. The standard error bars on the experimental results are included. In both graphs, the trend lines are mostly within the standard error bars of the data points.

Despite the consistency, the numerically determined hydraulic efficiency tends to give higher values especially for runner speeds exceeding best efficiency point, thus indicating that numerical analysis over-predicts the performance. One of possible the reasons for over-prediction of efficiency is that the mechanical efficiency of the turbine is not considered in the numerical calculations. The mechanical efficiency increases with runner speed. Hence at higher speeds, the contribution of mechanical losses is increased which lowers the shaftpower and widens the difference between the efficiencies.

It is inevitable for the discrepancies (both qualitative and quantitative) in numerical and experimental analyses to be present due to several reasons. Numerical analyses are associated with numerical errors and it is difficult to describe mathematically some flow processes. Numerical errors arise from the numerical procedure, mesh, data processing and geometric complexity of the turbine. Further, assumptions on the boundary conditions and turbulence model are major sources of numerical errors and may generate unrealistic simulation results.

In this numerical study, the numerical approach in the ANSYS CFX homogeneous free surface model does not support buoyancy and Coanda effect. Therefore, segregation between phases that occurs due to the gravitational action on the flow has not been modelled. Further, the deflection of the jet due to Coanda effect has not been modelled. However, due to complex nature of the flow and geometry of flow domain, numerical approach has proven to be superior in the design of 'partial' turbines such as the Crossflow.

7. Conclusions and recommendations

The conclusions that can be drawn from this numerical study of the flow profile and performance of Crossflow turbine are as follows:

1. The numerically determined performance results compare favorably with the experimental results. This shows that the CFD is an important tool for the design of 'partial' turbines such as the Crossflow.
2. The study shows that Crossflow turbine, at best efficiency point, operates using both impulse and reaction principles.
3. Incidence losses are greatly reduced when the jet does not strike the shaft in the runner. This shows that avoidance of jet from striking the shaft is important in the design of a high efficient Crossflow turbine.
4. Some pockets of negative pressures were identified in the flow in the runner. This signals the turbine runner being susceptible to cavitation; a problem which has been identified in practice.
5. At constant head and valve opening, the flow pattern in the runner depends on runner speed. At lower speeds, the flow in the runner strikes the shaft leading to increased incidence losses. At higher speeds, back flow in the runner was observed, again reducing the efficiency of the turbine.

On improving the efficiency over a range of operational runner speeds, it is recommended to consider designing the Crossflow turbine that does not have the shaft inside of the runner where crossed-flow jet passes through. Since the shaft provides an additional structural integrity to the turbine runner (together with the blades that are welded to the discs), it is important to carry out another numerical study, using finite element analysis, to investigate the effect of removing the shaft on performance and structural strength of the runner.

Acknowledgement

The authors appreciate the support from NORAD under the Energy and Petroleum (EnPe) Program for funding this research at Waterpower Laboratory, Department of Energy and Process Engineering, Norwegian University of Science and Technology.

References

[1] Mockmore C, Merryfield F. The Banki water-turbine. Engineering Bulletin, Series Number 25, Oregon State University, Corvallis, United States of America, 1949.

[2] Nakase Y, Fukutomi J, Watanabe T. A study of the Crossflow turbine, effects of nozzle shape on its performance. Journal of small hydropower fluid machinery, American Society of Mechanical Engineers (ASME), 1982, pp: 29-133.

[3] Johnson W, Ely R, White F. Design and test of an inexpensive Crossflow turbine. In Proceedings of American Society of Mechanical Engineers (ASME) annual symposium on small hydropower fluid machinery, New York, USA, 1982.

[4] Durgin W, Fay W. Some fluid flow characteristics of a Crossflow type hydraulic turbine. In Proceedings of American Society of Mechanical Engineers (ASME), Winter Annual Meeting on small hydropower fluid machinery, New Orleans, USA, 1984.
[5] Khosrowpanah S. Experimental study of the Crossflow turbine. PhD dissertation, Colorado State University, USA, 1984.
[6] Hothersall R.J. A review of the Crossflow turbine. In Proceedings of American Society of Civil Engineers (ASCE) on Waterpower, New York, N.Y., 1985, vol. 2, page 914.
[7] Ott R, Chappel J. Design and efficiency testing of a Crossflow turbine. In Proceedings of American Society of Civil Engineers (ASCE) on Waterpower, New York, N.Y., 1989, 1534-1539.
[8] Desai V, Aziz N. Parametric evaluation of Crossflow turbine performance. Journal of Energy Engineering, 1994, vol. 120, no. 1. pp: 17-34. American Society of Civil Engineers.
[9] Olgun H. Investigation of the performance of the Crossflow turbine. International Journal of Energy Research, issue 22, pp 953 – 964, 1998. John Wiley & Sons, Ltd.
[10] Andrade J, Curiel C, Kenyery F, Aguillon O, Vasquez A, Asuaje M. Numerical investigation of the internal flow in a Banki turbine. International Journal of Rotating Machinery, 2011, vol. 2011, article ID 841214, 12 pages.
[11] Zoppe B, Pellone C, Maitre T, Leroy P. Flow analysis inside a Pelton turbine bucket. Journal of Turbomachinery, 2006, vol. 128, issue 3, pp 500-511.doi: 10.1115/1.2184350
[12] Fukutomi J, Senoo Y, Nakase Y. A numerical method of flow through a Crossflow runner. JSME International Journal, 1991, series II, vol. 34, no. 1, pp 44-51.
[13] Fukutomi J, Nakase Y, Ichimiya M, Ebisu H. Unsteady fluid forces on a blade in a Crossflow turbine. JSME International Journal, 1995, series II, vol. 38, no. 3, pp 404-410.
[14] Pereira N.H, Borges J.E. Study of the nozzle flow in a Crossflow turbine. International Journal of Mechanical Sciences, 1996, vol 38, n. 3, pp 283-302.
[15] Arzola F, Rodriquez C, Martin J, De Andrade J, Vasquez A, Asuaje M. Technical assessment for overhaul project in small hydropower plant. In Proceedings of the 24th Symposium on hydraulic machinery and systems, Iguassu, Brazil, October 2008.
[16] Choi Y, Lim J, Kim Y, Lee Y. Performance and internal flow characteristics of a Crossflow hydro turbine by the shapes of nozzle and runner blades. Journal of Fluid Science and Technology, 2008, vol.3, no. 3, pp. 398 – 409.
[17] Khosrowpanah S, Fiuzat A, Albertson M. Experimental study of Crossflow turbine. Journal of Hydraulic Engineering, 1988, vol. 114, no. 3 pp: 299-314. American Society of Civil Engineers (ASCE).
[18] JLA & Co. Limited. Crossflow turbine Information. Product information by the JLA & Co, a turbine manufacturing company in B-4520 Moha, Belgium, 2010. Available online at <http://www.jlahydro.be/en/jla-cross-flow-turbines> [12 July 2013].
[19] Ossberger Limited. Ossberger turbine information. Ossberger Limited, Weissenburg/Bavaria Germany, 2013. Available <http://www.ossberger.de/cms/pt/hydro/ossberger-turbine/> [20 August 2013]
[20] Joshi B, Seshadri V, Singh S. Parametric study on performance of Crossflow turbine. Journal of Energy Engineering, 1995, vol.121, no. 1, pp: 28-45. American Society of Civil Engineer (ASCE).
[21] Haimerl L. A. The Crossflow turbine. In Conference proceedings on Water Power, London (UK), 1960, vol. 12 pp: 5-23.
[22] ANSYS. ANSYS CFX Solver Modelling Guide. Published by the ANSYS, Inc. in United States of America, 2010, Release 13.
[23] Vargaftik, N. B.; Volkov, B. N.; Voljak, L. D. J., 1983. International tables of the surface tension of water. Journal of Physical and Chemical Reference Data, vol.12, No. 3, pp: 817-820. Published by American Institute of Physics (AIP) Publishing.
[24] ANSYS CFX, 2009. Best Practices Guide for Turbomachinery. ANSYS CFX – 12 Reference guide. Published by the ANSYS, Inc. in United States of America, 2009, Release 12.
[25] Rodi W. Turbulence models and their application in hydraulics: A state of the art review. Taylor and Francis, 1993, Third Edition, ISBN 9789054101505.
[26] Varga J. Tests with the Banki water turbine. Journal of the Hungarian Academy of Sciences, 1959, vol. pp: 779-102.

2

Modeling the importance of biomass qualities in biomass supply chains for bioenergy production

T.P. Upadhyay, J. H. Greibrokk

University of Tromsø –Arctic University of Norway, School of Business and Economics, Campus Alta, Follumsvei 31, N-9509 Alta, Norway.

Abstract

A tactical-operational level quantitative model can be an important decision support tool for bioenergy producers. Goal programming approach can help analyze the costs and volume implications of various competing goals in terms of biomass characteristics on part of the bioenergy producers. One cost and six quality characteristics goals, namely moisture and ash contents, and thermal values of two types of biomass (forest harvest residue and un/under-utilized species) are selected for the four bioenergy producers in northwestern, Ontario, Canada. We run four models cenarios: i) benchmark total cost and ceilings of mean values of six biomass qualities (Initial Goals), ii)relaxing the quality goals by 10% from the Initial Goals scenario, iii) increasing the conversion efficiency by 10%, and iv) all goals as in Initial Goals except the Atikokan Generating Station (AGS)being supplied with only un/under-utilized biomass. The smaller power plants have relatively less per unit biomass procurement cost. While per unit procurement costs increased, the total costs and biomass volume required to produce the same amount of bioenergy for each power plant decreased in all scenarios compared to the benchmark costs. The goal programming approach, and the results thereof are found to be useful in making effective decisions in the biomass supply chains for bioenergy production.

Keywords: Combined heat and power (CHP); Decisions support system; Forest harvest residue; Northwestern Ontario; Supply chain modeling; Woody biomass.

1. Introduction

Globally forest land-use has significant potential to reduce greenhouse gas (GHG) emissions if managed and used properly. However, it can manifest differently in different regions of the globe depending on the level of development. For example, halting deforestation and forest/soil degradation and enhancement of carbon pool in managed forest ecosystems in developing countries can reduce the GHGs emissions to a large extent [1-3]. Similarly, use of forest residues and un-merchantable trees (mainly under-utilized hardwood species) from sustainably managed forests for bioenergy production in the developed world can replace the fossil fuel-based energy usages, thereby helping reduce the present level of GHGs emissions. Further, use of woody biomass for bioenergy production has many environmental and socioeconomic benefits –this being sustainable renewable and CO_2-neutral resource, reducing risk of forest fire events, increased rural employment and income, etc. [4, 5]. Nonetheless, bioenergy production faces many emerging challenges that include uncertainty of biomass feedstock supply due to its sparse distribution over space and time, not yet fully developed biomass and bioenergy markets, relatively

higher costs of production if the environmental and other benefits not taken into account which is the case at present [5, 6]. Moreover, as more biomass-based power plants come into operation in the near future there would be higher volume demand for biomass feedstock that results in an increase in transportation distance, and hence the higher biomass procurement costs [6-8]. In order to analyze decisional problems of such supply/value chains strategic, tactical and operational level planning models are used depending on the scales in terms of planning horizons in general. In this context, a tactical-operational level goal programming (GP) model can be an important decision tool for bioenergy producers in general. The modelling approach in this study, therefore, exemplifies the biomass procurement cost structures under various resource constraints with varying biomass quality, and procurement cost goals on part of the power plants under various model scenarios.
Globally, fossil fuels burning and land-use changes are the major contributors for the GHG emissions [2]. In this context, various international climate policies have exerted pressure on reducing GHG emissions for developed countries in which the Canadian government has also made a commitment to reduce GHG emissions from its major industrial sectors. The forest products industry of Canada, which requires a huge amount of energy to operate, has been a leader in utilizing bioenergy (e.g., burning black liquor and hog fuel for a major part of their energy needs). Beyond the forest industry, several independent power plants generate electricity from forest biomass. For example, utilization of wood biomass for bioenergy production has recently increased dramatically in northwestern Ontario (NWO) with four major energy plants with estimated biomass demand of about 2.21 million green tonnes [9]. Currently, three major combined heat and power(CHP) plant developments in NOW have the potential to use forest biomass feedstock for bioenergy production. These include Resolute Forest Products Thunder Bay CHP Plant (ABTB), Resolute Forest Products Fort Frances CHP Plant (ABFF) and Domtar Dryden CHP Plant (DDPP) with different levels of electrical and thermal power production capacities. The Atikokan Generating Station (AGS), another power plant in NWO, is currently being converted to use forest biomass feedstock instead of coal. Its power generating capacity is 230 MW_e, with a plan to run at 10% capacity [10]. With the development of these biomass based bioenergy plants, the entire bioenergy system in the NWO will generate major socio-environmental consequences in terms of reducing GHG emissions and stabilizing the economy of many small rural communities.
The two types of woody biomass used for this study are: FHR - forest harvest residue which includes tops and branches and wood left after stand harvesting; and UUW - un/under-utilized wood which includes un-harvested tree species that are not commercially important for timber. These biomass sources have variable costs and qualities in terms of thermal value, moisture content and ash content. A power plant manager can have various biomass quality goals as well as the cost target so that the plant can be run cost effectively. This kind of decision problem in a biomass supply chain can best be handled by using GP modeling technique. However, we found very few studies on modeling the wood biomass for bioenergy supply chains in Canada that consider multiple goals in terms of biomass qualities. Most of the existing studies focus mainly on optimizing harvesting and transportation of raw material for forest products industries from forest management units (FMUs) to the processing facilities [11, 12].
Our previous study [13] used the goal programming approach to model the cost and quality goals of varied sources of wood biomass. However, that study did not used the engineering equations to endogenize the biomass requirements once the biomass qualities change. The biomass requirements for each of the four plants were fixed instead of biomass being the function of amount of energy productions. This was modeled such as the information available at the time was constrained. The main objective of this study is, therefore, to improve upon our previous paper [13] with updated database in order to make the model more practical and policy relevant at operational level for the bioenergy producers. We model the biomass supply chains by using the engineering equation to first decide the amount of biomass (with various physical characteristics) required as function of amount of energies to be produced by each power plant. And then the GP model optimize the amount of 'right' type of biomass to be harvested from the forest cells. We analyze different scenarios relating to various biomass quality goals and technical efficiency change, and their impacts on procurement costs per green tonne and for total biomass for each of the power plant. The variations in quality characteristics (thermal value, moisture and ash contents) of two types of biomass distributed over the productive forest cells derive the total biomass requirement to produce bioenergy at the capacity levels of each power plant. The end results of the entire modelling process are to get to the estimates of costs structures with respect to different sets of goals/targets by different model scenarios. Though the modelling technique in this study is developed for optimizing the biomass supply chains pertinent to the Ontario bioenergy producers, this

can be easily adapted to the similar cases elsewhere. Hence, the importance of this work is of global nature in the era of emerging sustainable energy development around the globe in order to mitigate the GHGs emissions.

2. Data and method

2.1 Study area and data

The study area consists of 18 forest management units (FMU) west of Lake Nipigon in NWO where four power plants are running with biomass as their feedstock (Figure 1). The NWO study area is 167,184 km^2, with an annual average harvest of 60,867 ha (2002-2009) which is 0.61% of the productive forest area per year. GIS data related to forest areas and depleted forest for the period 2002-2009 were collected from Land Information Ontario, Sustainable Forest Licence (SFL) holders, and consultant companies in the formats of Shapefile and Geodatabase. The entire modeling system undergoes into two steps. First a database relating to the logistics costs to transport the biomass from each forest cells to the four power plants are estimated by road network optimization model [9]. Then the database thus obtained is used in GP model to estimate the optimal costs structures for biomass supply chains for each of the power plant.

Figure 1. Map of study area (Source: [14])

In the first step, the original vector data is converted to raster and finally to spatial database text files for the entire research area using Arc GIS software. Three main spatial layers (land use, forest depletion and cost layers) were prepared on a raster grid size of 1 km×1 km (1 km^2), and this study examines 20,315 productive forest cells where the timber harvesting activities occurred from 2002 to 2009. The detailed methodology for estimating forest harvest residue and un/under-utilized biomass availability for all 20,315 forest depletion cells is described in Alam et al.[9].

The important estimates of techno-economics parameters used in the GP model are as mentioned in Table 1.

Table 1. Estimates of parameters used in the model

Descriptions	Unit	Estimates	Remarks
Harvesting and processing costs (FHR)	CAD/gt	26	[15]
Harvesting and processing costs (UUW)	CAD/gt	31	[15]
Fixed cost due to load/unload overhead	CAD/gt	4.85	[9]
Electrical Capacity of ABTB Power plant	MWe	61	Power Plant data
Thermal Capacity of ABTB Power plant	MWth	16	Power Plant data
Electrical Capacity of ABFF Power plant	MWe	50	Power Plant data
Thermal Capacity of ABFF Power plant	MWth	61	Power Plant data
Electrical Capacity of DDPP Power plant	MWe	30	Power Plant data
Thermal Capacity of DDPP Power plant	MWth	37	Power Plant data
Electrical Capacity of AGS Power plant	MWe	23	Power Plant data
Harvesting factor*	% of BM	67	[4]
Number of forest depletion cells **	No	20,315	New estimate

Note: CAD = Canadian Dollar, BM = Biomass, gt = green tonne, yr = year
**The percentage of total biomass that can be extracted from the given area*
***1kmX1km grid of depletion cells in harvesting sites*

Descriptive statistics of biomass qualities for all 20,315 forest depletion cells, which are updated version from our previous study, are as depicted in Table 2. This helps to get the initial target levels for each of the quality goals. We select six quality characteristic related goals, namely moisture and ash contents of both forest biomass types (four goals) and thermal value of each forest biomass type (two goals) that give us a fairly good account of biomass quality information to feed into the GP model. Although estimating the values of all these parameters for the 20,315 individual forest depletion cells is a daunting task, we use [16] and Hosegood [17] to approximate the estimates of these parameters.

Table 2. Descriptive statistics of biomass quality and target levels by scenarios (n=20,315)

	Moisture content FHR (% gw basis)	Thermal value FHR (GJ/ODt)	Ash content FHR (%)	Moisture content UUW (% gw basis)	Thermal value UUW (GJ/ODt)	Ash content UUW (%)
Mean	**33.48**	**18.92**	**1.59**	**39.89**	**16.99**	**1.87**
Minimum	31.97	18.50	1.30	30.15	15.30	1.00
Maximum	39.96	20.00	3.00	55.34	18.50	2.50
Standard Deviation	2.23	0.47	0.56	9.66	1.29	0.45
Initial goals	34.00	19.00	2.00	40.00	17.00	2.00
10 % relaxation	37.40	17.10	2.20	44.00	15.30	2.20
10 % increase in efficiency	34.00	19.00	2.00	40.00	17.00	2.00
UUWAGS	No FHR for AGS only, but other plants are using it.			40.00	17.00	2.00

Note: FHR = Forest Harvest Residue, UUW = Un/under-utilized wood biomass, gw= green weight

2.2 GP model for biomass procurement

In the past, the multi-criteria decision making models, which is a common name given to all relevant models of multi-objective decision model (MODM) techniques and other related simulation models, have been applied to solve complex production and management problems in natural resources management fields including forestry [1, 18]. The goal programming model, a variant of MODM, is found to be more useful in production systems analysis because it can handle continuous problems that involve the optimisation of several simultaneous objectives. A brief sketch of GP model has been presented in [13].

The GP model is specified as minimizing the sum of positive and negative deviations from the target levels as appropriate depending upon the problem being studied. In our model, we have minimized the positive deviations of cost and heat values of FHR and UUW for each forest depletion cell and negative deviations of moisture and ash contents of two types of biomass – FHR and UUW. The formal GP model is specified as below.

$$\textbf{Minimize } Z = p1 + \sum_{j=1}^{20315} p_{2j} + \sum_{j=1}^{20315} p_{3j} + \sum_{j=1}^{20315} n_{1j} + \sum_{j=1}^{20315} n_{2j} + \sum_{j=1}^{20315} n_{3j} + \sum_{j=1}^{20315} n_{4j} \quad (1)$$

Subject to

$$\sum_{i=1}^{4}\sum_{j=1}^{20315}(XRES_{ij}(PRC + TC_{ij})) + \sum_{i=1}^{4}\sum_{j=1}^{20315}(XUNB_{ij}(PUC + TC_{ij})) - p1 \le C \quad (2)$$

$$XRES_{ij}*(MCR_j+n_{1j}) \le g_1*XRES_{ij} \quad (3)$$

$$XUNB_{ij}*(MCU_j+n_{2j}) \le g_2*XUNB_{ij} \quad (4)$$

$$XRES_{ij}*(AshR_j+n_{3j}) \le g_3*XRES_{ij} \quad (5)$$

$$XUNB_{ij}*(AshU_j+n_{4j}) \le g_4*XUNB_{ij} \quad (6)$$

$$XRES_{ij}*(TVR_j-p_{2j}) \ge g_5*XRES_{ij} \quad (7)$$

$$XUNB_{ij}*(TVU_j-p_{3j}) \ge g_6*XUNB_{ij} \quad (8)$$

$$\sum_{i=1}^{4} XRES_{ij} \le ARES_j \quad (9)$$

$$\sum_{i=1}^{4} XUNB_{ij} \le AUNB_j \quad (10)$$

$$THE_l = TCAP_i * 1000 * 350 * 24 * 3.6 \quad (11)$$

$$EE_l = ECAP_i * 1000 * 350 * 24 * 3.6 \quad (12)$$

$$TE_l = THE_i + EE_i = DB_i \quad (13)$$

$$XTB_i = \sum_{j=1}^{20315} XRES_{i,j} * (TVR_j - 0.2164 * MCR_j) + \sum_{j=1}^{20315} XUNB_{i,j} * (TVU_j - 0.2164 * MCU_j) \quad (14)$$

$$XTB_i \ge DB_i \quad (15)$$

$$XRES_{ij}, XUNB_{ij} \text{ and deviations} \ge 0 \quad (16)$$

where, p_1, p_{2j} and p_{3j} are the positive deviations from the total cost target (C) thermal values target of FHR (g_5) and UUW (g_6) for each forest depletion cells, respectively;
n_{1j}, n_{2j} and n_{3j} and n_{4j} are negative deviations from target/goal levels of moisture contents of FHR (g_1) and UUW (g_2), and ash contents of FHR (g_3) and UUW (g_4) for each forest depletion cell j, respectively;
PRC is processing (harvesting and grinding/chipping) cost ($\$\cdot gt^{-1}$) of FHR at roadside;
PUC is processing (harvesting and grinding/chipping) cost ($\$\cdot gt^{-1}$) of UUW at roadside;
DB_i is annual forest biomass demand converted to energy units (GJ) of power planti ;
$ARES_j$ is annual technical availability (gt) of FHR in forest depletion cell j;
$AUNB_j$ is annual technical availability (gt) of UUW in forest depletion cell j;
$TC_{ij}$is biomass transportation cost ($\$\cdot gt^{-1}$) from the jth forest depletion cell to the ith power plant including loading/unloading overhead;
$XRES_{ij}$ is amount of annual FHR harvested (gt) from the jth forest depletion cell for the ith power plant;
$XUNB_{ij}$ is amount of annual UUW harvested (gt) from the jth forest depletion cell for the ith power plant;
and XTB_i is annual forest biomass converted into energy units (GJ) to be brought in the ith power plant.

THE_i and EE_i are the total annual thermal and electrical energy to be produced by each plant and $TCAP_i$ and $ECAP_i$ are the thermal and electrical capacity of each plant (cf. Table 2). We assume to run the power plant for 350 days in a year and 24 hours a day. The value of 3.6 in eqns. (11, 12) is the amount of GJ energy in one kilo watt hour of power production. The parameter and equation to estimate the energy content of each type of biomass from each forest cells, which is a function of thermal value and the moisture contents (that vary among the biomass types in each depletion cells) are taken from [19].
In the above model specification, equations 11-14 are the engineering equations, which are new to this paper and that make the model more practical. Equations 2-8 are the goal constraint equations where the right hand side scalars are chosen as the goal or target, each representing the decision maker's objectives to be met with some relevant deviations by selecting the optimal choice variables $XRES_{ij}$ and $XUNB_{ij}$. The cost target is selected based on the total cost obtained from the linear programming (LP) model without putting any goal constraints with the same technical constraints, **benchmark scenario**. Different quality targets are selected based on four goal set scenarios as mentioned above. The Initial Goals scenario selects the ceiling of mean values of each of the six biomass quality characteristics as shown in Table 2. Equations (9) and (10) represent harvesting constraints, suggesting the annual harvest for each type of biomass should not exceed the available biomass in each forest depletion cell. Equation (15) constrains the total amount of energy (GJ per annum) derived from the optimal level of forest biomass harvested for i^{th} power plant should at least meet the energy production (GJ per annum) level of that plant. The general algebraic modeling system (GAMS) optimization software has been used to solve this complex problem.

2.3 Model scenario

Four different model scenarios in terms of goals sets and technical efficiency are used in order to test the various situations of multiple objectives of biomass procurement decisions for the four power plants. Before running the GP models, we ran the benchmark LP model with total cost (sum of the costs for all four plants) minimization objective with usual constraints without any quality targets (no target level set on moisture contents, thermal values and ash contents). Results of the benchmark LP model gave an idea of cost goal and results for comparisons of the biomass procurement costs for different scenarios in the GP model.
The first scenario, **Initial Goals,** includes a goal set with biomass qualities (MC and ash contents, and thermal values) having ceilings of mean values of the corresponding variables (Table 2). This scenario is set to establish a baseline goals relating to the quality of biomass, where the power plant manager may want to have higher quality feedstock as defined by these threshold goals/targets. This goal set will introduce constraints in the model as the biomass to be harvested should have MC and ash contents not more than the targets and the thermal value should be at least equal to the target. The second scenario with **10% Relaxation** deals with goal sets that relax target values for biomass qualities (the target values are as shown in Table 2) from the Initial Goals scenario by 10%. These two scenarios are run to test the sensitivity of the changes in goal levels (targets) to the costs structures of biomass procurement problems, which exemplify the importance of biomass qualities in biomass supply chains for bioenergy production. The third scenario, **10 % increase in conversion efficiency**, tests the sensitivity of costs and biomass volume to be harvested under new technological era. The fourth scenario is to use only un/under-utilized biomass for the AGS (**UUWAGS** scenario) power plant as it is planning to use UUW to produce pellets for power production in the future. Due to strict ash content requirements (<1% ash) of high quality pellets, only un/under-utilized biomass can be used for this purpose as logging residues would result in excessive bark and thus more ash content. In the last two scenarios, other quality goals remain same as in the **Initial Goals** scenario.

3. Results and discussion

First, the results obtained from the benchmark LP model scenario suggest for an optimal solution for supplying forest biomass feedstock (both FHR and UUW) from forest depletion cells in the case study area to the four power plants on an annual basis by minimizing the total annual harvesting, processing and transportation costs of biomass subject to the availability of forest biomass in each depleted forest cell. The LP model selects the optimal harvest cells for each power plant as depicted in Figure 2. From the figure we can see that Dryden CHP plant is located near the middle of the research area, and there are many forest cells closer to this power plant along with a denser network of higher class and straighter roads that imply a lower per gt procurement cost, location matters in terms of per unit biomass procurement cost (cf. Table 3).On the other hand, Abitibi Bowater Thunder Bay CHP plant is located in the southeastern part of research area, where there is no competing power plant on its northern, eastern and southern sides in the research area. The other two power plants (ABFF and AGS) need to compete more for the forest biomass as they are located close to each other in the research area (Figure 1). Here the biomass procurement costs per gt depend on two factors, namely, location and competition as per the total volume requirements for each nearby power plants. A final note on the results of both the LP and GP models with various scenarios - this modeling approach gives the optimal network of selected forest cells to harvest both FHR and UUW biomass to meet the biomass requirements for the four power plants.

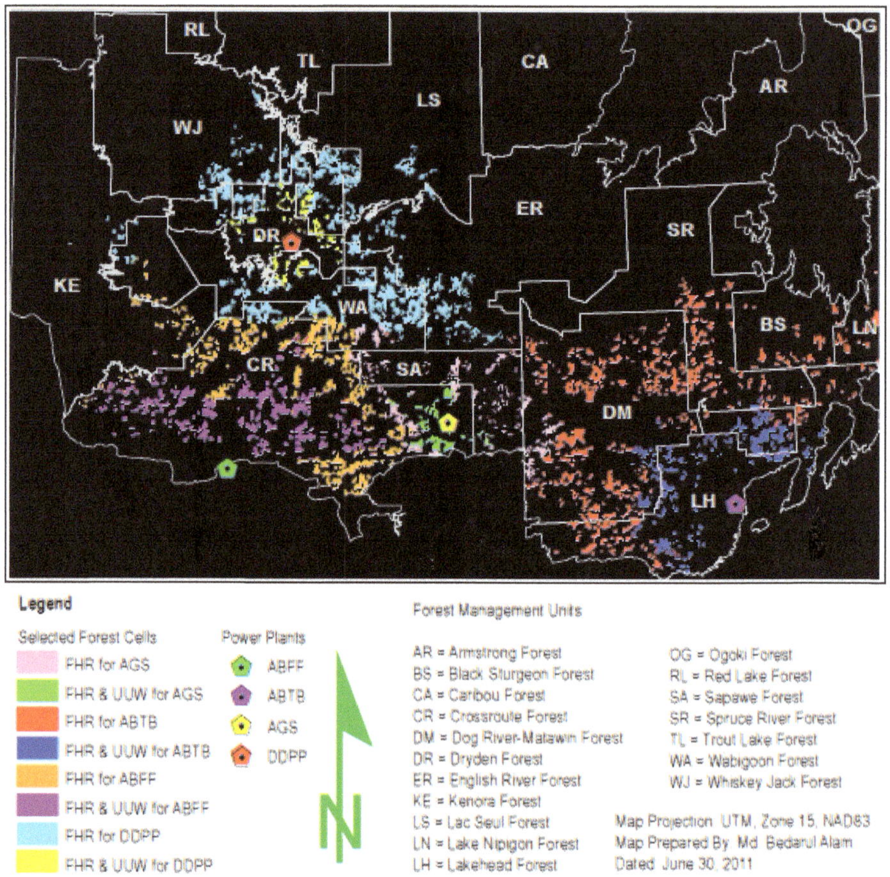

Figure 2. Optimal harvest of FHR and UUW biomass for four power plants with benchmark LP model

The models choose the most efficient forest cells out of 20,315 cells to meet the demands for all the four plants developing different networks of distribution of forest harvest cells along with the volume harvested for each scenario. We have developed a map producing application in Visual Basic, which takes the optimal cells generated by either LP or GP model as an input, and produce the visual map as shown in Figure 2. These results are very important in developing a decision support system for biomass supply chain for bioenergy productions anywhere in the world, where the similar datasets are available.

Table 3 presents the average cost estimates of per green tonne harvestin relation to various goals set and technical efficiency increase scenarios by power plants. While comparing the per gt cost in each scenario with respect to the benchmark results, we can see that per unit costs have increased significantly as we are looking for high quality biomass, thus end up paying higher per unit prices. However, the per gt costs have decreased in 10% relaxation scenario (column 3, Table 3) when compared with the Initial Goals scenario (column 3, Table 3) as this scenario allows relatively lower quality biomass than the Initial goals scenario. We can also observe that DDPP being suitably located around the dense forest areas with no nearby competing power plant has the lowest per gt cost in all scenarios. On the other hand, AGS power plant which has the lowest total volume of biomass requirement has the second least cost option to produce bioenergy, except in the UUWAGS scenario.

One interesting result to ponder is that the per gt cost for AGS in the UUWAGS scenario is significantly higher compared with other scenario because the goal of the AGS plant manager is only to procure the hardwood biomass for better quality pellet production in this case. Thus, Atikokan plant has to pay more per gt, while at the same time other competing plants have an option to grab the low cost FHR biomass previously used by AGS, thereby reducing their unit procurement costs (compare the costs with Initial Goals scenario in Table 3). These results can be useful for power plant managers to evaluate the effect of trade-offs between qualities and the costs, as well as the impacts on costs of the strategy of other competing plants. The 10% increase in conversion efficiency scenario shows the lower per gt cost as compared to the Initial goals scenario. The modeling approach developed for this study can accommodate other relevant goals or technical scenarios as well as need be to analyze the sensitivity of changing parameters on the procurement costs and forest cells network as shown in Figure 2.

Table 3. Biomass procurement costs in each scenario by power plants (CAD/gt)

Power plants	Benchmark	Initial Goals	10% Relaxation	10% increase in efficiency	UUWAGS
ABTB	38.11	41.16	38.41	40.85	41.15
ABFF	38.77	40.21	38.99	39.88	39.99
DDPP	35.74	37.57	36.00	37.34	37.54
AGS	36.13	37.46	35.96	37.21	39.11

The more important results of this improved modelling work can be observed while looking at the total volume and thereby total costs required to produce the same amount of bioenergy as discussed in our previous paper [13]. The total cost estimates are not presented here as it is obvious from the total biomass volume, which is directly proportional to the costs. Once engineering equations are introduced in the model, and with the introduction of goals constraints in the biomass qualities, we require the less total volume of biomass to produce the same amount of bioeenrgy. Table 4 summarizes the results in terms of total biomass volume required in each scenario for each power plant. We can see the drastic reduction of volume of biomass with respect to the benchmark LP scenario as we are using higher quality biomass in our GP model scenarios. The figures in the parentheses in Table 4 show the percentage reduction of total biomass requirement from the benchmark case for each power plant by model scenarios. These are important results of this study while compared with the results of [13] where total biomass volumes were kept constant for each power plant in all the model scenarios. The most drastic reduction of total biomass requirement is found in the technological improvement scenario, where about 33% reduction is observed. In rest of the scenarios, we can observe the reduction in biomass harvest in the range of 25%-27% with respect to the benchmark case.

Table 4. Total volume of biomass procurement in each scenario by power plants (gt/yr)

Power plants	Benchmark	Initial Goals	10% Relaxation	10% increase in efficiency	UUWAGS
ABTB	730,000	536,642 (-26.49)	541,051 (-25.88)	487,026 (-33.28)	536,404 (-26.52)
ABFF	800,000	586,713 (-26.66)	589,268 (-26.34)	532,440 (-33.45)	584,889 (-26.89)
DDPP	480,000	351,669 (-26.74)	355,162 (-26.01)	319,276 (-33.48)	351,548 (-26.76)
AGS	200,000	147,163 (-26.42)	148,204 (-25.90)	133,653 (-33.17)	149,361 (-25.32)

With the extension and improvement upon [13], the present GP model accounts for impacts of higher quality biomass goals and potential improvement in technology on procurement cost structures both in terms of per gt and total volume by model scenarios. By using higher quality biomass as set in GP scenario, it is found that total wood biomass harvested is reduced significantly in the bioenergy supply chains, even though the delivered per unit costs are higher. The trade-off between quality and cost has been demonstrated in this model effort, where the total biomass demand for each power plant has been endogenized through engineering process equations. The impacts of changing goals are found to be inter-dependent between plants; changing a goal in one plant affects costs at other plants.

There are many important ongoing discussions about increasing the share of renewable energy into the entire energy portfolio due to various sustainability implications in the recent past. Lund [3] brings three aspects of sustainable energy development strategy, namely, energy saving on the demand side, efficiency improvements in the supply side, and replacement of fossil-based fuels by renewable energies. Our findings can shed light on the energy efficiency scenario, and how cost-effectively renewables with biomass sources can be produced in order to fulfill the rural energy demands. This study further aids in the literature of biomass supply chain modelling with multi criteria decision-making approach by explicitly capturing the effects of biomass qualities as goals sets on the procurement costs for multiple power plants in NWO region of Ontario. This differs with the past studies in this area with similar paradigm [8, 11, 12]. Regarding technological development, we would expect an improved efficiency of CHP and power only plants with the use of higher quality biomass (as discussed in this study) in the future [20]. The trade-off between increase in conversion efficiency and the supply chain costs can be modeled by using this approach.

4. Summary and conclusions

Biomass procurement problems for four bio-energy power plants in NWO have been modelled using GP method with engineering equations and improved dataset. The approach taken in this study can greatly help in day to day operational planning problems of biomass supply chain management for bioenergy production in general. The biomass currently utilized is mainly mill and logging residues, but in future there will be a need to utilize under-utilized species and un-merchantable standing trees to meet the growing demands for biomass. All of the biomass sources have variable costs and qualities, and potential impact on other wood users (e.g., utilizing standing trees for energy would compete with other users). In this study, six quality characteristics goals, namely moisture and ash contents and thermal values of both forest biomass types are taken into account, which give a fairly good account of biomass quality information to feed into the GP model. After establishing the cost and physical quality goals characteristics we ran four different scenarios with different sets of quality goals and technological improvement situation in order to analyze the sensitivity of these scenarios on the procurement costs. The results are contrasted with the benchmark LP model, which give very good policy and operational indications for biomass supply chain management on part of the bioenergy producers.

This work represents an extension and improvement upon Upadhyay et al. [13]that helps us to find the practical real time solutions for impacts of higher quality biomass goals and potential improvement in technology on biomass procurement cost structures both in terms of per gt and total volume by model scenarios. It is found that total wood biomass harvested is reduced significantly in the bioenergy supply chains, even though the delivered per unit costs are higher. The trade-off between quality and cost has been demonstrated in this modeling effort, where the total biomass demand for each power plant has been endogenized through engineering process equations. The impacts of changing goals are found to be inter-dependent between plants; changing a goal in one plant affects costs at other plants

Conflict of interests disclosure

The authors declare that there is no conflict of interests regarding the publication of this article.

References

[1] Upadhyay, T.P., B. Solberg, and P.L. Sankhayan, Use of models to analyse land-use changes, forest/soil degradation and carbon sequestration with special reference to Himalayan region: A review and analysis. Forest Policy and Economics, 2006.9: p. 349-371.

[2] IPCC, Climate Change 2007: Mitigation. Contribution of Working Group III to the Fourth Assessment Report of the Intergovernmental Panel on Climate Change [B. Metz, O.R. Davidson, P.R. Bosch, R. Dave, L.A. Meyer (eds)], 2007.Cambridge University Press, Cambridge, United Kingdom and New York, NY, USA., XXX pp.

[3] Lund, H., Renewable energy strategies for sustainable development. Energy, 2007. 32: p. 912-919.

[4] Gan, J., and C.T. Smith, Availability of logging residues and potential for electricity production and carbon displacement in the USA. Biomass and Bioenergy, 2006.30: p. 1011-1020.

[5] Ballarin, A., D. Vecchiato, T. Tempesta, F. Marangon, and T. Troiano, Biomass energy production in agriculture: A weighted goal programming analysis. Energy Policy, 2011.39: p. 1123-1131.

[6] Rauch, P., and M. Gronalt, The Terminal Location Problem in the Forest Fuels Supply Network. International Journal of Forest Engineering,2010.21: p. 32-40.

[7] Tahvanainen, T., and A. Perttu, Supply chain cost analysis of long-distance transportation of energy wood in Finland. Biomass and Bioenergy, 2011. 35: p. 3360-3375.

[8] Tenerelli, P., and S. Carver, Multi-criteria, multi-objective and uncertainty analysis for agro-energy spatial modeling. Applied Geography, 2012. 32: p. 724-736.

[9] Alam, M.B., R. Pulkki and C. Shahi, Road network optimization model for supplying wood biomass feedstock for energy production in northwestern Ontario. The open forest science Journal, 2012. 5:p. 1-14

[10] Ontario Power Generation Inc. (OPG), Atikokan Generation Station Biomass Repowering Project. Ontario Power Generation Inc.). 2010.http://www.opg.com/power/thermal/AtikokanFactSheet1009.pdf, [Accessed May 25, 2013].

[11] Pulkki, R. Role of supply chain management in the wise use of wood resources, Southern African Journal, 2001.191:p. 89-95.

[12] Chauhan, S.S., J.M. Frayret, and L. LeBel, Multi-commodity supply network planning in the forest supply chain. European Journal of Operational Research, 2009. 96: p. 688–696.

[13] Upadhyay, T.P., M.B. Alam, R. Pulkki, C. Shahi and M. Leitch, Integrating goal programming in biomass procurement for bio-energy value chains J-FOR, 2011. vol. 1, no. 2, pp. 45–53.

[14] Ontario Ministry of Natural Resources (OMNR). Map of Ministry office locations. http://www.mnr.gov.on.ca/en/ContactUs/1ColumnSubPage/STEL02_179001.html, 2014. [Accessed May 25, 2014].

[15] Ontario Ministry of Energy (OME). An Assessment of the Viability of Exploiting Bio-Energy Resources Accessible to the Atikokan Generating Station in Northwestern Ontario. 2006.A consultancy report prepared by Forest BioProducts Inc. Sault Ste. Marie, Ontario, Canada.

[16] Gautam, S., Assessment of Fuel Quality Changes during Storage of Biofibre and Its Effect on Cost. M.Sc. Thesis, 2010. Lakehead University, Thunder Bay, Ontario.

[17] Hosegood, S. Fuel Characteristics of Seven Northwestern Ontario Tree Species. M. Sc. Thesis, 2010.Lakehead University, Thunder Bay, Ontario.

[18] Romero, C., and T. Rehman, Multiple Criteria Analysis for Agricultural Decisions. 1989. Amsterdam: Elsevier.

[19] Woodenergy.net. http://www.woodenergy.ie/frequentlyaskedquestions/, 2013.[Accessed May 27, 2013].

[20] U. S. Environmental Protection Agency (EPA). Biomass Combined Heat and Power Catalog of Technologies (www.epa.gov/chp). 2007.USA.

Linear irreversible thermodynamic performance analyses for a generalized irreversible thermal Brownian refrigerator

Zemin Ding[1,2,3], Lingen Chen[1,2,3], Yanlin Ge[1,2,3], Fengrui Sun[1,2,3]

[1] Institute of Thermal Science and Power Engineering, Naval University of Engineering, Wuhan 430033, China.
[2] Military Key Laboratory for Naval Ship Power Engineering, Naval University of Engineering, Wuhan 430033, China.
[3] College of Power Engineering, Naval University of Engineering, Wuhan 430033, China.

Abstract
On the basis of a generalized model of irreversible thermal Brownian refrigerator, the Onsager coefficients and the analytical expressions for maximum coefficient of performance (COP) and the COP at maximum cooling load are derived by using the theory of linear irreversible thermodynamics (LIT). The influences of heat leakage and the heat flow via the kinetic energy change of the particles on the LIT performance of the refrigerator are analyzed. It is shown that when the two kinds of irreversible heat flows are ignored, the Brownian refrigerator is built with the condition of tight coupling between fluxes and forces and it will operate in a reversible regime with zero entropy generation. Moreover, the results obtained by using the LIT theory are compared with those obtained by using the theory of finite time thermodynamics (FTT). It is found that connection between the LIT and FTT performances of the refrigerator can be interpreted by the coupling strength, and the theory of LIT and FTT can be used in a complementary way to analyze in detail the performance of the irreversible thermal Brownian refrigerators. Due to the consideration of several irreversibilities in the model, the results obtained about the Brownian refrigerator are of general significance and can be used to analyze the performance of several different kinds of Brownian refrigerators.

Keywords: Linear irreversible thermodynamics; Generalized model; Irreversible thermal Brownian refrigerator; COP.

1. Introduction

In macroscopic systems, thermal fluctuations are not directly observable and their influences on the system can be ignored. However, when the system is small enough, thermal fluctuations become the major driving force of the system and can no longer be ignored. Brownian motor is a typical device which can rectify thermal fluctuations to produce directed motion [1-4]. Nowadays, people are trying to invent miniature and nanoscale devices which help to utilize energy resources in the microscopic scale. And the Brownian motor systems have attracted much interest due to their importance in achieving microscopic energy conversion. Actually, as to the Brownian motors, there are a variety of nonequilibrium driving forces besides the thermal fluctuations, such as external modulation of an underlying potential [5, 6], external force [7-9], chemical potential differences [10, 11] and so on. So far, thermal Brownian motor is the most extensively studied one among the different kinds of Brownian motors.

In the analyses of thermal Brownian motors, the thermodynamics performance is an important factor which has been analyzed by many authors [1, 12-15]. And the central issues as to the thermodynamics performance are the mechanism and efficiency of energy conversion of the Brownian motor systems. By noting the fact that the strict thermodynamic definition of efficiency is external load-dependent and is not adequate for microscopic energy conversion systems, Derényi *et al.* [16] proposed a load-independent new definition of generalized efficiency for the microscopic engines and analyzed its application to a Brownian heat engine. Meanwhile, many researchers are focusing on the efficiency performance of Brownian motors following the classical thermodynamics theory [13, 17-22].

In the past decades, the theory of finite time thermodynamics (FTT) has made tremendous progresses in the performance analyses of conventional macroscopic and quantum energy conversion systems [23-34]. Optimum performance and the transmission losses between the heat reservoirs in energy conversion systems are two major consideration factors in FTT. Parrondo and de Cisneros [35] pointed out that the strategies and principles developed in FTT theory are also valuable for the studies of Brownian motors. So far, the FTT theory has already been applied to analyze performance of Brownian motor systems, such as thermal Brownian heat engines, refrigerators and heat pumps [14, 36-42], and many significant results have been obtained.

Linear irreversible thermodynamics (LIT) is a powerful tool for studying the performance of linear processes and coupled phenomena, such as thermodiffusion, thermoelectric and thermomagnetic effects [43, 44]. In a long time, the LIT theory is limited to study the performance of isothermal energy conversion systems. Recently, Van den Broeck [45] derived the efficiency at maximum power of a heat engine using the LIT theory, and found that the efficiency at maximum power is equal to Novikov-Chambadal-Curzon-Ahlborn (NCCA) efficiency which is one of the most important results obtained in FTT [46-48]. In the derivation of NCCA efficiency in FTT, an endoreversible approximation was used. However, Van de Broeck had shown that in the frame of LIT theory, NCCA efficiency is a fundamental result obtained without approximation. Van den Broeck's work [45] also paves the way for analyzing the nonisothermal heat engines using the theory of LIT. Later, Jiménez de Cisneros *et al.* [49] extended the proposal of Van den Broeck [45] to refrigeration cycle and derived the coefficient of performance (COP) at maximum cooling load which could be equivalent to the NCCA efficiency by using the theory of LIT. At the same time, some research work has been carried out for conventional energy conversion systems within the realm of LIT, e.g., see Refs. [50-52].

Recently, due to its great significance in revealing the performance characteristic of energy conversion systems, the LIT theory has already been extended to the studies of Brownian motor systems. Van den Broeck and Kawai [53] first calculated the heat flow for an exactly solvable microscopic Brownian refrigerator model by using LIT and compared it with the results of molecular dynamics simulations. Gomez-Marin and Sancho [54] analyzed the tight coupling in a thermal Brownian motor and discussed the model acting as a refrigerator. They calculated the Onsager coefficients and showed how the reciprocity relation holds and that the determinant of the Onsager matrix vanishes. Gao *et al.* [55] calculated the Onsager coefficients and generalized efficiency of a thermal Brownian motor and discussed the influences of the main parameters on the performance of the system. Gao and Chen [56] later derived the Onsager coefficients and calculated the efficiency at maximum power of an irreversible thermally driven Brownian motor.

However, so far the LIT performance analysis for the Brownian motor systems mainly focuses on the system operating as a heat engine and the LIT performance of irreversible thermal Brownian refrigerators have been rarely investigated. Therefore, in this paper, a further step will be taken to analyze in detail the LIT performance of a thermal Brownian refrigerator. On the basis of a generalized irreversible thermal Brownian refrigerator model [42], the Onsager coefficients are derived, and the maximum COP as well as the COP at maximum cooling load of the refrigerator are analytically calculated. It is found that the heat leakage and the heat flow via the kinetic energy change have great influences on the performance of the refrigerator and when the two kinds of heat flows are not considered, the refrigerator becomes a perfectly coupled system. Moreover, the LIT performance of the refrigerator are compared with the FTT performance, and it is shown that theory of LIT and FTT can be used in a complementary way to analyze in detail the performance of the irreversible thermal Brownian refrigerators.

2. Performance characteristics and parametric optimum criteria of a Brownian [42]

A model of a generalized irreversible thermal Brownian refrigerator is shown in Figure 1 [42]. The refrigerator is modeled as moving Brownian particles in a viscous medium which is alternately in contact with a hot heat reservoir (at temperature T_H) and a cold heat reservoir (at temperature T_C) along the space coordinate. Additionally, a periodic sawtooth potential and an external force F are applied to the particles. In the figure, x is the horizontal axis of the coordinate, $\dot{N}_+$ and $\dot{N}_-$ are the numbers of forward and backward jumps per unit time, L_1 and L_2 are the widths of the left and right parts of the potential, and U_0 is the barrier height of the potential.

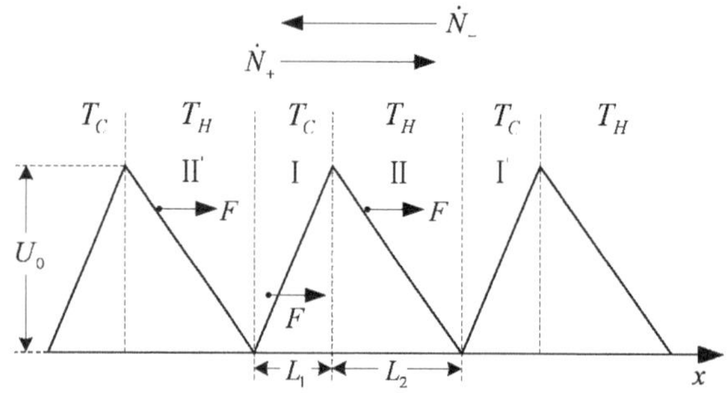

Figure 1. Schematic diagram of a thermal Brownian refrigerator

In the present model, both the irreversibility of heat leakage between two heat reservoirs and the irreversible heat flow via the change of kinetic energy of particles are considered. According to Refs. [42, 57], the rates of total heat absorbed from the cold reservoir ($\dot{Q}_C$) and released to the hot reservoir ($\dot{Q}_H$) of the Brownian refrigerator can be given by

$$\dot{Q}_C = (\dot{N}_+ - \dot{N}_-)(U_0 - FL_1) - k_B(\dot{N}_+ + \dot{N}_-)(T_H - T_C)/2 - C_i(T_H - T_C) \tag{1}$$

$$\dot{Q}_H = (\dot{N}_+ - \dot{N}_-)(U_0 + FL_2) - k_B(\dot{N}_+ + \dot{N}_-)(T_H - T_C)/2 - C_i(T_H - T_C) \tag{2}$$

where k_B is the Boltzmann's constant and it is taken to be unity for simplicity in the following calculations, C_i is the coefficient of heat leakage, $\dot{N}_+ = (1/t)\exp[-(U_0 - FL_1)/(k_B T_C)]$ and $\dot{N}_- = (1/t)\exp[-(U_0 + FL_2)/(k_B T_H)]$ are the numbers of forward and backward jumps for the Brownian particles per unit time with t a proportionality constant. The derivation of $\dot{N}_+$ and $\dot{N}_-$ is based on the assumption that the system is in a stable flow state and the rates of both forward and backward jumps are proportional to the corresponding Arrhenius' factor [57].

The heat flows between the two heat reservoirs defined by Eqs. (1) and (2) consist of three parts, respectively. The first is the heat flow caused by the particles' moving through the potential barrier, as shown by the first item in the right hand side of Eqs. (1) and(2). The second is the heat flow via the change of kinetic energy due to the particles' recrossing the boundary between the two regions, as shown by the second item in the right hand side of Eqs. (1) and (2). The influence of this kind of heat flow on the performance of Brownian motor systems was first considered by Derényi and Astumian [17] and Hondou and Sekimoto [18], and was later analyzed by many authors [14, 20-22, 38-41]. The last kind is the heat leakage between the two reservoirs, which is similar to the bypass heat leakage in conventional macroscopic heat engines and refrigerators [58, 59]. Velasco *et al.* [57] first considered the heat leakage in a Feynman's ratchet. Later this factor was extended to the studies of several kinds of thermal Brownian motors [41, 42, 55, 56].

In order to show more clearly the configuration of the system, the thermodynamic representation for the generalized model of irreversible thermal Brownian refrigerator is shown Figure 2, where P is the power input into the system, $\dot{Q}_L$ is the heat leakage between the two heat reservoirs, $\dot{Q}_1$ and $\dot{Q}_2$ are, respectively, the rates of heat absorbed from the cold reservoir and released to the hot reservoir by the system and are defined as $\dot{Q}_1 = (\dot{N}_+ - \dot{N}_-)(U_0 + FL_2) - k_B(\dot{N}_+ + \dot{N}_-)(T_H - T_C)/2$ and $\dot{Q}_2 = (\dot{N}_+ - \dot{N}_-)(U_0 - FL_1) - k_B(\dot{N}_+ + \dot{N}_-)(T_H - T_C)/2$.

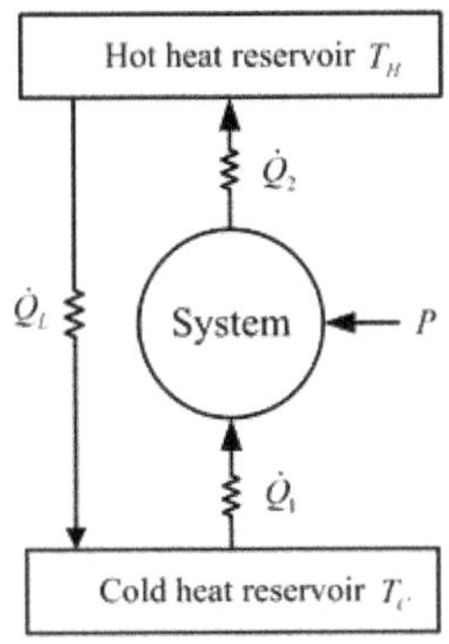

Figure 2. Thermodynamic representation for the generalized irreversible thermal Brownian refrigerator model

3. Onsager coefficients for the thermal Brownian refrigerator

According to the second law of thermodynamics, the entropy generation rate of the system can be expressed as

$$\sigma = \dot{Q}_H/T_H - \dot{Q}_C/T_C \tag{3}$$

Under the external force F, the Brownian particles will move from the cold part to the hot part and the refrigerator provides a cooling load $\dot{Q}_C$ with the power input P. In LIT theory, the sole requirement for the definition of thermodynamic forces and associated fluxes is $\sigma \geq 0$. And in the system, the external force F is the source of input power. Thus, one can consider a driving force $X_1 = F/T_H$ and a thermodynamic flux $J_1 = \dot{x}$, where x is a thermodynamically conjugate variable and the dot refers to the time derivative [45, 49], so that the power input is $P = F\dot{x} = J_1 X_1 T_H$. In the cold reservoir, the rate of heat $\dot{Q}_C$ is pumped at the cost of the input power P. Thus the thermodynamic force can be chosen as $X_2 = 1/T_H - 1/T_C$ with the corresponding flux $J_2 = \dot{Q}_C$. In the system, it is assumed that the temperature difference $\Delta T = T_H - T_C$ is small compared to T_H or T_C so that the driven force X_2 can be written as $X_2 = -\Delta T/T_H^2$.

In linear response regime, by following the LIT theory, the entropy generation rate can be expressed as

$$\sigma = (X_1, X_2)\begin{pmatrix} L_{11} & L_{12} \\ L_{21} & L_{22} \end{pmatrix}\begin{pmatrix} X_1 \\ X_2 \end{pmatrix} \tag{4}$$

where L_{ij} $(i, j = 1, 2)$ are the Onsager coefficients. Substituting Eqs. (1) and (2) into Eq. (3) and making some simplification following the rules in steady state ($F \to 0$ and $\Delta T \to 0$) gives

$$\begin{aligned} \sigma = {} & e^{-U_0/(k_B T_H)}(F/T_H)^2 (L_1 + L_2)^2/(k_B t) + (\Delta T/T_H^2)^2 [e^{-U_0/(k_B T_H)} U_0^2/(k_B t) + e^{-U_0/(k_B T_H)} k_B T_H^2/t \\ & + C_i T_H^2] - 2e^{-U_0/(k_B T_H)}(F/T_H)(\Delta T/T_H^2)(L_1 + L_2)U_0/(k_B t) \end{aligned} \tag{5}$$

One can obtain the Onsager coefficients for the Brownian refrigerator by comparing Eq. (5) with Eq. (4)

$$L_{11} = e^{-U_0/(k_B T_H)}(L_1 + L_2)^2/(k_B t) \tag{6}$$

$$L_{22} = e^{-U_0/(k_B T_H)} U_0^2/(k_B t) + e^{-U_0/(k_B T_H)} k_B T_H^2/t + C_i T_H^2 \tag{7}$$

$$L_{12} = L_{21} = -e^{-U_0/(k_B T_H)}(L_1 + L_2)U_0/(k_B t) \tag{8}$$

The Onsager coefficients offer a lot of information about the non-equilibrium thermodynamic properties of the Brownian refrigerator. It is easily found from Eqs. (6)-(8) that the reciprocity relation $L_{12} = L_{21}$ is fulfilled and the diagonal coefficients L_{11} and L_{22} are always positive. The coefficients L_{11} and $L_{12} = L_{21}$ are independent of the heat leakage and the heat flow via the kinetic energy change of the particle; while

L_{22} is closely dependent on the two kinds of irreversible heat flows. Especially, one can find that the relation $L_{11}L_{22} > L_{12}^2$ holds. This implies that the Brownian refrigerator model is inherently irreversible and there exists an entropy generation due to the existence of heat leakage and the heat flow via the kinetic energy change. Similar analyses for Brownian heat engine have been carried out in Refs. [54-56].

A dimensionless coupling strength q defined by Van den Broeck [45] can be introduced to analyze the non-equilibrium thermodynamics performance of the refrigerator

$$q = \frac{L_{12}}{\sqrt{L_{11}L_{22}}} \tag{9}$$

By substituting Eqs. (6)-(8) into Eq.(9), one can find that the absolute value of coupling strength $|q|$ is always smaller than unity.

If the heat flow via the kinetic energy change is ignored, the coefficient L_{22} becomes

$$L_{22} = e^{-U_0/(k_B T_H)} U_0^2 / (k_B t) + C_i T_H^2 \tag{10}$$

Eqs.(6), (8) and (10) can be used to study the performance of a Brownian refrigerator only considering the heat leakage. It is similar to the Brownian heat engine model where only heat leakage is considered [57]. In this condition, the coupling strength $|q|$ is smaller than unity. Similarly, if the heat leakage is ignored, the coefficient L_{22} becomes

$$L_{22} = e^{-U_0/(k_B T_H)} U_0^2 / (k_B t) + e^{-U_0/(k_B T_H)} k_B T_H^2 / t \tag{11}$$

Eqs.(6), (8) and (11) can be used to study the performance of a Brownian refrigerator only considering the heat flow via the kinetic energy change of the particle, which is just the model studied by Lin and Chen [38] and Ai *et al.* [20]. In this condition, the coupling strength $|q|$ is also smaller than unity.

Moreover, if both the heat leakage and the heat flow via the kinetic energy change of the particles are ignored, the coefficient L_{22} becomes

$$L_{22} = e^{-U_0/(k_B T_H)} U_0^2 / (k_B t) \tag{12}$$

Eqs.(6), (8) and (12) can be used to study the performance of a Brownian refrigerator considering neither the heat leakage nor the heat flow via the kinetic energy change of the particle, which is just the model considered by Asfaw and Bekele [19] and Gomez-Marin [54]. In this condition, the coupling strength $|q| = 1$, which implies that the relevant relation $L_{11}L_{22} = L_{12}^2$ is fulfilled and the refrigerator is built with the condition of tight coupling between fluxes and forces. The refrigerator operates in a reversible regime with zero entropy generation.

4. COP performance analyses

The LIT theory is based on the assumption of local equilibrium with the following linear relation between the fluxes and forces [43, 45]

$$\begin{aligned} J_1 &= L_{11}X_1 + L_{12}X_2 \\ &= [e^{-U_0/(k_B T_H)}(L_1 + L_2)^2 F / T_H + e^{-U_0/(k_B T_H)}(L_1 + L_2)U_0 \Delta T / T_H^2] / (k_B t) \\ &= e^{-U_0/(k_B T_H)}(L_1 + L_2)[(L_1 + L_2)F / T_H + U_0 \Delta T / T_H^2] / (k_B t) \end{aligned} \tag{13}$$

$$\begin{aligned} J_2 &= L_{21}X_1 + L_{22}X_2 \\ &= -e^{-U_0/(k_B T_H)}(L_1 + L_2)U_0 F / (T_H k_B t) - [e^{-U_0/(k_B T_H)} U_0^2 / (k_B t) \\ &\quad + e^{-U_0/(k_B T_H)} k_B T_H^2 / t + C_i T_H^2] \Delta T / T_H^2 \end{aligned} \tag{14}$$

The physical meaning of the diagonal coefficients can be obtained from the above two equations [45]. For $X_2 = 0$, i.e., $T_C = T_H$ and $\Delta T = 0$, one can find that $\dot{x} = J_1 = L_{11}F/T_H \ = e^{-U_0/(k_B T_H)}(L_1 + L_2)^2 F / (T_H k_B t)$, so that

L_{11}/T_H is the mobility of the refrigerator system in response to the external force F. For $X_1 = 0$, i.e., $F = 0$, one obtains $\dot{Q}_C = J_2 = -L_{22}\Delta T/T_H^2 = -[e^{-U_0/(k_B T_H)}U_0^2/(k_B t) + e^{-U_0/(k_B T_H)}k_B T_H^2/t + C_i T_H^2]\Delta T/T_H^2$, so that L_{22}/T_H^2 is a coefficient of thermal conductivity. The reciprocity relation $L_{12} = L_{21}$ describes the cross coupling of the system, which has been analyzed in many well-documented cases, such as the Seebeck, Peltier, and Thomson effects [43, 44].

If the motion of the system halts, i.e., $J_1 = \dot{x} = 0$, one has

$$X_1 = -\frac{L_{12}X_2}{L_{11}} = -\frac{e^{-U_0/(k_B T_H)}(L_1 + L_2)U_0\Delta T/(T_H^2 k_B t)}{e^{-U_0/(k_B T_H)}(L_1 + L_2)^2/(k_B t)} = \frac{U_0\Delta T}{T_H^2(L_1 + L_2)} \equiv X_1^{stop} \tag{15}$$

where X_1^{stop} is the stopping force. The external force F^{stop} corresponding to the stopping force is then

$$F^{stop} = X_1^{stop}T_H = U_0\Delta T/[T_H(L_1 + L_2)] \tag{16}$$

Using Eqs. (13) and (14), the power input (P) and COP (ε) of the Brownian refrigerator can be given, repectively by

$$P = J_1 X_1 T_H = L_{11}X_1^2 T_H + L_{12}X_1 X_2 T_H \tag{17}$$

$$\varepsilon = \frac{\dot{Q}_C}{P} = -\frac{T_C}{T_H - T_C}\frac{L_{12}X_1X_2 + L_{22}X_2^2}{L_{11}X_1^2 + L_{12}X_1X_2} \tag{18}$$

One may note that the expressions for the power input and COP for the Brownian refrigerator are the same as those for a conventional microscopic refrigerator [49]. Therefore, it is concluded that in the frame of LIT theory, the COP of different refrigerators have a unified expression while the expressions for the Onsager coefficients of the refrigerators may be different from each other.

4.1 Maximum COP

In Eq. (18), for fixed X_2, maximizing the COP with respect to X_1 by setting $d\varepsilon/dX_1 = 0$ yields [49]

$$X_1 = -X_2\sqrt{L_{22}/L_{11}}\,(1 + \sqrt{1 - q^2})/q \tag{19}$$

And the maximum COP is

$$\varepsilon_{max} = \frac{T_C}{T_H - T_C}\frac{q^2}{2\sqrt{1 - q^2} - q^2 + 2} = \frac{\varepsilon_C q^2}{2\sqrt{1 - q^2} - q^2 + 2} \tag{20}$$

Substituting Eqs. (6)-(8) into Eq. (20) yields the analytical expression of maximum COP for the generalized irreversible thermal Brownian refrigerator. One may note that when $|q| \to 0$, $\varepsilon_{max} \to 0$; and when $|q| \to 1$, i.e., both the heat leakage and the heat flow via the kinetic energy change of the particles are ignored, the maximum COP ε_{max} can attain the Carnot value ε_C.

4.2 COP at maximum cooling load

The efficiency at maximum power (for a heat engine), or the COP at maximum cooling load (for a refrigerator), is the most important parameter considered in FTT theory. The parameters can also be derived using the LIT. The efficiency at maximum power output for a Brownian heat engine in LIT has been analyzed in Ref. [56]. The COP at maximum cooling load for the irreversible Brownian refrigerator will be discussed in this section. Jiménez de Cisneros *et al.* [49] showed that in LIT theory the function J_2X_2 for a refrigerator is equivalent to $J_1X_1 \propto P$ for a heat engine. Therefore, the COP at maximum cooling load is equivalent to the COP when J_2X_2 is maximized.

For the Brownian refrigerator, maximizing $J_2X_2 = L_{12}X_1X_2 + L_{22}X_2^2$ with respect to X_2 for fixed X_1 by setting $d(J_2X_2)/dX_2 = 0$ gives

$$X_2 = -L_{12}X_1/(2L_{22}) \tag{21}$$

Substituting Eq. (21) into Eq. (18) yields the corresponding COP at maximum cooling load

$$\varepsilon_{J_2X_2} = \frac{T_C}{T_H - T_C}\frac{q^2}{2(2-q^2)} = \frac{\varepsilon_C q^2}{2(2-q^2)} \tag{22}$$

$\varepsilon_{J_2X_2}$ is equal to half of the Carnot COP times a q-dependent factor $q^2/(2-q^2)$. One may further note that Eq. (22) shares the same form as the COP at maximum cooling load for a conventional cascade refrigerator [49]; and the factor is the same as that for a Brownian heat engine optimized at maximum power condition. In the case of tight coupling, i.e., $|q| \to 1$, the COP at maximum cooling load is exactly equal to half of the Carnot COP.

4.3 Discussions

Comparing Eq. (22) with Eq.(20), one can find that the COP ($\varepsilon_{J_2X_2}$) at maximum cooling load is always smaller than the maximum COP (ε_{max}).

The theory of LIT and FTT can be used in a complementary way to analyze in detail the performance of the irreversible thermal Brownian refrigerators. The FTT performance of the generalized irreversible thermal Brownian refrigerator has already been extensively analyzed in Ref. [42]. The connection between the LIT performance and the FTT performance of the thermal refrigerator can be interpreted by the coupling strength q.

For conventional macroscopic refrigerator, if $|q| = 1$, the refrigerator becomes a perfectly coupled system in LIT, and meanwhile the coupled system corresponds to an endoreversible refrigerator in FTT where the sole irreversibility comes from the finite rate heat transfer [23-27, 60-62]; while for the thermal Brownian refrigerator, the perfectly coupled system with $|q| = 1$ corresponds to an refrigerator without considering the heat leakage and the heat flow via kinetic energy change in FTT where the sole irreversibility comes from the particle transport process.

For conventional macroscopic refrigerator, if $|q| < 1$, the refrigerator in LIT corresponds to an irreversible refrigerator with internal irreversibility and heat leakage besides the irreversibility of finite rate heat transfer; while for the thermal Brownian system, the Brownian refrigerator in LIT with $|q| < 1$ corresponds to an refrigerator considering the heat leakage or the heat flow via kinetic energy change, or both of them in FTT besides the irreversibility in the process of particle transport.

5. Conclusions

Based on a generalized irreversible thermally driven Brownian refrigerator model built in Ref. [42], the Onsager coefficients and the analytical expressions for maximum COP and the COP at maximum cooling load are derived by using the theory of linear irreversible thermodynamics in this paper. The COP performance of the refrigerator are analyzed and it is found that in the frame of LIT, the expressions of cooling load and COP of the refrigerator share the same forms as those for a conventional macroscopic irreversible refrigerator. The influences of heat leakage and the heat flow via the kinetic energy change on the LIT performance of the refrigerator are further analyzed and it is shown that they affect not only the COP performance but also the Onsager coefficients of the refrigerator. When the two kinds of irreversible heat flow are ignored, the Brownian refrigerator becomes a perfectly coupled system. Moreover, the results obtained by LIT theory are compared with those obtained by using the FTT theory. It is found that connection of the LIT and FTT performances for the refrigerator can be interpreted by the defined parameter, i.e., the coupling strength, and the theory of LIT and FTT can be used in a complementary way to analyze in detail the performance of the irreversible thermal Brownian refrigerators. The results obtained about the irreversible model are general and can be used to analyze the performance of several different kinds of Brownian refrigerators.

Acknowledgements

This paper is supported by The National Natural Science Foundation of P. R. China (Project Nos. 51306206 and 10905093) and the Natural Science Foundation of Naval University of Engineering (Project No. HGDQNJJ12013).

Nomenclature

C_i	coefficient of heat leakage (W/K)	x	direction of the coordinate
F	external force	*Greek symbols*	
J_1, J_2	thermodynamic fluxes	ΔT	temperature difference (K)
k_B	Boltzmann's constant (J/K)	ε	coefficient of performance (COP)
L_1, L_2	lengths of the left and right part of the potential	ε_C	Carnot COP
L_{ij} ($i, j = 1,2$)	Onsager coefficients	σ	entropy generation rate (W/K)
P	power input (W)	I, I'	cold regions of the ratchet
$\dot{Q}$	rate of heat flow (W)	II, II'	hot regions of the ratchet
q	dimensionless coupling strength	*Superscripts*	
R	cooling load (W)	*stop*	stopping force
T	temperature (K)	C	cold electron reservoir
t	a proportionality constant with a time dimension	H	hot electron reservoir
U_0	barrier height of the potential	*max*	maximum value
X_1, X_2	thermodynamic forces	+, −	forward and backward jumps of Brownian particle

References

[1] Astumian R D. Thermodynamics and kinetics of a Brownian motor. Science 1997;276(5314): 917-22.

[2] Astumian R D, Hänggi P. Brownian motors. Phys Today 2002;55(11):33-9.

[3] Reimann P. Brownian motors: noisy transport far from equilibrium. Phys Rep 2002;361(2/4): 57-265.

[4] Hänggi P, Marchesoni F. Artificial Brownian motors: controlling transport on the nanoscale. Rev Mod Phys 2009;81(1):387-442.

[5] Faucheux L P, Bourdieu L S, Kaplan P D, Libchaber A J. Optical thermal ratchet. Phys Rev Lett 1995;74(99):1504-7.

[6] Dinis L, Parrondo J M R, Cao F J. Closed-loop strategy with improved current for a flashing ratchet. Europhys Lett 2005;71(4):536-41.

[7] Marcelo M O. Forced thermal ratchets. Phys Rev Lett 1993;71(10):1477-81.

[8] Mateos J L. Chaotic transport and current reversal in deterministic ratchets. Phys Rev Lett 2000;84(2):258-61.

[9] Savel'ev S, Marchesoni F, Hänggi P, Nori F. Nonlinear signal mixing in a ratchet device. Europhys. Lett., 2004, 67(2): 179-185.

[10] Astumian R D, Bier M. Fluctuation driven ratchets: Molecular motors. Phys Rev Lett 1994; 72(11):1766-69.

[11] Astumian R D, Derényi I. A chemically reversible Brownian motor: application to kinesin and Ncd. Biophys J 1999;77(2):993-1002.

[12] Parrondo J M R, Blanco J M, Cao F J, Brito R. Efficiency of Brownian motors. Europhys Lett 1998;43(3):248-54.

[13] Sumithra K, Sintes T. Efficiency optimization in forced ratchets due to thermal fluctuations. Physica A 2001;297(1/2):1-12.

[14] Ai B Q, Xie H Z, Wen D H, Liu X M, Liu L G. Heat flow and efficiency in a microscopic engine. Eur Phys J B 2005;48(1):101-6.

[15] Boksenbojm E, Wynants B. The entropy and efficiency of a molecular motor model. J Phys A Math Theor 2009;42(44):445003.

[16] Derényi I, Bier M, Astumian R D. Generalized efficiency and its application to microscopic engines. Phys Rev Lett 1999;83(5):903-6.

[17] Derényi I, Astumian R D. Efficiency of Brownian heat engines. Phys Rev E 1999;59(6): R6219-22.

[18] Hondou T, Sekimoto K. Unattainability of Carnot efficiency in the Brownian heat engine. Phys Rev E 2000;62(5):6021-5.

[19] Asfaw M, Bekele M. Current, maximum power and optimized efficiency of a Brownian heat engine. Eur Phys J B 2004;38(3):457-61.

[20] Ai B Q, Wang L, Liu L G. Brownian micro-engines and refrigerators in a spatially periodic temperature field: heat flow and performances. Phys Lett A 2006;352(4/5):286-90.

[21] Asfaw M. Modeling an efficient Brownian heat engine. Eur Phys J B 2008;65(1):109-16.

[22] Zhang Y, He J. Thermodynamic performance characteristics of an irreversible micro- Brownian heat engine driven by temperature diference. Chin Phys Lett 2010;29(9):090502.
[23] Andresen B, Berry R S, Ondrechen M J, Salamon P. Thermodynamics for processes in finite time. Acc Chem Res 1984;17(8):266-271.
[24] Bejan A. Entropy generation minimization: The new thermodynamics of finite-size devices and finite-time processes. J Appl Phys 1996;79(3):1191-218.
[25] Hoffmann K H, Burzler J M, Schubert S. Endoreversible thermodynamics. J Non-Equilib Thermodyn 1997;22(4):311-55.
[26] Chen L, Wu C, Sun F. Finite time thermodynamic optimization of entropy generation minimization of energy systems. J Non-Equilib Thermodyn 1999;24(4):327-59.
[27] Salamon P, Nulton J D, Siragusa G, Andresen T R, Limon A. Principles of control thermodynamics. Energy 2001;26(3):307-19.
[28] Chen L, Sun F. Advances in Finite Time Thermodynamics: Analysis and Optimization. New York: Nova Science Publishers, 2004.
[29] Chen L. Finite-Time Thermodynamic Analysis of Irreversible Processes and Cycles. Higher Education Press, Beijing, 2005.
[30] Andresen B. Current trends in finite-time thermodynamics. Angewandte Chemie International Edition 2011; 50(12) : 2690-2704.
[31] Chen L. Progress in entransy theory and its applications. Chinese Science Bulletin 2012; 57(34): 4404-4426.
[32] Li J, Chen L, Ge Y, Sun F. Progress in the study on finite time thermodynamic optimization for direct and reverse two-heat-reservoir thermodynamic cycles. Acta Physica Sinica 2013; 62(13): 130501.
[33] Sieniutycz S, Jezowski J. Energy Optimization in Process Systems and Fuel Cells. 2013, Oxford, UK: Elsevier.
[34] Chen L. Progress in optimization of mass transfer processes based on mass entransy dissipation extremum principle. Science China: Technological Sciences 2014; 57(12): 2305-2327.
[35] Parrondo J M R, de Cisneros B J. Energetics of Brownian motors: A Review. Appl Phys A 2002;75(2):179-91.
[36] Schmiedl T, Seifert U. Efficiency at maximum power: An analytically solvable model for stochastic heat engines. Europhys Lett 2008;81(2):20003.
[37] Tu Z C. Efficiency at maximum power of Feynman's ratchet as a heat engine. J Phys A Math Theor 2008;41(31):312003.
[38] Lin B, Chen J. Performance characteristics and parametric optimum criteria of a Brownian micro-refrigerator in a spatially periodic temperature field. J Phys A Math Theor 2009;42(7): 075006.
[39] Ding Z, Chen L, Sun F. Thermodynamic characteristic of a Brownian heat pump in a spatially periodic temperature field. Sci China Phys Mech Astron 2010;53(5):876-85.
[40] Zhang Y, He J, Ouyang H, Qian X. Performance characteristics of a micro-Brownian refrigerator in a one-dimensional lattice. Phys Scr 2010;82(5): 055005.
[41] Ding Z, Chen L, Sun F. Generalized model and optimum performance of an irreversible thermal Brownian microscopic heat pump. Math Comput Model 2011;53(5/6):780-792.
[42] Chen L, Ding Z, Sun F. A generalized model of an irreversible thermal Brownian refrigerator and its performance. Appl Math Model 2011;35(6):2945-58.
[43] De Groot S R, Mazur P. Non-equilibrium Thermodynamics. New York: Dover, 1984.
[44] Callen H B. Thermodynamics and an Introduction to Thermostatistics, 2nd Edition. New York: Wiley, 1985.
[45] Van den Broeck C. Thermodynamic efficiency at maximum power. Phys Rev Lett 2005; 95(19):190602.
[46] Novikov I. The efficiency of atomic power stations (A review). Atommaya Energiya 3 1957; 11: 409.
[47] Chambadal P. Les Centrales Nucleaires. Pairs: Armand Colin, 1957.
[48] Curzon F L, Ahlborn B. Efficiency of a Carnot engine at maximum power output. Am J Phys 1975;43(1):22-4.
[49] Jiménez de Cisneros B, Arias-Hernández L A, Calvo Hernández A. Linear irreversible thermodynamics and coefficient of performance. Phys Rev E 2006;73(5):057103.
[50] Jiménez de Cisneros B, Calvo Hernández A. Collective working regimes for coupled heat engines. Phys Rev Lett 2007;98(13):130602.

[51] Jiménez de Cisneros B, Calvo Hernández A. Coupled heat devices in linear irreversible thermodynamics. Phys Rev E 2008;77(4):041127.
[52] Wang X. Irreversible cycle in linear irreversible thermodynamics. J Phys A Math Theor 2010; 43(42):425003.
[53] Van den Broeck C, Kawai R. Brownian refrigerator. Phys Rev Lett 2006;96(21):210601.
[54] Gomez-Marin A, Sancho J M. Tight coupling in thermal Brownian motors. Phys Rev E 2006; 74(6):062102.
[55] Gao T, Zhang Y, Chen J. The Onsager reciprocity relation and generalized efficiency of a thermal Brownian motor. Chin Phys B 2009;18(8):3279-86.
[56] Gao T, Chen J. Non-equilibrium thermodynamic analysis on the performance of an irreversible thermally driven Brownian motor. Mod Phys Lett B 2010;24(3):325-33.
[57] Velasco S, Roco J M M, Medina A, Calvo Hernández A. Feynman's ratchet optimization: maximum power and maximum efficiency regimes. J Phys D Appl Phys 2001;34(6): 1000-6.
[58] Bejan A. Theory of heat transfer-irreversible power plant. Int J Heat Mass Transfer 1988; 31(6):1211-9.
[59] Bejan A. Theory of heat transfer-irreversible refrigeration plants. Int J Heat Mass Transfer 1989;32(9):1631-9.
[60] Goth Y, Feidt M. Optimum COP for endoreversible heat pump or refrigerating machine. C R Acad Sc Pairs 1986;303(1):19-24.
[61] Chen L, Sun F, Chen W. Optimization of the specific rate of refrigeration in combined refrigeration cycles. Energy 1995;20(10):1049-53.
[62] Chen L, Wu C, Sun F. Heat transfer effect on the specific cooling load of refrigerator. Appl Therm Eng 1996;16(12):989-97.

4

Biogas purification using membrane micro-aeration: A mass transfer analysis

Wathsala Perera[1], Deshai Botheju[2], Rune Bakke[1,3]

[1] Telemark University College, Porsgrunn, Norway.
[2] Agility Group AS, Sandefjord, Norway.
[3] Tel-Tek, Porsgrunn, Norway.

Abstract
When sulfur containing organic feedstocks undergo anaerobic digestion, sulfides are formed due to the biological activities of sulfur reducing bacteria. Presence of hydrogen sulfide (H_2S) negatively affects the usage of biogas and needs to be reduced to levels that depend on the intended biogas application. Conversion of sulfide to its oxidized forms can be carried out by aerobic chemolithotrophic bacteria consuming oxygen as the electron acceptor. Membrane micro-aeration is a recently developed reliable method of safely supplying oxygen into anaerobic digesters. In this study, mass transfer models are developed to represent diffusion and back diffusion of gases through tubular polydimethylsiloxane (PDMS) membranes. The models are utilized to determine the required membrane area and length in order to supply the stoichiometric amount of oxygen for biologically oxidizing a given amount of sulfide feed into elemental sulfur. Penetration of oxygen and nitrogen into the digester and transfer of methane, carbon dioxide and hydrogen sulfide back into the membrane tube are analyzed using these mass transfer models. Circulating air or aerated water inside the membrane tube is considered as two alternatives for supplying micro-aeration to the digester. Literature digester performance and sulfide data are used for example calculations. The required membrane length depends on circulating water flow rates and dissolved oxygen concentrations when water is used inside the membrane. A considerable fraction of CO_2 can also be removed from the biogas in this case. Circulating air inside the membrane is, however, more promising solution as it requires much less membrane area and thereby also causes insignificant methane loss. The proposed membrane micro-aeration technique cuts N_2 biogas dilution in half compared to direct air purging for in-situ sulfide oxidation.

Keywords: Biogas; Hydrogen sulfide; Mass transfer; Micro-aeration; PDMS membrane; Sulfide oxidation.

1. Introduction

Significant amounts of hydrogen sulfide can be formed inside anaerobic digesters when sulfur containing organic substrates such as, paper mill effluents [1], seafood processing wastes [2], animal manures [3], food wastes [4], are fed into digesters. The reduction of S containing compounds is performed by sulfur reducing bacteria growing inside the digester. Depending on the pH these reduced sulfides can be present in three different forms, i.e. HS^-, S^{2-} and H_2S [5]. H_2S transfers into the gas phase as part of the biogas, restricting the direct use of raw biogas as a fuel. Accelerated corrosion of utilities (combustors, compressors, engines, boilers, etc.) and reduced lifespan of pipe work and other installations are among the major impacts of H_2S presence in biogas. Presence of high sulfide concentrations in digesters can also lead to inhibition of the methanogenesis stage of digestion [6]. Heat production boilers require H_2S concentration to be less than 1000 ppmv and electricity production using internal combustion engines require it to be less than 100 ppmv [7]. Cleaning of raw biogas is thus often essential to achieve such purity levels.

Chemical, physicochemical and biological methods can be used to remove sulfides from biogas [8, 9]. Biological methods are assumed to have the advantages of low cost and the environmental sustainability. Photoautotrophic or chemolithotrophic microorganisms are involved in biological sulfide removal methods [10]. Photoautotrophs use CO_2 as the terminal electron acceptor in an anaerobic process carried out by purple and green sulfur bacteria [10]. Biological sulfide oxidation is most commonly applied with colorless chemolithotrophic bacteria such as Thiobascillus sp. Reduced inorganic sulfur compounds like S^{2-}, S^o, $S_2O_3^{2-}$ and organic sulfur compounds like methanethiol are suitable substrates for these types of bacteria [10]. Aerobic chemolithotrophic species use oxygen as the terminal electron acceptor and anaerobic species can use nitrate and nitrite as electron acceptors [10]. According to [10], most of the chemolithotrophic bacteria thrive under mesophilic conditions while Thiobascillus genera can survive both in thermophilic and mesophilic conditions.

Aerobic chemolithotrophs obtain energy by the following reactions where the final product depends on the amount of oxygen available [10].

$$H_2S + 1/2O_2 \rightarrow H_2O + S^0 \qquad \Delta G^0\text{=-209.4kJ/reaction} \qquad (1)$$

$$S^0 + 2H_2O + 3/2O_2 \rightarrow SO_4^{2-} + 2H^+ \qquad \Delta G^0\text{=-587.1kJ/reaction} \qquad (2)$$

$$H_2S + 2O_2 \rightarrow SO_4^{2-} + 2H^+ \qquad \Delta G^0\text{=-798.2kJ/reaction} \qquad (3)$$

$$S_2O_3^{2-} + H_2O + 2O_2 \rightarrow SO_4^{2-} + 2H^+ \qquad \Delta G^0\text{=-818.3kJ/reaction} \qquad (4)$$

Anaerobic digesters can be provided with limited amounts of oxygen (micro-aeration) to have beneficial effects (e.g. enhanced hydrolysis) without inhibiting the anaerobic biochemical pathways leading to methane generation [8, 11, 12]. Micro-aeration can be applied either to the head space or to the liquid phase of the anaerobic digester. Supply of pure oxygen for this purpose, in order to avoid biogas dilution by nitrogen in air, can appear to be advantageous but air as an oxygen source is much less expensive. The direct introduction of air or oxygen into anaerobic digesters is, however, not approved by the safety regulations in Scandinavian countries, due to the explosion risk of methane and oxygen mixtures. Oxygen transfer using dense membrane tubes can be considered as a safer and more controllable mean of supplying micro-aeration into anaerobic digesters, to comply with such regulation restrictions.

In this analysis, micro-aeration circulating aerated water or air in a dense tubular poly-dimethylsiloxane (PDMS) membrane placed inside the headspace of the anaerobic digester is studied. O_2 and N_2 diffuse into the headspace while CH_4, CO_2, and H_2S can diffuse the other way into the membrane tube from the biogas containing headspace. The aim is to quantify and model these transports in order to evaluate the practical potential of such solutions. The analysis is based primarily on the knowledge that transport of gases through dense polymeric membranes can be described by the solution-diffusion mechanism [13]. First, the gas is dissolved into the polymer membrane from the feed side and then diffuses through the membrane according to the direction of the concentration driving force. The rate of gas transfer across the membrane depends on the mass transfer resistance and the extent of driving force. Mass transfer resistance is caused by the membrane material itself and the liquid and/or gas films on the membrane surfaces.

2. Model development

Two models are developed to analyze the mass transfer of gases across a tubular PDMS membrane. One is when water is circulated inside the membrane and the other is when air is circulated. The membrane is placed in the biogas containing digester headspace in both cases.

2.1 Water circulation

Mass transfer of gases into and out of the PDMS tube is derived conceptualizing the resistances in series model. Accordingly, water flows inside the membrane tube and biogas exists outside the membrane tube within the headspace. Figure 1 illustrates the suggested placement of the membrane tube loop inside the headspace of the reactor and one way of aerating the water that flows inside the membrane tube. Both the outward diffusion and the backward diffusion are analyzed here. The following assumptions are made for the model development.

- Water inside the membrane is completely mixed because of the high water circulation rate.
- Flow inside the membrane tube is a fully developed laminar flow.
- Biogas in the head space of the anaerobic digester is completely mixed.
- Diffusion through the membrane follows Fick's first law of diffusion.
- Gas phase temperature is constant and uniform; and equal to the reactor liquid phase temperature.
- Biogas behaves as an ideal gas under the given conditions.

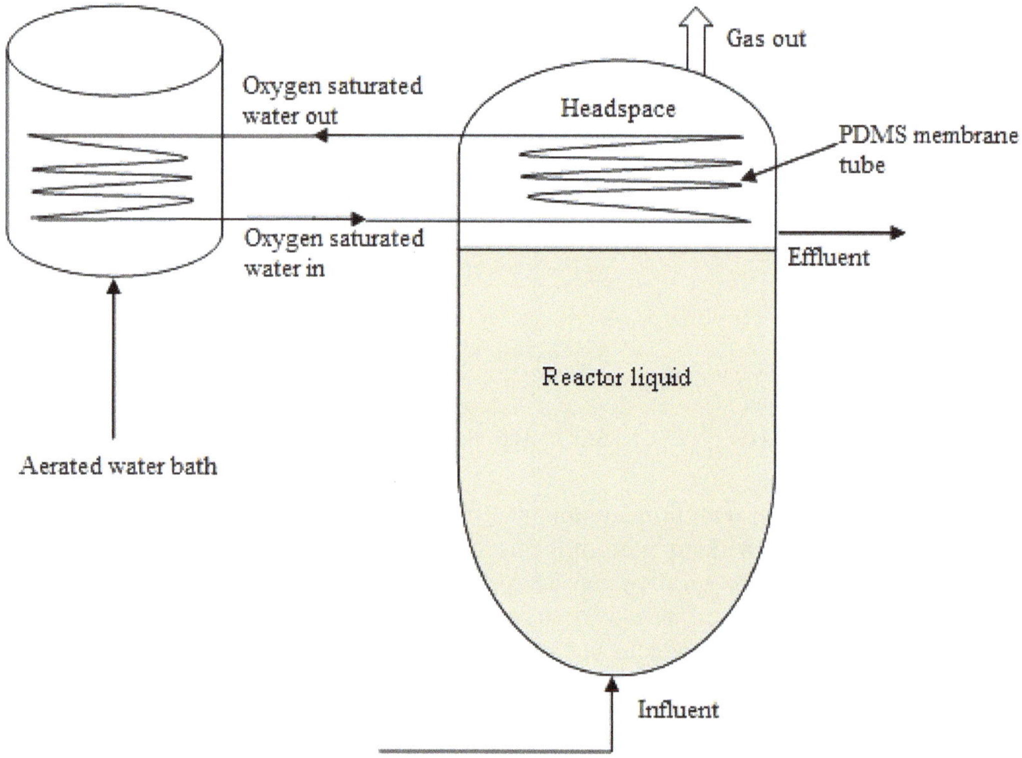

Figure 1. Arrangement of the mass transfer membrane loop inside the digester using an aerated water bath to supply oxygen to the circulating water

Outward Diffusion

Dissolved oxygen and nitrogen gases are diffused into the head space of the anaerobic digester. Figure 2 represents the distribution of concentration gradients.

Mass transfer rate of oxygen and nitrogen across the water film boundary layer inside the membrane tube is given by Eq. 5.

$$J_i = k_{i,in} 2\pi r_i L (C_{i,in} - C_{i,in,eq}) \tag{5}$$

Transfer of gases across the membrane is given by the Fick's first law (Eq.s 6 and 7).

$$J_i = -D_{i,m} A_{lm} \frac{dc}{dr} \tag{6}$$

$$J_i = D_{i,m} 2\pi L \frac{(C_{i,m,in} - C_{i,m,out})}{In\left(\frac{r_o}{r_i}\right)} \tag{7}$$

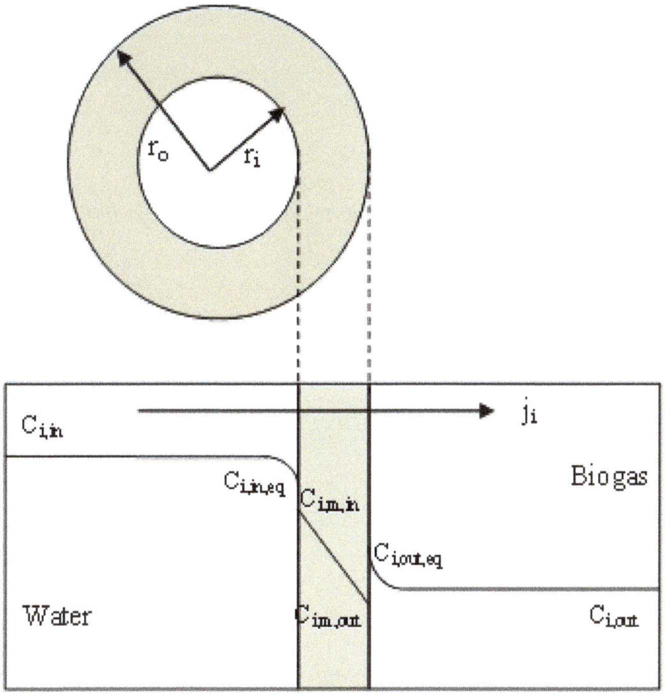

Figure 2. Distribution of oxygen and nitrogen concentration gradients

Equilibrium partition coefficient for water side is given by $S_{i,mle}$ (Eq. 8).

$$\frac{C_{i,m,in}}{C_{i,in,eq}} = S_{i,mle} \tag{8}$$

Equilibrium partition coefficient for gas side is given by $S_{i,mge}$ (Eq. 9).

$$\frac{C_{i,m,out}}{C_{i,out,eq}} = S_{i,mge} \tag{9}$$

The concentrations on the membrane can be substituted with the water and gas phase partition coefficients as given by Eq.s 8 and 9 to obtain Eq. 10.

$$J_i = \frac{D_{i,m} 2\pi L S_{i,mle}\left(C_{i,in,eq} - C_{i,out,eq}\frac{S_{i,mge}}{S_{i,mle}}\right)}{\ln\left(\frac{r_o}{r_i}\right)} \tag{10}$$

Mass transfer rate across the boundary layer outside the membrane tube is given by Eq. 11.

$$J_i = k_{i,out} 2\pi r_o L (C_{i,out,eq} - C_{i,out}) \tag{11}$$

Equation 12 is the mass transfer rate of oxygen and nitrogen from water side to gas side and is obtained by solving equation 5, 10 and 11.

$$J_i = \frac{\left(C_{i,in} - \frac{S_{i,mge}}{S_{i,mle}} C_{i,out}\right)}{\frac{1}{k_{i,in} 2\pi r_i L} + \frac{\ln(r_o / r_i)}{D_{i,m} 2\pi L S_{i,mle}} + \frac{S_{i,mge} / S_{i,mle}}{k_{i,out} 2\pi r_o L}} \tag{12}$$

Back diffusion

Methane, carbon dioxide and hydrogen sulfide gases diffuse from the biogas headspace into the membrane tube as illustrated in Figure 3.

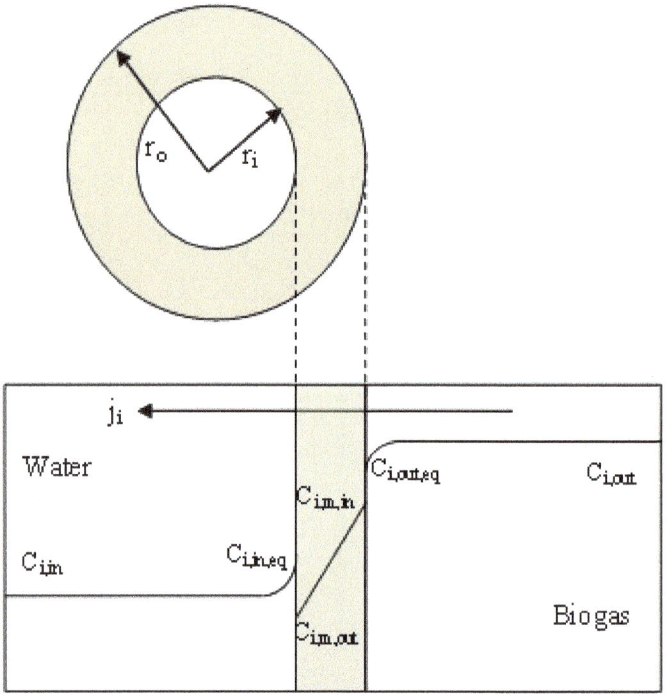

Figure 3. Distribution of methane, carbon dioxide and hydrogen sulfide concentration gradients

Mass transfer rate across the boundary layer in the gas side is given by Eq. 13.

$$J_i = k_{i,out} 2\pi r_o L (C_{i,out} - C_{i,out,eq}) \tag{13}$$

Back diffusion of gases across the membrane is, again, given by the Fick's first law (Eq. 14).

$$J_i = -D_{i,m} 2\pi L \frac{(C_{i,m,in} - C_{i,m,out})}{\ln\left(\frac{r_o}{r_i}\right)} \tag{14}$$

Substitution of membrane concentrations with equilibrium partition coefficients gives Eq. 15.

$$J_i = \frac{-D_{i,m} 2\pi L S_{i,mge} (C_{i,in,eq} \frac{S_{i,mle}}{S_{i,mge}} - C_{i,out,eq})}{\ln(r_o / r_i)} \tag{15}$$

Mass transfer rate of methane, carbon dioxide, and hydrogen sulfide across the water side boundary layer inside the membrane tube is given by Eq 16.

$$J_i = k_{i,in} 2\pi r_i L(C_{i,in,eq} - C_{i,in}) \tag{16}$$

Solving Eq.s 13, 15 and 16 lead to Eq. 17 which describes the back diffusion of gases.

$$J_i = \frac{\left(C_{i,out} - \dfrac{S_{i,mle} C_{i,in}}{S_{i,mge}}\right)}{\dfrac{S_{i,mle}}{k_{i,in} 2\pi r_i L S_{i,mge}} + \dfrac{\ln(r_o / r_i)}{D_{i,m} 2\pi L S_{i,mge}} + \dfrac{1}{k_{i,out} 2\pi r_o L}} \tag{17}$$

2.1.1 Equilibrium Partition coefficient

Partition coefficient is the ratio of concentrations of a compound in the two phases of a mixture of two immiscible solvents at equilibrium. Relations for equilibrium partition coefficients are derived using Eq.s 18 - 21.

Solubility of a gas in a liquid can be expressed as in Eq. 18.

$$C_{i,liquid} = S_{i,liquid} P_{i,gas} \tag{18}$$

Applying Henry's law leads to Eq.s 19 and 20.

$$P_{i,gas} = H_{i,membrane} C_{i,membrane} \tag{19}$$

$$P_{i,gas} = H_{i,liquid} C_{i,liquid} \tag{20}$$

Applying ideal gas law leads to Eq. 21.

$$P_{i,gas} = C_{i,gas} RT \tag{21}$$

Now, considering the gas side of the membrane, we can express,

$$C_{i,out} = \frac{P_{i,out}}{RT} \tag{22}$$

Using Eq. 19:

$$C_{i,out} = \frac{H_{i,mg} C_{i,m,out}}{RT} \tag{23}$$

$$\frac{C_{i,m,out}}{C_{i,out}} = \frac{RT}{H_{i,mg}} \tag{24}$$

$C_{i,out}$ is replaced with $C_{i,out,eq}$ and the relation for $S_{i,mge}$ can be written as Eq. 25.

$$S_{i,mge} = \frac{RT}{H_{i,mg}} \tag{25}$$

Relation for $S_{i,mle}$ is obtained in a similar manner from Eq.s 18 and 20 to yield Eq.s 26-28.

$$C_{i,in} = S_{i,in}P_{i,in} \tag{26}$$

$$P_{i,in} = H_{i,gl}C_{i,in} \tag{27}$$

$$S_{i,in} = \frac{1}{H_{i,gl}} \tag{28}$$

Applying Henry's law according to Eq.19 leads to Eq. 29.

$$P_{i,in} = H_{i,mg}C_{i,m,in} \tag{29}$$

Eq.s 26 and 28 give Eq. 30 and 31.

$$\frac{C_{i,in}}{S_{i,in}} = H_{i,mg}C_{i,m,in} \tag{30}$$

$$\frac{C_{i,m,in}}{C_{i,in}} = \frac{1}{S_{i,in}H_{i,mg}} \tag{31}$$

By substituting $C_{i,in}$ with $C_{i,in,eq}$ in Eq. 28 leads to Eq. 32.

$$S_{i,mle} = \frac{H_{i,gl}}{H_{i,mg}} \tag{32}$$

2.1.2 Mass transfer coefficients

Mass transfer coefficients of the inside liquid and outside gas films are calculated using Sherwood number (Sh) correlations.

$$Sh = \frac{k_{i,in}d_h}{D_{i,l}} \tag{33}$$

For inside water film:

$$Sh = 3.66 + \frac{0.0668(\frac{Pe.d_h}{L})}{(1+0.04(Pe.d_h/L)^{2/3})} \tag{34}$$

This is valid $\frac{x/d_h}{\mathrm{Re}.Sc} < 0.10$, and $10^4 > \frac{\pi d_h}{4x}\mathrm{Re}.Sc > 10$ for a fully developed parabolic velocity profile and laminar flow conditions in a tube [14].

Water side mass transfer coefficient can be calculated using Eq.s 33 and 34.

$$k_{i,in} = \left[3.66 + \frac{0.0668(\frac{Pe.d_h}{L})}{(1+0.04(Pe.d_h/L)^{2/3})} \right] \frac{D_{i,l}}{d_h} \tag{35}$$

For outside gas film:
According to the derivations presented by [15], if the membrane tube can be assumed to be in a region of free convection, Equation 36 is found to fit with mass transfer data for tubular rings within the range $5.5\times10^{5} < Sc.Gr < 9.4\times10^{8}$. Deviations from the single ring data at the outer surface of helical coils depend on the number of turns per coil. The maximum deviation is found to be 12 %.

$$Sh = 0.55(Sc.Gr)^{0.25} \tag{36}$$

Mass transfer coefficient is given by:

$$k_{i,out} = 0.55\left[\frac{\mu}{\rho D_{i,g}}\frac{g\beta\rho^{2}(C_{i,out,eq}-C_{i,out})L^{3}}{\mu^{2}}\right]^{0.25}\frac{D_{i,g}}{d_h} \tag{37}$$

For the mass transfer rate calculations below, resistance to mass transfer in the gas film is neglected since its contribution to the total resistance is very low.

2.1.3 Parameters

Table 1 summarizes the values of diffusivity, solubility, Henry's coefficients and partition coefficients related to the gases of relevance here. Parameters are evaluated assuming a water temperature of 20^{0}C. The anaerobic digester is, however, considered to operate under mesophilic conditions (close to 35 ^{0}C). Diffusion coefficients of gases in water are estimated using Eq. 38 [16].

$$D = \frac{7.4x10^{-8}(\phi \overline{M}_{H2O})^{1/2}T}{\mu_{H2O}\overline{V}^{0.6}_{gas}} \tag{38}$$

D is the diffusion coefficient in cm^2/sec. V_{gas} can be calculated using Eq. 39 [17].

$$\overline{V}_{gas} = 0.285V_c^{1.048} \tag{39}$$

According to [18], "solubility and diffusivity in polydimethylsiloxane membranes show very small variations in the temperature range 25-65 ^{0}C". Therefore, the diffusion coefficients and the Henry's law coefficients used in this analysis are assumed approximately constant for the temperature range of 20-55 ^{0}C.

Table 1. Parameters related to gas transfer

Parameter	Oxygen	Nitrogen	Methane	Carbon dioxide	Hydrogen sulfide	Reference
$D_{i,m}$(m^2/s)	1.6E-09	1.5E-09	1.3E-09	1.1E-09	1.55E-09	[19] [20]
$D_{i,l}$(m^2/s)	2.1E-09	1.88E-09	1.77E-09	1.81E-09	1.91E-09	Estimated
$^{1}H_{mg}$(Pa.m^3/mol)	7321.7	15131.5	3982	1031.7	^{2}2694	[19]
H_{gL}(Pa.m^3/mol)	74879.7	146561.8	68527.2	2590.8	881.5	[21]
$S_{i,mle}$	10.23	9.68	17.21	2.51	0.33	Estimated
$S_{i,mge}$	0.33	0.16	0.64	2.48	0.95	Estimated

^{1}Estimated from solubility data, ^{2} [22]

Mass transfer rates of the gases are calculated for 3 different sizes of silicone membrane tubes commonly available. Different water velocities in the laminar flow range are considered. Several dissolved oxygen concentrations, saturated and sub-saturated, are also tested in calculations. Table 2 gives data on different sizes of membrane tubes used in the calculations.

Table 2. Three membrane tube sizes used in the analysis

r_i (mm)	1	1.5	4
r_o (mm)	1.5	2.5	5
Log mean radius, r_{lm} (mm)	1.2	1.9	4.5

2.2 Air circulation

The case of air instead of water circulating inside the membrane is analyzed as a simplified version of the water case described above. The gas film mass transfer resistance (Eq.37) is determined to be insignificant compared to the membrane resistance so that both inside and outside gas film resistances are assumed negligible for this case. Eq. 40 therefore describes the outward diffusion of gases and Eq. 41 describes the back diffusion of gases in this case.

$$J_i = \frac{S_{i,mge,in}C_{i,in} - S_{i,mge,out}C_{i,out}}{\dfrac{\ln\left(r_o/r_i\right)}{D_{i,m}2\pi L}} \quad (40)$$

$$J_i = \frac{S_{i,mge,out}C_{i,out} - S_{i,mge,in}C_{i,in}}{\dfrac{\ln\left(r_o/r_i\right)}{D_{i,m}2\pi L}} \quad (41)$$

3. Results and discussion

The above developed mass transfer models are used to calculate the required PDMS membrane area and also the length of the membrane tube for a specific level of sulfide removal in a given case. Two published AD studies, representing high and moderate sulfide situations, were selected as case studies for our analysis. Table 3 summarizes some parameters characterizing these studies. Membrane requirements and mass transfer are first calculated for the option of using water inside the tube and next using air.

Table 3. Operating conditions for the two case studies selected

Case no.	1	2
Reference	[12]	[23]
Feed source	Waste water +Sodium sulfate	Municipal organic waste
Working volume of the reactor (m^3)	0.2	0.538
Operating temperature (K)	308.15	296.15
Avg. total sulfur input (mg-S/day)	7890	Not given
Avg. inlet COD Concentration (g/l)	71	Not given
Avg. daily biogas production (l/day)	200	*960
Avg. H_2S conc. in biogas under anaerobic conditions (ppmv)	14400	1100
Avg. CH_4 concentration in biogas (% v/v)	62	65
*Required O_2 flow rate to oxidize total sulfide(kg/s)	2.11E-08	8.05E-09

* Estimated from given data

3.1 Use of water inside the membrane

For the above two cases, membrane lengths required to induce the necessary amount of oxygen are calculated using the developed mass transfer models given by Eq.s 12 and 17. Table 4 summarizes the results for different membrane sizes.

Table 4. Estimated membrane requirement for different membrane sizes in the two cases

Case	r_{lm} (mm)	U (m/s)	$C_{O2,in}$ (mg/l) (Saturated)	Liquid film resistance to O_2 transfer (s/m^3)	Membrane resistance to O_2 transfer (s/m^3)	L (m)	A_{lm} (m^2)
1	1.2	0.016	7.26	308330	34970	90	0.7
	1.9	0.028	7.26	299910	43350	91.5	1.1
	4.5	0.1	7.26	315285	27990	62	1.7
2	1.2	0.016	8.67	984770	93020	40.5	0.3
	1.9	0.028	8.67	959375	118300	40	0.5
	4.5	0.1	8.67	960150	117560	17.5	0.5

Case 1: Reactor temperature = 35^{0}C and water temperature = 35^{0}C
Case 2: Reactor temperature = 23^{0}C and water temperature = 23^{0}C

The required length of the membrane is higher in Case 1 due to higher biogas H_2S concentration compared to the other case. Increase in tube diameter decreases the required length while the required area increases. There is a small effect of membrane thickness on the mass transfer. Most of the resistance to mass transfer is in the liquid film (~90 %) and the membrane resistance only accounts for 10 % of the total resistance. A turbulent water flow would decrease the liquid film resistance but may not be practical.

Gas penetration rates across the membrane calculated for both cases are given in Table 5. Oxygen and nitrogen diffuse into the headspace of the anaerobic digester while other gaseous components back diffuse into the membrane tube from the headspace. Diffusion rate of carbon dioxide is considerably higher compared to the other gases. Some of the hydrogen sulfide back diffuses into the membrane tube, so that the actual oxygen requirement will be lower than the calculated value. This is a significant factor for case 1 because it has a high H_2S concentration. Back diffusion of some methane can negatively impact the process performance.

Table 5. Mass flow rates through the membrane (r_{lm} = 4.481 mm)

Gas	Case 1 Water Temperature: 35^{0}C Reactor Temperature: 35^{0}C		Case 2 Water Temperature: 23^{0}C Reactor Temperature: 23^{0}C	
	Mass flow rate (mg/day)	Gas flow rate (l/day)	Mass flow rate (mg/day)	Gas flow rate (l/day)
Oxygen	1832	1.44	696	0.53
Nitrogen	2851	2.57	1080	0.94
Methane	2670	4.22	1071	1.69
Carbon dioxide	73958	42.48	24538	14.12
Hydrogen sulfide	3024	2.25	73	0.05

Membrane lengths required for different water velocities are estimated considering the largest tube diameter assuming that oxygen concentration is kept at saturation level (Figure 4). The observed relation is close to a second order polynomial showing that less membrane is required at higher velocities.

Calculated required membrane length for different concentrations of oxygen in the circulating water (for the same conditions as in Figure 4) are shown in Figure 5. Oxygen saturated condition demands a less membrane length, as can be expected, compared to sub-saturated cases since the highest driving force is available when the water is saturated with oxygen.

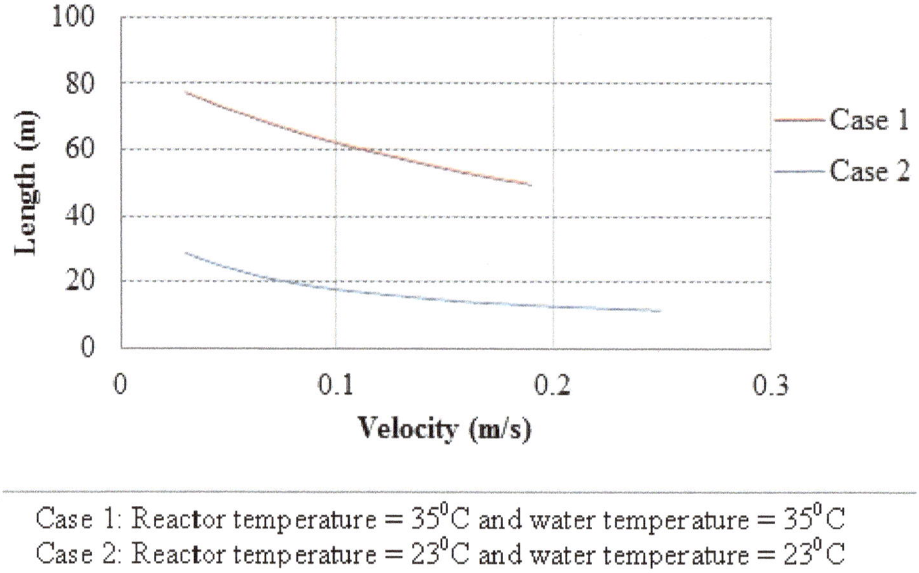

Figure 4. Calculated required membrane length vs. Velocity for the two cases (r_{lm} = 4.5 mm)

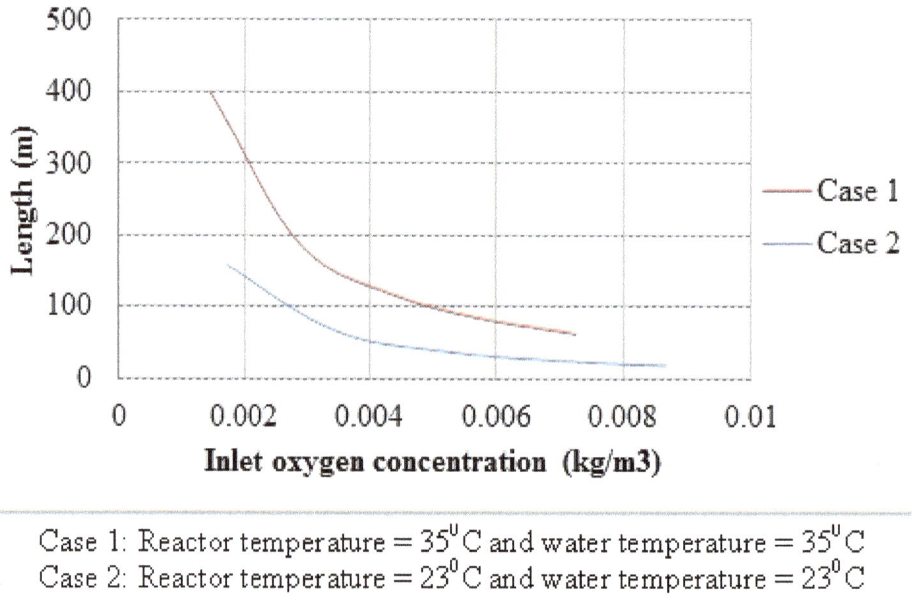

Figure 5. Calculated required membrane length vs. Oxygen concentration (r_{lm} = 4.5 mm)

Digester temperatures and inlet water temperature can also influence the required membrane length, but, according to Table 6, the temperature influence is not that significant. Membrane tube log mean radius is 4.5 mm. Increasing the digester temperature decreases the required memebrane length, mainly due to the higher values of diffusion coefficients at increased temperatures.

Table 6. Effect of temperature on the required membrane length

Case 1			Case 2		
Reactor temperature (K)	Water temperature (K)	Length of the membrane (m)	Reactor temperature (K)	Water temperature (K)	Length of the membrane (m)
35^0C	20^0C	66.5	23^0C	23^0C	17.6
35^0C	35^0C	61.8	35^0C	35^0C	16.4
55^0C	55^0C	58.2	55^0C	55^0C	16.3

3.1.1 Different configurations of membrane micro-aeration for water circulation

Water bath vs. open air aeration of the membrane tube

Water flowing through the membrane tube mainly carries oxygen and nitrogen into the anaerobic digester and absorbs methane, carbon dioxide and hydrogen sulfide from the biogas. These absorbed gases need to be released from the circulating water. Two options for this purpose evaluated are:

1. Submerging the membrane section outside the anaerobic digester in an aerated water bath (as illustrated in Figure 1).
2. Placing the membrane section outside the anaerobic digester in open air.

Reduced nitrogen transfer into the membrane tube and less methane loss from the biogas are advantages of the first option. Low solubility of nitrogen in water decreases the diffusion of nitrogen into the membrane and hence penetration of nitrogen into the digester is kept at a low level. Methane also has low solubility in water and accordingly the amount of methane diffused into the water bath through the membrane tube is low. Therefore most of the methane, diffused from the biogas to the membrane remains inside the membrane tube which leads to a smaller methane concentration gradient between anaerobic digester gas and the membrane tube; so that the methane loss is reduced (compared to results in Table 5).

Nitrogen diffusion into the biogas headspace will be higher and the methane loss from the biogas will be higher in option 2 than the first option. Option 2 is, however an easier and more efficient way to supply the required oxygen, thus the required membrane length outside the AD will be shorter.

3.2 Use of air inside the membrane

Air consists of 78.1 % nitrogen, 20.9 % oxygen, 0.033 % carbon dioxide, and other trace gases including argon. Effect of trace gases such as argon is not considered in the following calculations. Both mass transfer resistances inside and outside of the membrane is considered insignificant, according to argument in Chapter 2.1.2 that gas phase mass transfer resistance is much less than liquid film and membrane resistance. Only the membrane resistance is therefore accounted for in the calculations for this case. The calculated required lengths of the different sizes of membranes for Cases 1 and 2 for both the water and air circulation scenarios are compared in Table 7. Water as circulating fluid requires more than one hundred times larger membrane than air as oxygen supply gas.

Table 7. Comparison of required membrane lengths when water or air is circulated inside the membrane tube

Case		1						2					
Circulating fluid		Water			Air			Water			Air		
Temperature in the reactor & fluid (^{0}C)		23	35	55	23	35	55	23	35	55	23	35	55
L (m)	r_{lm} 1.2 mm	100.66	88.35	77.94	0.30	0.30	0.30	38.20	32.55	29.00	0.11	0.11	0.11
	r_{lm} 1.9 mm	102.12	88.50	78.90	0.37	0.37	0.37	36.25	31.50	28.50	0.14	0.14	0.14
	r_{lm} 4.5 mm	65.73	61.86	58.16	0.16	0.16	0.16	17.60	16.36	16.27	0.06	0.06	0.06

No effect of temperature on the membrane length is noticeable when air is circulated instead of water. Increase in tube diameter reduces the required membrane length, as for the water flow case. A larger effect from membrane thickness on the required membrane length is observed for the air than for the water option. E.g.: The membrane tube having a log mean radius of 1.9 mm and a thickness of 1 mm requires a longer membrane than the 1.2 mm option with a thickness of 0.5 mm.

The estimated gas penetration rates through the membrane for air circulation cases are shown in Table 8. For a fixed amount of oxygen supply; penetration of methane, carbon dioxide and hydrogen sulfide is less compared to when water is used inside the membrane because of the reduced membrane length. It was hypothesized that water was advantageous to avoid loss of methane due to low methane solubility in water, but these calculations refute this hypothesis. Nitrogen, similarly, that has a higher diffusion rate per length through the membrane for the air circulation case, which can be understood based on the water solubility of N_2 and the higher concentration of nitrogen in air, but the overall transfer is less than in the water case. The nitrogen to oxygen ratio supplied (~2) is, thereby, half of the ratio in air (~4), implying another significant advantage compared to directly adding air to headspace.

Back diffusion of gases is minimal for this air circulation case, due to the far shorter membrane length compared to water circulation cases. Hydrogen sulfide diffusion is negligible and methane also has a very low back diffusion, making this option potentially very attractive. The only possible disadvantage of air compared to water circulation is the reduced capability of carbon dioxide removal from biogas.

Table 8. Gas penetration rates (mg/day) with air inside the membrane

Case		1			2		
Circulating fluid		Air			Air		
Temperature in the reactor & fluid		23	35	55	23	35	55
r_{lm} 1.2 mm	O_2	1833	1833	1833	696	696	696
	N_2	3099	3099	3099	1177	1177	1177
	CH_4	130	135	144	51	53	57
	CO_2	686	714	762	243	253	270
	H_2S	11	11	12	0	0	0
r_{lm} 1.9 mm	O_2	1829	1829	1829	700	700	700
	N_2	3094	3094	3094	1184	1184	1184
	CH_4	130	135	144	52	54	57
	CO_2	685	713	760	244	254	271
	H_2S	11	11	12	0	0	0
r_{lm} 4.5 mm	O_2	1829	1829	1829	700	700	700
	N_2	3092	3092	3092	1183	1183	1183
	CH_4	130	135	144	52	54	57
	CO_2	684	713	760	244	254	271
	H_2S	11	11	12	0	0	0

3.3 Membrane aeration vs. direct air purging

Composition of the biogas for the two cases where (i) the reactor is membrane aerated and (ii) the reactor is directly air purged, are presented in Table 9. The following assumptions were made for the calculations.

All the oxygen supplied to the reactor is consumed.

1. Hydrogen sulfide is totally removed from biogas due to oxidation and back diffusion.

Approximately the same amounts of nitrogen gas are diffused into the reactor headspace in the air and water circulation scenario, which is approximately half the amount of nitrogen dilution caused by direct air purging. The membrane aeration techniques suggested here maintain the biogas nitrogen dilution within the limits for vehicle fuel quality as opposed to the direct air injection method for high sulfide situations (Case 1).

Table 9. Calculated composition of biogas before and after aeration by membrane and by air purging (r_{lm}= 4.5 mm)

Description	Case 1 (35^0C)				Case 2 (23^0C)			
	Without aeration	After Membrane Aeration (Water)	After Membrane Aeration (Air)	After Air purging	Without aeration	After Membrane Aeration (Water)	After Membrane Aeration (Air)	After Air purging
Daily biogas production (l/day)	200	200	200	200	960	960	960	960
O_2 (%)	0	0	0	0	0	0	0	0
N_2 (%)	0	1.7	1.4	2.7	0	0.1	0.1	0.2
CH_4 (%)	62.5	78.9	61.7	61.7	65	65.9	64.9	65
H_2S (%)	1.5	0	0	0	0.1	0	0	0
CO_2 (%)	36	19.4	36.8	35.6	34.9	34	36.8	34.8
CH_4/CO_2 Ratio	1.73	4	1.67	1.73	1.86	1.94	1.67	1.86

4. Conclusions

Membrane micro-aeration appears as a sound technique for oxygen supply to AD headspace for in situ aerobic microbial oxidation of hydrogen sulfide in raw biogas. Diffusion through a dense membrane gives the ability of supplying a precise amount of oxygen in a safe manner without mixing air and biogas directly.

Preferential transfer of relevant gasses (Methane, oxygen, hydrogen sulfide, nitrogen and carbon dioxide) for alternative scenarios can be studied and understood based on the mass transfer models presented. The mass transfer analysis conducted here can be utilized to design and evaluate appropriate practical configurations for experimental and on-field anaerobic digesters.

Comparing the use of air and water inside the membrane tube to supply the required oxygen shows that air is the most promising option because it requires little membrane area and cause insignificant loss of methane. The main advantage of using water is that it can remove a considerable fraction of carbon dioxide from the biogas.

The main mass transfer advantage of the proposed membrane micro-aeration technique compared to direct air purging for in-situ sulfide oxidation is the reduced biogas dilution by N2, to a level that is acceptable for vehicle fuel quality, even in the high sulfide case studied.

Nomenclature

J_i	Flow rate of solute i (kg/s)
r_i	Inside radius of membrane tube (m)
r_o	Outside radius of membrane tube (m)
L	Length of the membrane tube (m)
A_{lm}	Log mean area of the membrane (m^2)
$C_{i,in}$	Concentration of solute in water (kg/m^3)
$C_{i,out}$	Concentration of solute in biogas (kg/m^3)
$C_{i,in,eq}$	Equilibrium solute concentration in water side (kg/m^3)
$C_{i,out,eq}$	Equilibrium solute concentration in gas side (kg/m^3)
$C_{i,m,in}$	Concentration of solute in membrane in liquid side (kg/m^3)
$C_{i,m,out}$	Concentration of solute in membrane in gas side (kg/m^3)
$C_{i,liquid}$	Concentration of solute in a liquid (kg/m^3)
$C_{i,gas}$	Concentration of solute in a gas (kg/m^3)
$P_{i,gas}$	Partial pressure of i in the gas (Pa)
$S_{i,liquid}$	Solubility of i in a liquid ($kg/m^3.Pa$)
$D_{i,m}$	Diffusion coefficient for diffusion of solute in membrane (m^2/s)
$S_{i,mle}$	Equilibrium partition coefficient at the liquid membrane interface (-)
$S_{i,mg}$	Equilibrium partition coefficient at the gas membrane interface (-)
$S_{i,mge,in}$	Equilibrium partition coefficient at the gas membrane interface inside the membrane when air is circulated (-)

$S_{i,mge,in}$	Equilibrium partition coefficient at the gas membrane interface outside the membrane when air is circulated (-)
$H_{i,gl}$	Henry's Law constant for solute in gas liquid system ($Pa.m^3/mol$)
$H_{i,mg}$	Henry's Law constant for solute in gas-membrane system ($Pa.m^3/mol$)
$H_{i,membrane}$	Henry's Law constant for solute in membrane ($Pa.m^3/m$
$H_{i,liquid}$	Henry's Law constant for solute in liquid ($Pa.m^3/mol$)
R	Gas constant ($Pa.m^3.K^{-1}\ mol^{-1}$)
T	Temperature (K)
$k_{i,in}$	Mass transfer coefficient water side (m/s)
$k_{i,out}$	Mass transfer coefficientgas side (m/s)
x	Distance from entrance (m)
d_h	Hydraulic diameter(m)
Re	Reynolds number (-)
Sc	Schmidt number (-)
Sh	Sherwood number (-)
Pe	Peclet number (-)
$D_{i,l}$	Diffusion coefficient for diffusion of solute in liquid (m^2/s)
μ	Dynamic viscosity of water (Pa.s)
ρ	Density of fluid (kg/m^3)
u	Velocity of water inside the membrane tube (m/s)
Gr	Grashof Number (-)
g	Acceleration due to gravity (m/s^2)
β	Volumetric thermal expansion coefficient (equal to approximately 1/T, for ideal fluids, where T is absolute temperature)
$D_{i,g}$	Diffusion coefficient for diffusion of solute in gas (m^2/s)
Φ	Empirical parameter (For water it is 2.6)
$\widetilde{M}_{H_2O}$	Molecular weight of water (Daltons)
μ_{H_2O}	Viscosity of water (Centipoise)
$\widetilde{V}_{gas}$	Molecular volume of gas (cm^3/mol)

References

[1] Rintala, J.A. and J.A. Puhakka, Anaerobic treatment in pulp- and paper-mill waste management: A review. Bioresource Technology, 1994. 47(1): p. 1-18.

[2] Omil, F., R. Méndez, and J.M. Lema, Anaerobic treatment of saline wastewaters under high sulphide and ammonia content. Bioresource Technology, 1995. 54(3): p. 269-278.

[3] Rasi, S., J. Läntelä, and J. Rintala, Trace compounds affecting biogas energy utilization – A review. Energy Conversion and Management, 2011. 52(12): p. 3369-3375.

[4] Oyarzún, P., et al., Biofiltration of high concentration of hydrogen sulphide using Thiobacillus thioparus. Process Biochemistry, 2003. 39(2): p. 165-170.

[5] Duranceau, S.J., et al., Innovative Hydrogen Sulfide Treatment Methods: Moving Beyond Packed Tower Aeration. FLORIDA WATER RESOURCES JOURNAL, 2010.

[6] Sarti, A., et al., The treatment of sulfate-rich wastewater using an anaerobic sequencing batch biofilm pilot-scale reactor. Desalination, 2009. 249(1): p. 241-246.

[7] VILNIŠKIS1, R., et al., Research and assessment of biogas evolved during anaerobic digestion of biodegradable agricultural waste. Ecological Chemistry and Engineering S, 2011. 18(4): p. 409-427.

[8] Zee, F.P.v.d., et al., Sulfide removal by moderate oxygenation of anaerobic sludge environments. Bioresource Technology, 2007. 98(3): p. 518-524.

[9] Janssen, A.J.H., G. Lettinga, and A.d. Keizer, Removal of hydrogen sulphide from wastewater and waste gases by biological conversion to elemental sulphur: Colloidal and interfacial aspects of biologically produced sulphur particles. Colloids and Surfaces A: Physicochemical and Engineering Aspects, 1999. 151(1-2): p. 389-397.

[10] Tang, K., V. Baskaran, and M. Nemati, Bacteria of the sulphur cycle: An overview of microbiology, biokinetics and their role in petroleum and mining industries. Biochemical Engineering Journal, 2009. 44(1): p. 73-94.

[11] Botheju, D. and R. Bakke, Oxygen Effects in Anaerobic Digestion – A Review. The Open Waste Management Journal, 2011. 4: p. 1-19.

[12] Diaz, I., et al., Effect of oxygen dosing point and mixing on the microaerobic removal of hydrogen sulphide in sludge digesters. Bioresour Technol, 2011. 102(4): p. 3768-75.

[13] Mulder, M., Basic principles of membrane technology. 1996: Kluwer Academic Publications.

[14] Perry, R.H. and D.W. Green, Perry's chemical engineers' handbook. 2008, New York: McGraw-Hill.

[15] Naphon, P. and S. Wongwises, A review of flow and heat transfer characteristics in curved tubes. Renewable and Sustainable Energy Reviews, 2006. 10(5): p. 463-490.

[16] Cussler, E.L., Diffusion: mass transfer in fluid systems. 2009, Cambridge: Cambridge University Press.

[17] Reid, R.C., J.M. Prausnitz, and T.K. Sherwood, The properties of gases and liquids. Vol. 3rd Edition. 1977: McGraw-Hill.

[18] Rathnasiri, P.G., Anaerobic digestion process using membrane integrated microaeration, 2009, Norwegian university of science and technology: Trondheim.

[19] Zhang, H. and A. CLoud. The Permeability Characteristics of Silicone. in Global Advances in Materials and Process Engineering. 2006. Dallas.

[20] Turk, B.S., et al., Novel technologies for gaseous contaminants control, 2001, Research Triangle Institute.

[21] Tchobanoglous, G., F.L. Burton, and H.D. Stensel, Wastewater engineering: treatment and reuse. 2003, Boston: McGraw-Hill. XXVIII, 1819 s. : ill.

[22] Merkel, T.C., et al., Mixed-gas permeation of syngas components in poly(dimethylsiloxane) and poly(1-trimethylsilyl-1-propyne) at elevated temperatures. Journal of Membrane Science, 2001. 191(1–2): p. 85-94.

[23] Soreanu, G., et al., Laboratory pilot scale study for H2S removal from biogas in an anoxic biotrickling filter. Water Sci Technol, 2008. 57(2): p. 201-7.

5

Phisical and chemical properties and thermobaric conditions of occurrence of hard-to-recover oils

I.G. Yashchenko, Y.M. Polishchuk

Institute of Petroleum Chemistry, Siberian Branch of Russian Academy of Sciences, 4, Akademichesky Ave., 634021, Tomsk, Russia.

Abstract
The criteria necessary to classify oils as hard-to-recover oil reserves are determined. The distribution of viscosity, heavy, paraffin and highly resin oils is analyzed in terms of their reserves. The features of the physic-chemical properties of these oils are studied under various conditions of their occurrence and formation temperatures. The results obtained could be used to solve practical issues in the oil sector.

Keywords: Hard-to-recover oils; High-viscosity oils; Heavy oils; Paraffin oils; Highly resin oils; Physical and chemical properties of oils; Conditions of occurrence; Formation temperature; Reservoir pressure; Gas content in oil.

1. Introduction

Regularities of spatial distribution and characteristics of physical and chemical properties of oils were considered in [1-4]. Currently, the world has witnessed the growth of the share of hard-to-recover oils in the total balance due to the depletion of readily available oils. The oils in complex geological formations and deposits or those containing low-mobility oils (e.g., high-viscosity, high-paraffin oils) are classified as hard-to-recover reserves. They are characterized by relatively low flow rates due to low reservoir productivities, unfavorable conditions of oil occurrence (gas and oil deposits at the depths more than 4,500 meters, etc.) or anomalous physical and chemical properties. Currently, however, systematic materials on the properties of such oils in the literature is extremely insufficient. In this regard, the aim of our work was the consideration of physical and chemical properties of hard-to-recover oils and features of the conditions of their occurrence.

2. About hard-to-recover oils

The most reasonable approach to the differentiation of hard-to-recover reserves was proposed by Khalimov in 1987. Purtova and coauthors of [5] made a list of the main criteria for differentiation of hard-to-recover reserves, according to which the such oil reserves are characterized by their specific properties and conditions of their occurrence, namely:

(1) oil reserves possessing anomalous physical and chemical properties (high viscosity and density and high content of paraffins, asphaltenes and resins);
(2) oil reserves in low permeability reservoirs and oil-water and oil-gas zones;
(3) oil reserves with high gas saturation (more than $200 m^3/m$) or with dissolved and/or free gas containing aggressive components (hydrogen sulfide, carbon dioxide) in the amounts requiring the use of special equipment for well-drilling and oil production;

(4) oil reserves occurring at great depths (below 4,500m);
(5) oil reserves with the reservoir temperature of 100°C and higher or less than 20°C (due to the little difference between the low reservoir temperature and the pour point of wax and resins);
(6) oil reserves with a high degree of water cut (to 75 - 80%).

Recent years have seen an incresed production of hard-to-recover oils with anomalous physical and chemical properties, i.e. paraffin, resin, viscosity and heavy oils. An investigation of these oils requires the development of the respective EOR methods, which in turn makes it necessary to identify patterns of their spatial distribution and to analyze the changes in their physico-chemical properties that depend on the geological and thermobaric conditions of their occurrence. The development of new technologies for production and transportation of oils with anomalous physical and chemical properties and improvement of the available processes require that new data on the properties and conditions of occurrence of hard-to-recover oils be obtained.

The reserves of heavy and viscosity oils are several times higher than those of readily available low-viscosity oils (810 and 162 billion tons, respectively) and represent a raw material source for the oil sector of Russia and other oil-producing countries of the world [6-10]. The development of deposits of viscosity oils is therefore receiving an increasing attention. To date, the average annual total production of these oils in the world is approaching 500 million tons and the cumulative production exceeds 14 billion tons. The most abundant resources of heavy and viscous oil are located in Canada and Venezuela. Mexico, the United States, Russia, Kuwait; China also possess significant heavy oil reserves. On the territory of Russia, the reserves of oil with a viscosity higher than 35 mm^2/s account for 7.3 billion tons oil and most of them are concentrated in the Komi Republic, Tatarstan, and in the Tyumen region. Paraffin crude oils [11, 12], whose viscosity and high pour point are controlled by the high paraffin content (more than 6%) make a considerable fraction in the total reserves of viscous oils. The percentage of paraffin oils is also significant and accounts for more than 25% of all world oils. In particular, 19% of oil fields on the territory of the West Siberian oil-gas basin contain paraffinic oils.

The investigations requiring an analysis of numerous literature sources need the data on physical properties and chemical composition of the global oils and on the geological characteristics of oil bearing basins. A large body of information has been accumulated in the data base on physical and chemical oil properties [13] of the Institute of Petroleum Chemistry, SB RAS (Tomsk). Therefore this data base was used as a source of information on hard-to-recover oils.

The data base includes more than 22,000 oil samples from 191 oil and gas basins of the world, most are those of Eurasian continental oils from 106 oil and gas basins and 4,064 oil fields located in 58 countries. The general characteristics of the data base are presented in Table 1. Being constantly developed and improved, the database is enriched with new data on the physical, chemical and geochemical properties of oils, their geographical location, geological modes of occurrence and formation temperature and pressure. Table 1 offers a description of the continental distribution of the database information.

Table 1. Data distribution by continents

Geography	Data base sample	Number of oil-gas bearing basins	Number of oil fields
Australia, New Zealand and Oceania	152	11	86
Africa	532	15	294
Eurasia	20,629	106	4,064
North and Central America	1,631	38	899
South America	428	21	256

The global database on physical and chemical properties of oils currently contains 5,104 heavy oil samples, 2,510 samples of viscous oil (with a viscosity higher than 35mm^2/s), 2,327 paraffin oil, and more than 2,000 resin oil samples (more than 13 % resin). More detailed characteristics of this information are presented in Table 2.

3. Generalized classification of oils by their physical and chemical parameters

To investigate the physical and chemical properties of hard-to-recover types of oils having anomalous physical and chemical properties and those with properties conditioned by anomalous modes of occurrence, we have developed a generalized classification of oils (Tables 3 and 4).

Table 2. Characteristics of database information on the oils with anomalous properties

Class of oil	Data base sample	Number of oil-gas bearing basins	Number of oil fields
Heavy (density > 0.88 g/cm^3)	5,104	117	1,637
Viscous (viscosity at 20 °C > 35 mm^2/s)	2,510	63	821
Resin (resin content >13 %)	2,030	52	684
Paraffin (paraffin content > 6 %)	2,327	58	800

Table 3. Classification of oils by their density and viscosity

Index	Class of oil		Limits of variation
Density (g/cm^3)	very light oil		< 0.80
	light oil		0.80÷0.84
	medium-density oil		0.84÷0.88
	heavy	higher-density oil	0.88÷0.92
		extra heavy crude oil	0.92÷0.96
		bituminous oil	> 0.96
Viscosity at 20 °C, (mm^2/s)	low-viscosity oil		< 10
	medium-viscosity oil		10 ÷35
	viscous	higher-viscosity oil	35÷100
		high-viscosity oil	100÷500
		extremely-high-viscosity oil	> 500

Table 4. Classification of oils by chemical components

Chemical component, wt%	Class of oil		Limits of variation
Sulfur content, wt%	low sulphur (sweet crude) oil		< 0.5
	medium sulphur oil		0.5÷1
	sulfurous oil		1÷3
	high-sulphur (sour) oil		> 3
Resin content, wt%	low-resin oil		< 8
	medium-resin oil		8÷13
	resin oil	moderately resin	13÷20
		highly resin	20÷30
		extremely resin	> 30
Asphaltene content, wt%	low-asphaltene oil		< 3
	medium-asphaltene oil		3 ÷10
	high-asphaltene oil		> 10
Paraffin content, wt%	low-paraffin oil		< 1.5
	medium-paraffin oil		1.5÷6
	paraffin oils	moderately paraffin	6÷ 10
		highly paraffin	10÷20
		extremely paraffin	> 20
Low-boiling fraction 200 °C, wt%	oils with low fraction-200		< 20
	oils with medium fraction-200		20 ÷30
	oils with high fraction-200		> 30
Low-boiling fraction 350 °C, wt%	oils with low fraction-200		< 25
	oils with medium fraction-350		25÷50
	oils with high fraction-350		50÷75
	oils with extremely high fraction-350		75÷100

4. Analysis of oil distribution by classes

Figure 1 shows the diagrams of distribution of oil types under study by classes in terms of their density, viscosity and content of resins and paraffin. As can be seen from Figures 1a and b, the amount of oil types (light, medium, and heavy oils and low-viscosity, medium-viscosity, and viscous oils) are almost equally distributed by classes, i.e. light, medium density, low viscosity and medium viscosity oils comprise more than 66% of the total sampling. Heavy and viscous oil account for slightly more than 33% of the total sample, and resinous and paraffin oils account for less than 30% in the total sample (Figures 1c and d).

Here we investigate the distribution of hard-to-recover oil reserves (heavy, viscosity, paraffin, and resin) in the Volga-Ural, West Siberian, and Timan-Pechora oil and gas bearing basins (in what follows referred to as Volga-Ural, West Siberian, and Timan-Pechora oil-gas-bearing basins, respectively) as the main Russian oil producing regions. Table 5 shows the distribution of information from the data base for these three basins. These data indicate that the most abundant reserves of viscous, heavy, paraffin, and resin oils are located in Western Siberia; the Volga-Ural basin possesses rich reserves of viscosity, heavy, and resin oils, while the Timan-Pechora basin has concentrated reserves of paraffin oils. The largest number of fields with hard-to-recover oils is located in the Volga-Ural oil-producing regions and the lowest – in the Timan-Pechora basin.

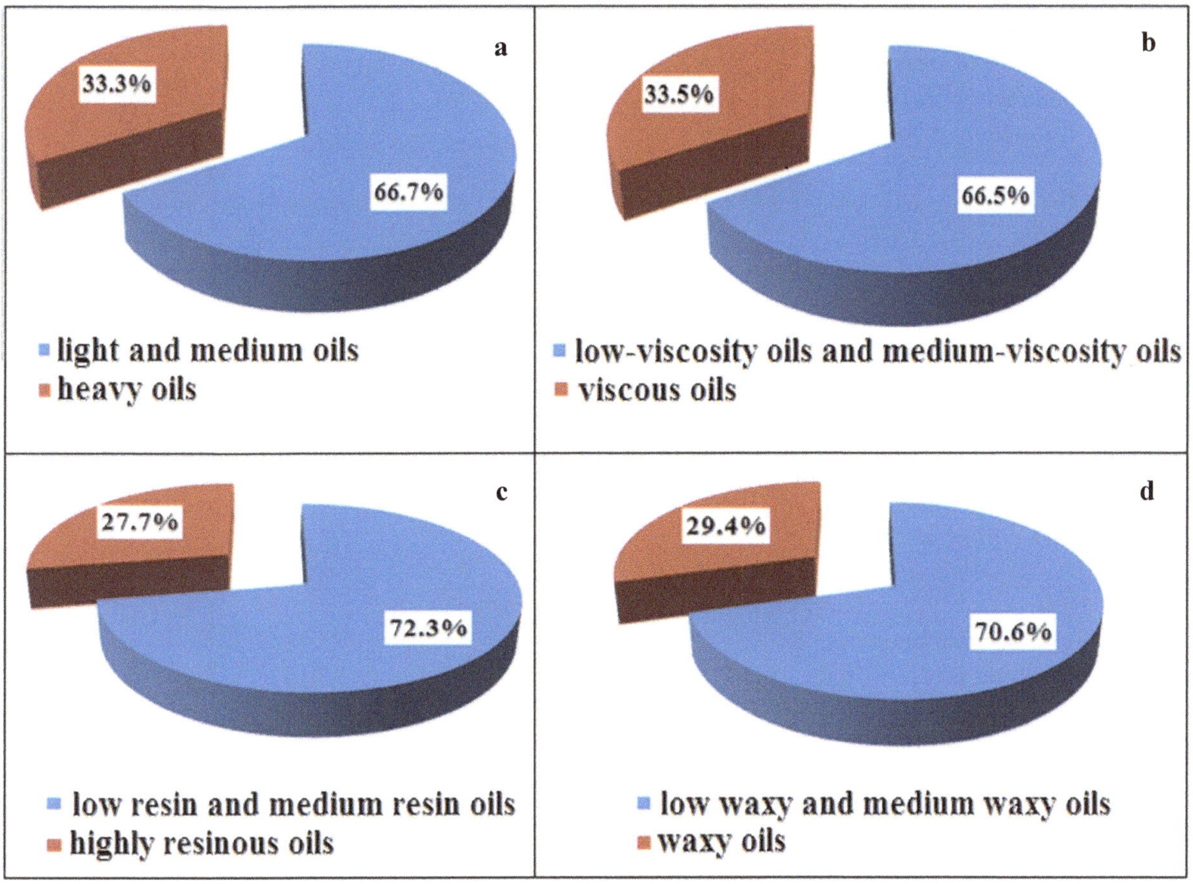

Figure 1. The distribution of the number of deposits by density (a), viscosity (b), content of resin (c) and paraffin (d)

5. Features of the physic-chemical properties of hard-to-recover oils

As already mentioned above, the hard-to-recover oils also include those with a gas factor above 200 m^3/t. The database contains 348 samples of these oils from 255 oil fields of 44 basins. They are distributed over the continents as follows: one – in Australia, 3 – in Africa, 20 – in America and the rest 20 – in Eurasia. Most deposits with high gas content are located in the Volga-Ural (81 deposits), 10 deposits – in the Lena-Tunguss, and 7 deposits each in Western Canadian, Pre-Caspian and Timan-Pechora basins. The physic-chemical properties of these oils are presented in Table 6. It is evident that, according to

Tables 3 and 4, these oils belong to the light, low viscosity, low sulfur, moderately paraffin, and low asphaltene oils.

Table 5. Hard-to-recover oils in the main oil-gas bearing basins of Russia

Statistical data	Oil-gas bearing basins		
	Volga-Ural	West Siberian	Timan-Pechora
	Viscous oils		
Number of samples	1,263	157	104
Number of oil fields	362	45	42
Russian oil reserves, %	33.46	37.06	14.30
	Heavy oils		
Number of samples	1,787	378	270
Number of oil fields	453	120	60
Russian oil reserves, %	38.35	40.70	9.85
	Paraffin oils		
Number of samples	322	449	95
Number of oil fields	140	136	34
Russian oil reserves, %	16.26	40.46	42.50
	Resin oils		
Number of samples	1,103	133	66
Number of oil fields	301	58	24
Russian oil reserves, %	45.87	36.46	9.79

Table 6. Physical and chemical properties of crude oils with high gas content (above 200m^3/t)

Physic-chemical parameters	Database sample	Mean value
Density g/cm^3	326	0.83
Viscosity at 20 °C, mm^2/s	148	7.21
Sulfur content, wt%	248	0.38
Paraffin content, wt%.	173	6.23
Resin content, wt%	104	5.30
Asphaltene content, wt%	126	0.82
Low-boiling fraction 200 °C, wt%	23	31.60
Low-boiling fraction 300 °C, wt%	23	48.28
Gas content in oil, m^3/t	348	346.19
Thermobaric conditions of occurrence		
Reservoir temperature, °C	241	85.88
Reservoir pressure, MPa	232	32.54

The presence of sulfur dioxide also impedes oil production due to the increased environmental hazard, which requires the use of special equipment for well drilling and oil production when the concentration of sulfur dioxide in the oil is higher than 5%. The database comprises 79 samples of oils containing sulfur dioxide gas from 55 oil fields of 19 basins. The physical and chemical properties of these oils are presented in Table 7. It is evident from Table 8 that the oils belong to medium density, high viscosity, sulfurous, medium paraffin, medium asphaltene and moderately resin classes.

The main criteria for differentiation of hard-to-recover reserves are the oil reservoir temperatures above 100°C or below 20°C. The database contains 838 samples of in-reservoir oils at the temperatures higher 100°C from 483 oil fields of 47 basins. Their physic-chemical properties are presented in Table 8. It implies that these oils belong to the light, medium viscosity, low sulfur, low resin and low asphaltene but moderately paraffin classes.

The data base contains 318 samples of oils occurring in low-temperature reservoirs (< 20 °C) from 181 deposits of 35 basins. The most of oil deposits characterized by low reservoir temperature are located in the Volga-Ural basin (81 deposits), the Lena-Tunguss basin (10 deposits) and in Western Canadian, Pre-Caspian and Timan-Pechora basins (7 deposits each). Their physical and chemical properties are presented in Table 9. It is evident from Table 9 that these oils belong to the high-density, extremely high-viscosity, sulfur, medium paraffin, moderately resin, medium asphaltene and low gas content oil classes.

Table 7. Physic-chemical properties of crude oils with high sulfur gas content (above 5%)

Physic-chemical parameters	Database sample	Mean value
Density, g/cm^3	23	0.8767
Viscosity at 20 °C, mm^2/s	10	49.33
Sulfur content, wt%	19	2.18
Paraffin content, wt%	13	4.05
Resin content, wt%	12	14.56
Asphaltene content, wt%	11	6.23
Gas content in oil, m^3/t	15	109.97
Thermobaric conditions of occurrence		
Reservoir temperature, °C	27	74.75
Reservoir pressure, MPa	30	31.81

Table 8. Physical and chemical properties of in-reservoir oils at the temperature above 100°C

Physic-chemical parameters	Database sample	Mean value
Density, g/cm^3	432	0.8251
Viscosity at 20 °C, mm^2/s	155	12.03
Sulfur content, wt%	237	0.36
Paraffin content, wt%	228	9.85
Resin content, wt%	214	5.36
Asphaltene content, wt%	203	1.27
Low-boiling fraction 200 °C, wt%.	75	29.41
Low-boiling fraction 300 °C, wt%.	64	49.10
Gas content in oil, m^3/t	117	163.63
Thermobaric conditions of occurrence		
Reservoir temperature, °C	838	119.40
Reservoir pressure, MPa	665	38.31

Table 9. Physical and chemical properties of in-reservoir oils at the temperature lower than 20 °C

Physic-chemical parameters	Database sample	Mean value
Density, g/cm^3	221	0.8855
Viscosity at 20 °C, mm^2/s	165	12,929.34
Sulfur content, wt%	137	1.46
Paraffin content, wt%	126	3.87
Resin content, wt%	112	14.52
Asphaltene content, wt%	106	3.98
Low-boiling fraction 200 °C, wt%.	24	18.75
Low-boiling fraction 300 °C, wt%.	24	33.13
Gas content in oil, m^3/t	104	36.82
Thermobaric conditions of occurrence		
Reservoir temperature, °C	318	15.37
Reservoir pressure, MPa	267	10.04

A comparison of the data from Tables 8 and 9 has shown that oil properties are strongly controlled by the formation temperature. The highest quality indicators are exhibited by the oils occurring in high-temperature reservoirs.

Deep-seated oils (below 4,500m) are also hard-to-recover reserves. There are 354 samples from 180 deposits of 24 basins in the data base. An analysis of their physical and chemical properties presented in Table 10 has revealed that in general they can be classified as light, high-viscosity, medium-sulphur, medium-paraffin, low-resin, low-asphaltene oils with a high content of low-boiling (200°C) fraction and average content of low-boiling (300°C) fraction. These oils are characterized by a high content of oil gas, which is also complicates oil production.

Table 10. Physical and chemical properties of oils from deep-lying strata (more than 4,500m)

Physic-chemical parameters	Database sample	Mean value
Density, g/cm^3	153	0.8377
Viscosity at 20 °C, mm^2/s	72	79.15
Sulfur content, wt%	104	0.52
Paraffin content, wt%	66	5.92
Resin content, wt%	59	5.91
Asphaltene content, wt%	62	1.81
Low-boiling fraction 200 °C, wt%.	47	32.14
Low-boiling fraction 300 °C, wt%.	32	46.57
Gas content in oil, m^3/t	9	263.24
Thermobaric conditions of occurrence		
Reservoir temperature, °C	77	117.48
Reservoir pressure, MPa	77	66.73

6. Conclusion

Currently, production of hard-to recover reserves increases due to depletion of readily available oils in the whole world. Here we have identified the criteria for classification of oils as hard-to-recover reserves. Using a global database on physical and chemical properties of oils, we have analyzed the distribution of viscous, heavy, paraffin and highly resin oils in terms of the volume of their reserves. In order to specify the physical and chemical properties of these oils as a function of the conditions and the depth of their occurrence and the reservoir-temperature differences, a generalized classification of oils by their density, viscosity and chemical composition has been developed. The results obtained can be used to develop new methods of enhanced oil recovery, to improve the available technologies, to optimize the transportation of oils with anomalous physical and chemical properties, and to solve a number of related oil-sector problems.

References

[1] Polishchuk Yu.M., Yashchenko I.G. Possible correlation between crude oil chemical composition and reservoir age. Journal of Petroleum Geology. 2006, Vol.29, 2, 189-194.

[2] Polishchuk Yu.M., Yashchenko I.G. Phisico-chemical properties of oils: statistical analysis of spatial and temporal changes. Novosibirsk: Izd-vo of SB RAS, Filial "Geo". 2004. – 109 p. [in Russian].

[3] Resource assessment of oil and gas in oil-gas-systems of the world / Eds. K.A. Kleschev, G.F. Ulmishek. - Moscow, VNIGNI. 2008. – 352 p. [in Russian].

[4] Lurier M., Schmidt F. Oil. On the discussion of the origin. Sulfur and metal contents as genetic characteristics. - Saarbrucken: LAP LAMBERT Academic Publishing GmbH & Co. KG, 2012. - 292 p. [in Russian].

[5] Purtova I.P., Varichenko A.I., Shpurov I.V. Hard-to-Recover Oil Reserves. Terminology. Problems and the State of their Development in Russia. Nauka and TEK. 2011, 6, 21-26 [in Russian].

[6] Yakutseni V.P., Petrova Yu.E., Sukhanov A.A. Dynamics of the Share of the Relative Content of Hard-to-Recover Oil Reserves in Balance Reserves. Neftegazovaya Geologiya. 2007, 2, 1-11. www.ngtp.ru [in Russian].

[7] Maksutov R., Orlov G., Osipov A. Development of High-Viscosity Oil Reserves in Russia. Technologies of TEK. 2005, 6, 36-40 [in Russian].

[8] Vartivarian D., Andrawis H. Delayed coking schemes are most econo-mical for heavy-oil upgrading. Oil & Gas Journal. 2006, 6, 52–56.

[9] James G. Speight. The Desulfurization of Heavy Oils and Residua, Second Edition, Revised and Expanded, Laramie, Wyoming, 1999. – 458 p.

[10] Polishchuk Yu.M., Yashchenko I.G. Spatial and temporal changes of density and chemical composition of heavy oils of Eurasia. International Journal of Energy and Environment. 2011, 2, Issue 4, 717-722.

[11] Yashchenko I.G., Polishchuk Yu.M. Physical, Chemical and Rheological Properties of Paraffin Oils. Oil and Gas (Almaty). 2011, 5(65), 59-68 [in Russian].

[12] Yashchenko I.G. Spatial Distribution of Paraffin Oils and their Physical and Chemical Properties. Vestnik TSK Rosnedra. 2011,1, 40-46 [in Russian].

[13] Polishchuk Yu.M., Yashchenko I.G., Kozin E.S., An V.V. Database on the composition and physic-chemical properties of oil and gas (DB OG). Official bulletin of the Russian Agency for Patents and Trademarks. 2001, 3, 340-341 [in Russian].

6

Numerical computation of solar heat storage in phase change material/concrete wall

Mustapha Faraji

Physics Department, LPMMAT Laboratory, Faculty of Sciences Ain Chock, Hassan II University, PO 5366- Maarif, Casablanca- Morocco.

Abstract
A one-dimensional mathematical model was developed in order to analyze and optimize the latent heat storage wall. Two layers of phase change material (PCM) are sandwiched within a concrete wall. The governing equations for energy transport were developed by using the enthalpy method and discretized with volume control scheme. A series of numerical investigations were conducted. The effect of the melting temperature on the possibility of increasing the energy performance of the proposed heating system was analyzed. Results are obtained for thermal gain and temperature fluctuation. The charging/discharging process was also presented and analyzed.

Keywords: Thermodynamics; Phase-change material; Building; Concrete; Latent heat; Thermal control.

1. Introduction

Thermal energy storage in PCMs has been a central topic in research for the last 20 years. Because of the high thermal mass of PCM walls, they are capable of minimizing the effect of large fluctuations in ambient temperature on the inside temperature of a building [1]. They can therefore be very effective in shifting the heating and cooling load to off peak electricity periods. Castell et al. [2] conducted an experimental set-up to test phase change materials with alveolar brick for Mediterranean construction in real conditions. Several cubicles were constructed and their thermal performance throughout the time was measured. For each construction material, RT-27 and SP-25 A8 PCM is added in one cubicle. The experiments show that the PCM can reduce the peak temperatures up to 18 °C and smooth out the daily fluctuations. A thermally activated panel with PCM for applications in lightweight buildings was developed [3]. The design for the panel exploits the properties of the PCM paraffin. Its high thermal storage capacity during the phase change enables the overall panel thickness to be limited to 5 cm. Studies of the thermal behavior of wallboards containing PCM have been carried out in order to build a light building envelope with less energy consumption. The thermal effectiveness of a building's roof with phase change material (PCM) was studied by Alawadhi et al. [4]. The considered model consists of a concrete slab with vertical holes filled with PCM. The thermal effectiveness of the proposed roof-PCM system is determined by comparing the heat flux at the indoor surface to a roof without the PCM during typical working hours. It was found that n-eicosane PCM shows the best performance among the examined PCMs. The results indicate, also, that the heat flux at the indoor surface of the roof can be reduced up to 39 % for a certain type of PCM. Latent heat thermal energy storage was integrated into cooling system in buildings to cool the rooms during peak period while charging the storage system during off peak period [5]. PCM can also be integrated into a building wall to acts like a heat reservoir to provide thermal comfort by reducing the cooling/heating loads. There is a significant increase in comfort even in the absence of an air conditional [6]. A PCM cold storage unit incorporated in air conditioning systems in buildings has been build and studied. The PCM is solidified during the night and used to cool the interior of the building during the hottest hours of the day. This concept is known as free cooling [7]. Berroug et al. [8] analyzed the thermal performance of a north wall made with hydrate salt phase change material as a storage medium in east-west oriented greenhouse. A numerical study has been developed to investigate the impact of the PCM on greenhouse temperature. Results show that, with an equivalent to 32.4 kg of PCM per square meter of the greenhouse area, temperature of inside air were found to be 6°C more at night in winter period with less fluctuations. In the present paper, a study of the thermal performance of a wall/phase change material system for building thermal energy management is analyzed and discussed. The thermal response to the realistic boundary conditions and using temperature ranges and material properties that are common in buildings are investigated. The effect of the PCM melting temperature was analyzed for the cold period (January). The thermal performance is compared to an ordinary wall.

2. Mathematical model

Figure 1 shows the physical system studied in the present work. It consists of a sandwiched wall made with five layers: The inner face is made with 1 cm mortar followed by two layers of PCM sandwiched by cement. The total thickness of the wall is L=30 cm and the PCM layer is located at constant positions x_{m1} and x_{m2}. The thickness of each PCM layers is kept constant e_m=2cm. The composite wall corresponding thermal properties are given in Table 1. The PCM wall was initially at a uniform temperature close to the thermal comfort requirement, $T_{ini} \sim T_c$= 22°C. The boundary conditions on the outer surface of wall are due to the combined effects of solar radiation and ambient air convection. Meteorological data for every hour servicing during January in Casablanca, Morocco (33°36'N, 7°36'W) are used (Figure 2).

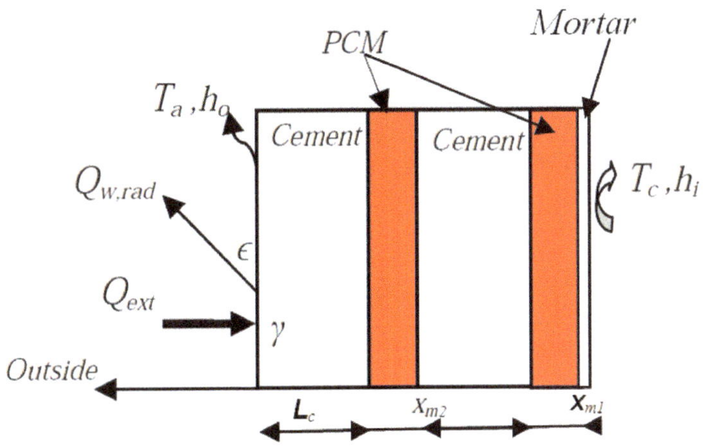

Figure 1. Physical model

Before presenting the equations that govern the heat transfer process occurring in the system, some assumptions were made in order to simplify its resolution. These assumptions are given as follows:

- thermal properties are constant and not varying with respect to temperature;
- end effects are neglected;
- interfacial resistances are negligible;
- heat transfer through the wall is assumed 1 D ;
- buoyancy induced flow in the melt PCM is neglected.

The energy transport in composite concrete/PCM may be written using the enthalpy formulation [9]:

$$\frac{\partial H}{\partial t} = \nabla\left(k\nabla T\right) \tag{1}$$

where,

$$H(T) = h(T) + \rho\lambda f \Delta H_f \tag{2}$$

And

$$h(T) = h(T_m) + \int_{T_m}^{T} \rho c_p dT \tag{3}$$

λ =1 in PCM and λ =0 in concrete layers.
Using sensible enthalpy h, Eq(1) is rewritten as:

$$\frac{\partial h}{\partial t} = \alpha \frac{\partial^2 h}{\partial x^2} - \rho\lambda \Delta H_f \frac{\partial f}{\partial t} \tag{4}$$

The liquid fraction *f* in the PCM layer is estimated as:

$$f = \begin{cases} 1 & if \;\; T > T_m \\ 0 & if \;\; T_m < T \\ 0 < f < 1 & if \; T = T_m \end{cases} \tag{5}$$

The continuity of the temperature and heat flux density are used at the interfaces concrete/PCM. The thermal properties at interfaces were obtained by the following relations:

$$k_i = \frac{k_+ k_- (\delta_- + \delta_+)}{k_+ \delta_- + k_- \delta_+}, \; k_m = f k_l + (1-f)\, k_s \,,$$
$$\rho c_p = f (\rho c_p)_l + (1-f)\, (\rho c_p)_s \tag{6}$$

where δ_+ is the distance between the interface and the first node inside the PCM region and δ_- is the distance between the interface and the first node inside the concrete layer.

- *Boundary conditions*
- at x=0 (indoor):

$$-k \left.\frac{\partial T}{\partial x}\right|_{x=0} = h_i (T_c - T) \tag{7}$$

- at $x=L$ *(outdoor)*:

$$-k \left.\frac{\partial T}{\partial x}\right|_{x=L} = h_o (T_a - T) + Q_{w,rad} + \gamma Q_{ext} \tag{8}$$

where,

$$Q_{w,rad} = \sigma\, \varepsilon\, F(T_{sky}^4 - T^4) = h_{sky}^r (T_a - T) \tag{9}$$

here, the sky is considered as black body at T_{sky} (°C) [10, 11]:

$$T_{sky} = 0.0552 \times (T_a + 273.15)^{1.5} - 273.15 \tag{10}$$

The irradiative heat transfer coefficient h_{sky}^r is given by:

$$h_{sky}^r = \frac{\sigma\, \varepsilon\, (T_{sky}^4 - T^4)}{T_a - T} \tag{11}$$

The shape factor F is assumed to be equal to one.
-Heat flux at interfaces, $x= x_{mi}$ and $x= x_{mi} +e_m$, between PCM and concrete is evaluated as:

$$q_i'' \approx \frac{k_+ k_-}{k_+ \delta_- + k_- \delta_+}(T_+ - T_-) \tag{12}$$

A self-Fortran computer code was developed in order to implement the numerical method. The convergence of the calculus was declared, at each time step, when a criterion based on the energy conservation principle was satisfied. A grid and time step refinement was carried out and showed that the time step of 2 s and a size grid of 50 nodes were found to be adequate for all the computations.

Table 1. Thermophysical properties [1]

Material	Organic PCM	Cement/mortar
T_m (°C)	16	-
ΔH_f (kJ/kg)	148.5	-
ρ (kg/m^3)	981 (Solid) 862 (Liquid)	2200
k (W/mK)	0.145	1.5
c_p (J/kg K)	1460 (Solid) 2130 (Liquid)	838
γ	-	0.9
ε	-	0.8

3. Results and discussion

Figure 2 shows the time wise variation of the ambient temperature and total solar radiations received by south faced wall, during a typical decade representing the cold period of the year (winter -January in Casablanca-Morocco, 33°36'N, 07°36'W). Data analysis shows the minimum temperatures are obtained during the night. On average, temperatures minima and maxima range between 5°C and 18°C, respectively, and the ambient temperature swing between these extremes. Radiations remain zero during the night and increase in the following day. Solar radiations, Q_{ext}, received by a south oriented vertical wall are zero during the first 6 hours every day and increases with the sunrise and the ambient temperature increases. Solar radiation reaches a maximum value (720 W/m²) between solar noon and 15h00 and falls to 0 W/m² at 18h00 (sunset and cancellation of solar radiations). Radiations remain zero during the night. Casablanca climate is characterized by significant temperature fluctuations with lower nocturnal values of temperature.

Figure 3 compares the thermal behavior of the ordinary wall and the composite wall with PCM, (T_m=16 °C -Table 1). Data analysis shows that the inside wall temperature increases from 16°C to 25°C for wall/PCM and from 14 °C to 34 °C for an ordinary concrete wall without PCM. Remember that ambient temperature varies between 5°C to 15°C (Figure 2), but the wall receives also a heat flux due to the solar radiations absorbed by the concrete (γ $=0.8$) combined with convective heat flux that increases the PCM temperature to the melting point (T_m= 16°C). Also, remark that sensible heat storage stops and latent heat process starts (Figure 4). As a result, a first layer of the liquid phase appears (f> 0) and the latent heat storage develops with the start up of the melting process. For the PCM wall case, the slop of T curve weakens during the phase change because the melting and solidification occur at a constant temperature. The increase of solar radiations accumulates sensible heat storage in concrete and latent heat in PCM layer. The composite PCM-Concrete wall can be considered as an important heat storage device. The excess of heat is stored in the wall with less temperature fluctuation until the full melting of PCM (f=1). The stored heat during a day is used for heating needs in the following night. Figure 3 shows also that PCM is totally melted at 11h00 (f=1) and the sensible heating continue and leads to the increase of wall face temperature between 11h00 and 17h00. This result is confirmed by Figure 4 which gives the latent, q_L, and sensible heat, q_s, variations. Analysis of such figure shows that, q_L, is more important than, q_s, and remains constant between 11h00 and 17h00 (f =1). The overheating of melted PCM layer is due to the raise of the sensible heat storage during that period and, therefore, leads to the increase of the wall temperature. This process continues until the decease of liquid fraction. The discharging process starts at

21h00 and the inner wall temperature falls after a certain delay due to the thermal inertia and to the weak value of its thermal conductivity. Note that minimum temperature reached in case of the wall with PCM is clearly greater than that achieved without PCM because the heat lost by the wall is shifted by the release of the important stored latent heat. Figure 4 shows, also, that PCM solidify and melts periodically and the stored latent heat, q_L, varies between 18.5MJ and 43MJ. Day/night cycles allow for charging/discharging of the PCM and wall temperature varies from 16°C to 25°C. The fluctuation intensity of the inner temperature (concrete with PCM case) decreases with the development of the latent heat of fusion. The accumulated heat during the melting processes added to the sensible overheating of the liquid PCM, q_s increases, lead to the enhancement of the inner temperature during the night. The latent heat stored in the wall/PCM shifts the fast decrease of the temperature during the night and the inside temperature of the PCM concrete wall is less fluctuating. Note that for wall without PCM sensible heat q_s decreases to negative values at night that explains the lower values of temperatures reached in case of ordinary wall.

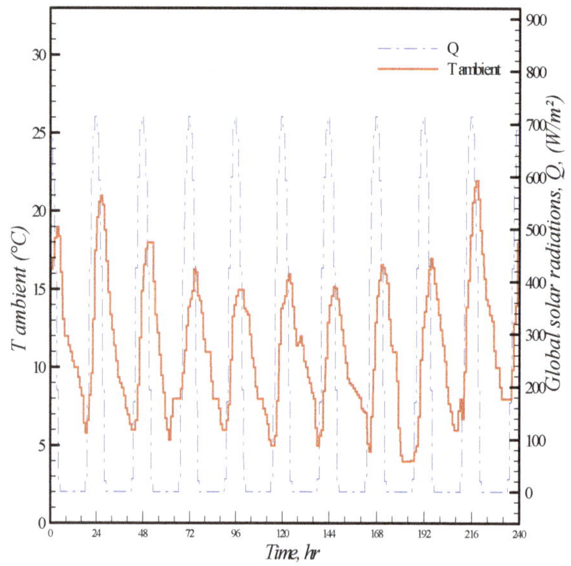

Figure 2. Time wise variations of the ambient temperature, T_a, and the global solar radiations, Q_{ext}, received by a south oriented vertical wall, (January- Casablanca, Morocco 33°36'N, 07°36'W)

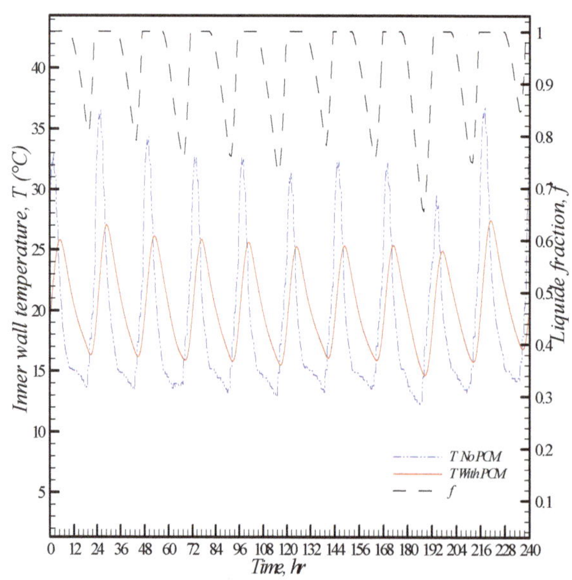

Figure 3. Variations of the inner wall temperature, $T_{1,in}$, and liquid fraction, f

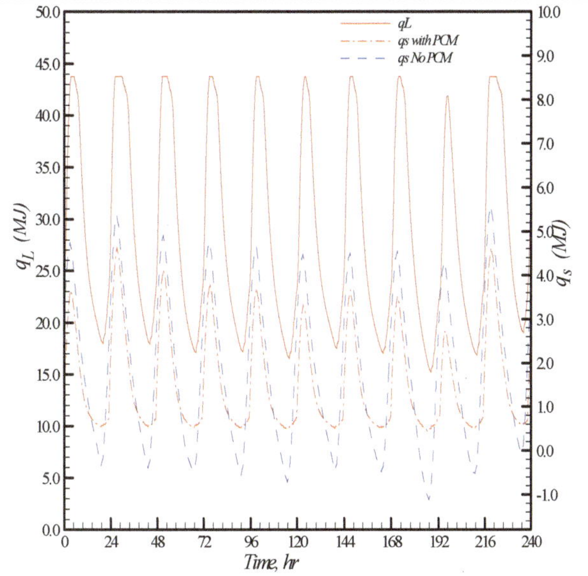

Figure 4. Variations of the latent heat, q_L, and sensible heat, q_s

In order to investigate the effect of the PCM melting temperature on its thermal performance, a study was made by modifying the parameter , T_m, then examining the effect on, $T_{1,in}$, q_L, q_s and temperature fluctuations. Figure 5 sketches the time wise variation of the concrete/PCM wall temperature for various melting temperature of the PCM layer: Values tested range within the interval less than, near and greater than the comfort temperature, T_c=22°C. The liquid fraction, f, is also presented. Data analysis shows that the inner temperature relative deviations $\frac{T_{max} - T_{min}}{T_c}$ are 0.53 , 0.19, 0.22 and 0.43 for melting temperature T_m equals to 10°C , 16 °C , 23 °C and 28 °C , respectively. So it's clear that when the melting temperature of the PCM layer is close to the comfort value, T_c=22°C, better are the living thermal conditions with less temperature fluctuations. For T_m equals to 10°C, PCM completely melts (f =1) and never solidifies and, so, latent heat is disabled. The same behavior was found when T_m equals to 28°C The PCM remains in solid state and a weak quantity of PCM is melted , f< 0.1, and latent heat doesn't work. In contrast when setting T_m near to the thermal comfort value 22°C an important amount of the PCM is discharged and solidified (f=0) at night and during the following day as a result good nocturnal heating of the wall and important diurnal melting of the PCM.

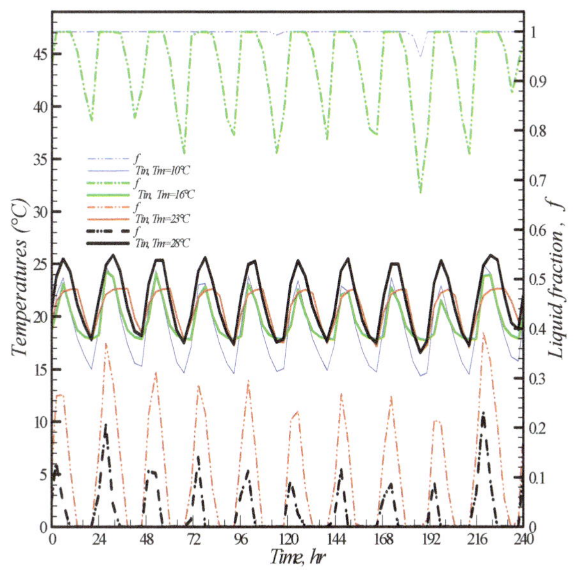

Figure 5. Variations of the inner temperature and liquid fraction for various value of T_m

The performance of the wall with PCM as a heating system for the building has been evaluated in terms of Thermal Load Leveling, *TLL*, and daily (24 hours) heating potential, Q_h, using the following equation [8]:

$$\begin{cases} TLL = \dfrac{\left(T_{i,\,in}\right)_{max} - \left(T_{i,\,in}\right)_{min}}{\left(T_{i,\,in}\right)_{max} + \left(T_{i,\,in}\right)_{min}} \\ Q_h = \sum_{t=1}^{24} c_p\left(T_{1,in}(t) - T_a(t) \right) \end{cases} \tag{13}$$

Thermal load leveling, *TLL*, puts a figure on the fluctuations of temperature inside the room. The less the fluctuations the better is the environment for the occupants. In winter, without heating arrangement for thermal comfort requirements, thermal load leveling should have lower value by incorporating heating method due to the increase of $\left(T_{i,\,in}\right)_{max} + \left(T_{i,\,in}\right)_{min}$) and the decrease of ($\left(T_{i,\,in}\right)_{max} - \left(T_{i,\,in}\right)_{min}$).

The results for daily variation of thermal load leveling for wall with and without PCM are summarized in Table 2. It can be seen that the value of *TLL* is maximum for the wall without PCM and it is reduced to about 20 % for composite wall with PCM with accurate melting temperature. The lower values of thermal load leveling indicate the decrease in the fluctuations of inner face and, thereby, there occurs an improvement of thermal comfort environment in the building. Also, the daily variations of total heating potential Q_h obtained from PCM wall for typical cold days were calculated and have been accumulated in Table 2. From the results, it is found that the heating potentials obtained in case with PCM wall were higher as compared to an ordinary wall.

Table 2. Thermal load leveling

T_m (°C)	$T_{in,min}$(°C)	$T_{in,max}$(°C)	*TLL*	Q_h (kJ/kg)
No PCM	14	34	0.45	157
10	14.7	26.3	0.28	159
16	16	25	0.17	160
23	17.8	22.5	0.10	163
28	17.4	26.7	0.21	161

4. Conclusion

A numerical investigation of the thermal performance of sandwiched concrete/PCM wall was performed. The results showed a significant reduction of indoor wall temperature due to absorption of solar gains in the composite wall in conjunction with melting of the PCM. It can therefore be concluded that PCM is effective for storage of solar gains and improvement of thermal comfort. Additional numerical simulations using different melting temperatures to control the inside temperature of the wall were conducted. The results showed that the wall thermal load leveling of the wall containing PCM with melting point near thermal comfort temperature was reduced about 25% compared to the wall without PCM with lower peak temperatures and more constant conditions, smoothing out the daily temperature fluctuations. This demonstrates the significant contribution and potential of the use of PCM in building envelopes for energy savings and thermal comfort.

Nomenclature

c_p	specific heat, (kJ/kg K)
f	liquid fraction
e_m	PCM layer thickness, (m)
h	enthalpy, (J)
h_i	inner wall convective heat transfer coefficient, (W/m^2 K)
h_o	outside wall convective heat transfer coefficient, (W/m^2 K)
h^r_{sky}	irradiative heat transfer coefficient, (W/m^2 K)
k	thermal conductivity, (W/m K)
L	total wall thickness, (m)

Indices

i	Interface
m	melting, PCM
c	Comfort
sky	Sky
in	inner wall face
out	outer wall face
+,-	right, left nodes
f	fusion

q_L	latent heat stored per square meter of the wall, (J)	***Greeks***	
q_S	sensible heat stored per square meter of the wall , (J)	ρ	density, (kg/m^3)
Q_{ext}	solar radiation, (W/m2)	γ	ABSORPTIVITY
$Q_{w,rad}$	heat lost to the ambient by radiations, (J)	α	thermal diffusivity, (m/s^2)
x	Cartesian coordinate, (m)	σ	Stefan Boltzmann constant
t	time, (s)	ε	concrete emissivity
T	temperature, (°C)	ΔH_f	latent heat of PCM (kJ/kg)

References

[1] Zalba, B., Marın, J.M., Cabeza, L.F., Mehling, H., 2003. Review on thermal energy storage with phase change: materials, heat transfer analysis and applications. Applied Thermal Engineering 23 (3), 251e283.

[2] A. Castell, I. Martorell, M. Medrano, G. Pe rez, L.F. Cabeza, Experimental study of using PCM in brick constructive solutions for passive cooling, Energy and Buildings 42 (2010) 534–540.

[3] Koschenz, M., Lehmann, B., 2004. Development of a thermally activated ceiling panel with PCM for application in lightweight and retrofitted buildings. Energy and Buildings 36 (6), 567e578.

[4] Esam M. Alawadhi , Hashem J. Alqallaf, Building roof with conical holes containing PCM to reduce the cooling load: Numerical study, Energy Conversion and Management 52 (2011) 2958–2964.

[5] F. Agyenim, N. Hewitt, The development of a finned phase change material (PCM) storage system to take advantage of off-peak electricity tariff for improvement in cost of heat pump operation, Energy and Buildings 42 (2010) 1552–1560.

[6] X. Fang, Z. Zhang, A novel montmorillonite-based composite phase change material and its applications in thermal storage building materials, Energy and Buildings 38 (2006) 377–380.

[7] V. Butala, U. Stritih, Experimental investigation of PCM cold storage, Energy and Buildings 41 (2009) 354–359.

[8] F. Berroug, E.K. Lakhal, M. El Omari, M. Faraji, H. El Qarnia, Thermal performance of a greenhouse with a phase change material north wall, Energy and Buildings 43 3027–3035 (2011).

[9] V.R. Voller and S.PENG An enthalpy formulation based on an arbitrarily mesh for solution of the stefan problem. Computational Mechanics, 14:492-502, 1994.

[10] W.C. Swinbank, Long-Wave radiation from clearskies, Quarterly Journal of the Royal Meteorological Society 381(89) 339-348, (1963).

[11] Patankar, S.V., Numerical Heat Transfer and Fluid Flow, Hemisphere, 1983.

A site-specific design of a fixed-pitch fixed-speed wind turbine blade with multiple airfoils as design variable

Arturo Del Valle-Carrasco[1], Delia J. Valles-Rosales[1], Luis C. Mendez[2], Alejandro Alvarado-Iniesta[2]

[1] New Mexico State University, Department of Industrial Engineering, MSC 4230/ECIII, PO Box 30001, Las Cruces NM, 88003-8001, USA.
[2] Autonomous University of Ciudad Juarez, Department of Industrial Engineering and Manufacturing Av. Del Charro, 450 Nte, Ciudad Juarez, Chihuahua, 32315, Mexico.

Abstract
This study seeks to optimize a fixed-pitch fixed speed (FPFS) wind turbine blade's performance using the chord, twist and the use of 3 different airfoils with varying relative thickness as design variables for the maximization of the Annual Energy Production for the wind profile of Roswell NM. A baseline design of the blade starts with a replica of the Phase VI blade utilized in a NASA-Ames experiment and a Matlab script utilizes the Blade Element Momentum Theory (BEM) for the aerodynamic analysis. The SQP method for Local Search are used to exploit the model utilizing the Phase VI design as a starting point which contains the S809 airfoil with a 0.21 relative thickness for the complete blade. Optimization results reduced the relative thickness of the airfoils to 0.17 and an increase of 36% in energy production was observed using this method.

Keywords: BEM; SQP; Optimization; HAWT; Wind energy.

1. Introduction

Global warming along with depletion of fossil fuels have created changes in energy policies in countries around the globe. Several countries in Western Europe are developing wind energy capacities aggressively such as Denmark having 30% of its electricity supplied by wind energy, Portugal 18%, Ireland 16%, Germany 10%, and the US presently consists of only 4.4%. However, the U.S. Department of Energy [1] stated a goal of reaching a 20% electricity production generated by wind resources by the year 2030.
There are two main types of wind turbines: the Vertical Axis Wind Turbine (VAWT) and the Horizontal Axis Wind Turbine (HAWT). Of the two topologies the most popular with manufacturers is the two or three-bladed HAWT due to its cost effectiveness in reliability, acoustics and efficiency [2]. Some HAWTs possess a blade pitching mechanism and therefore are known as a variable-pitch turbine, whereas their counterparts lacking this mechanism are known as fixed-pitch (FP) [3]. An additional variation is related to the speed at which the rotor spins: the two topologies are the fixed-speed (FS) and the variable-speed (VS) wind turbines. A variety of wind turbine which does not pitch its blades and has a fixed speed rotor is known as a fixed-pitch fixed-speed (FPFS) wind turbine. Compared to a FPVS turbine which use permanent magnet synchronous generators requiring complex electronics for grid

connections, the FPFS have the advantage of being robust, reliable and of lower cost than the variable-speed [4]. A FPFS will yield a maximum coefficient of power at only one given wind speed unlike a VS rotor, however research has shown that for high values of the Weibull's shape parameter of a wind profile, the FPFS turbines have competitive Annual Energy Production (AEP), to about 88% of its variable-speed version [5].

Wind turbine design optimization has been an ongoing research in academia and in the industrial practice during the last couple of decades. Previous optimization efforts have focused in an optimization of the twist and the chord of a given FPFS wind turbine with the objective to maximize the energy production [6-8]. Additional efforts have been performed in the area of airfoil optimization to improve the aerodynamic efficiency of the HAWT's airfoils for VS rotors [17-19]. For example Fuglsang (2004) designed the RISO family of airfoils to maximize the coefficient of torque and lift and minimize roughness [9], other studies modified the leading edge for roughness sensitivity [10, 11].

Some wind turbine airfoils are designed to have a high C_l / C_d and at the same time a high C_l for a given *alpha-design* angle of attack [12]. The optimizations of the C_l / C_d for a given *alpha-design* can be useful for an airfoil that is to be fitted in a VP or a VS turbine, since a control strategy can be implemented where the pitch or speed of the rotor is modified to maximize the exposure of the airfoil's optimum *alphas*. This is not the case in a FPFS which may not modify either parameter and hence its airfoils will be exposed at its optimum *alpha-design* for only one given wind speed, and consequently function at suboptimum levels for any other wind speed.

The Phase VI wind turbine will be utilized as a starting point design of the optimization process. In 2001, The U.S. National Renewable Energy Laboratory published experimental data [13] for the Phase VI wind turbine performed in the NASA-Ames wind tunnel located in Moffet Field, California. The wind turbine consists of 2-blades with 5.03-meter radius with chord and twist variable along the blade and a 3-degree pitch. The airfoil is the S809 and it is the same along the blade with a relative thickness of 0.21; the rotational velocity is 72 RPM [14].

The energy production of the rotor is simulated using BEM theory as in Manwell [15]; BEM Theory is widely used in the literature and in the industry to simulate the energy production of rotors for the energy optimization of wind turbines [7, 16-18]. The design proposed consists of a 2-bladed FPFS rotor with 3 different airfoils with variable relative thickness described by four Bezier curves; the chord and the twist are also modeled using one Bezier curve each. In particular the airfoils will have variable relative thickness with possible values ranging from 0.16 to 0.40 ratio *thickness/chord length (t/c)*. Since there are three airfoils, each one of them will be utilized for one third of the total length of the blade and will be referred as root, mid-section and tip airfoil.

The flow analysis of the airfoils is calculated with Drela's XFOIL [19] code to find their coefficients of lift and drag. XFOIL is a method used widely in the literature for the design of low speed airfoils [20-23]. Since BEM theory requires an extrapolation of the coefficients of lift and drag from -180 to 180 degrees the AERODAS method by Spera [24] was utilized to perform the extrapolation.

It is the aim of this study to investigate the impact of chord, twist with the addition of airfoil design and their relative thickness in the energy production of a FPFS wind turbine. The article is organized as follows: Section 2 explains the objective function to be maximized, Section 3 elaborates the design variables and constraints; Section 4 describes the methodology utilized for maximization of the objective. Finally, Section 5 and 6 close with results and conclusions of the study.

2. Objective function

The objective function to maximize will be the AEP using the wind profile for the location of Roswell NM. The data consists of hourly data wind speed obtained from the National Oceanic and Atmospheric Administration [25] station name Roswell Industrial Air Park, NM US with station ID GHCND:USW00023009 for the periods of January 1st 2010 to December 31st 2010. The data is characterized with a Weibull probability density function (*pdf*):

$$f_w(U) = (k / c) \cdot (U / c)^{k-1} \exp[-(U / c)^k] \tag{1}$$

where U represents the speed of the wind and k and c are the shape and scale parameters of the distribution. After a calculation of the Coefficient of Power (C_p) curve by BEM theory, the AEP can be obtained with the formula established by Hau [26]

$$AEP = 8760 \times \tfrac{1}{2}\eta\rho A \int_{cut-in}^{cut-out} U^3 C_p(U) \times f_w(U)dU \tag{2}$$

where η is the transmission efficiency of the wind turbine including the mechanical and electrical efficiency which is considered 85% in this study, ρ is the air density 1.062 kg/m^3, A=79.485m^2 corresponds to the area spanned by the rotor, the cut-in and cut-out speeds are 5 and 25 m/s.

3. Design variables and constraints

The shape of the blade is determined by the airfoil contour, the chord, and the twist distribution of each section of the blade. In order to achieve smooth distributions of the chord and the twist, two Bezier curves are used to describe them; for the airfoil shape four composite Bezier curves will be used to determine their shape.

The design proposed consists of a 2-bladed FPFS rotor with three different airfoils each described by four Bezier curves. Since three airfoils are proposed, each will be used for one third of the total length of the blade. In particular the first airfoil occupies the length from the root at 1.257 to 2.51m; the second airfoil from a length of 2.51 to 3.77m and the third airfoil from a length of 3.77m to the tip of the blade at 5.03m; they are referred as root, mid, and tip airfoils for simplicity.

3.1 Chord and twist distributions

Bezier curves require four control points each containing two coordinates, in this case the first coordinate represents the position of the point along the radius and the second represents either the chord or the twist values. Thus, the points that define the chord's Bezier are:

$$(r_{ci}, c_i) \quad \text{for} \quad i = 1,2,3,4 \tag{3}$$

And the twist's by

$$(r_{ci}, t_i) \quad \text{for} \quad i = 1,2,3,4 \tag{4}$$

The chord is constrained to be a decreasing curve for structural reasons, hence the constraint obtained is:

$$1.257 = r_{ch1} \le r_{ch2} \le r_{ch3} \le r_{ch4} = 5.03 \tag{5}$$

$$0.75 \ge c_1 \ge c_2 \ge c_3 \ge c_4 \ge 0 \tag{6}$$

In Eq. (6) a limit in the length of the chord is set at 0.75m, and each subsequent chord coordinate must be decreasing to achieve a blade that tapers down.

The Phase VI has twist values ranging from 25 to 3 degrees from the root to the tip respectively, in this study a larger set of values will be allowed for exploration resulting in the constraints

$$1.257 = r_{t1} \quad \text{and} \quad r_{t4} = 5.03 \tag{7}$$

$$1.257 \le r_{t2}, r_{t3} \le 5.03 \tag{8}$$

$$0 \le t_i \le 40 \quad \text{for} \quad i = 1,2,3,4 \tag{9}$$

An additional constraint is concerned with the distribution of the chord not being larger than the Phase VI's as it would be undesirable to create a design that produces more energy at the expense of a wider blade i.e. more costly; thus the chord distribution of the design will necessarily be constrained to be 'under' that of the Phase VI. This results in the constraint:

$$\int_{1.257}^{5.03} Bezier((r_{c1}, c_1), \ldots, ((r_{c4}, c_4))dr \le \int_{1.257}^{5.03} Bezier(PhaseVI - chord)dr = 2.06 \tag{10}$$

3.2 Airfoils

A variation in several of the dimension in the airfoils will be allowed in the study including relative thickness. As it was previously established by Grasso [12] a variation in relative thickness is recommended for the inner part of the blade to be larger than the outer part for structural requirements since they have a higher priority in the inside of the blade. In this study relative thickness values between 16% to 40% will be explored for the designed airfoil. The structural integrity of the blade is not considered in the scope of this study

For the parameterization of the airfoils, four composite Bezier curves are used as shown in Figure 1; The first curve is formed by points 1-4, the second curve by points 4-7, the third by 7-10, and the fourth curve is composed of points 10, 11, 12, and back to point 1. Points 1, 4, 7, and 10 are shared by adjacent curves. In addition, Point 1 and Point 7 are always fixed at (1,0) and (0,0) respectively.

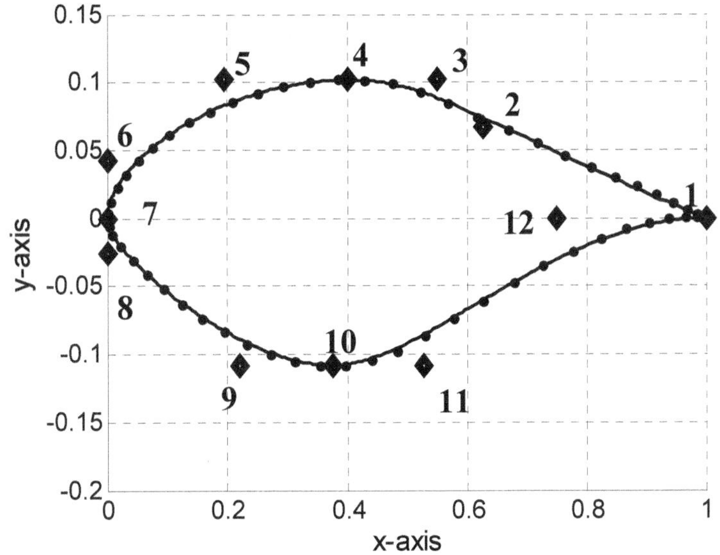

Figure 1. Parametrization of S809 with Beziers

Each *x* and *y-coordinates* of control points not fixed have upper and lower bound constraints. Table 1 sets the lower and upper limits for each coordinates of the airfoils' control points; *x-coordinates* in point 6 and 8 are fixed at zero to achieve a smooth transition between curves connecting at Point 7; the same case applies to Points 3 and 5 *y-coordinates* in which the values are set equal to Point 4, as this yields a smooth transition between curves connecting at Point 4.

Table 1. Control Points upper and lower bounds

	x-Coordinates		y-coordinates	
Point No.	min	max	min	max
Point 2	0.525	0.9	0.04	0.15
Point 3	0.52	0.675	Same as Pt. 4	
Point 4	Fixed at 0.4		0.08	0.2
Point 5	0.1075	0.2725	Same as Pt. 4	
Point 6	Fixed at 0		0.02	0.10
Point 8	Fixed at 0		-0.10	-0.01
Point 9	0.12	0.3	Same as Pt. 10	
Point 10	Fixed at 0.37425		-0.20	-0.08
Point 11	0.475	0.6375	Same as Pt. 10	
Point 12	0.65	0.85	-0.10	0.07

The problem yields six variables for the chord, six for the twist, and each airfoil requires the use of twelve variables for each of the three airfoils i.e. root, midsection and tip areas; a total of 48 variables are considered in the problem.

Airfoils performance is also affected by Reynolds number (Re) defined as [15]

$$Re = \rho U_{rel} c / \mu \tag{11}$$

where c is the chord length of the airfoil, U_{REL} is the relative wind velocity to the airfoil, and μ is the viscosity of the airfoil. An estimation of the Reynolds number in the blade's mid-section of the baseline Phase VI blade which has a chord length of 0.54m needs to be established in order to set a baseline of analysis

Considering that the most of the energy gathered from the wind profile in the area of Roswell are at wind speeds of 10m/s as shown in Figure 2 an analysis of the Reynolds number in the baseline blade will be made for 5, 7.5 and 10m/s.

In Table 2 the Reynold numbers for the midsection of the preliminary blade as estimated from the given wind speeds are calculated. This speed calculation is based in the triangle of speeds from Blade Element Theory. In order to obtain the quantity U_{REL} necessary for Eq. (1) the formula

$$U_{REL} = sqrt(U^2 + U_{TAN}^2) \tag{12}$$

Is used, in which

$$U_{TAN} = \Omega \cdot r \tag{13}$$

Since the rotor speed is constant at 72 RPM the Reynolds number of the HAWT do not vary significantly, which is in the range of 900,000 to 950,000. To simplify the optimization process the design Reynolds number is chosen as 925,000. However it must be clarified that this technique would not be appropriate for other topologies of wind turbines such as variable speed where these values vary significantly.

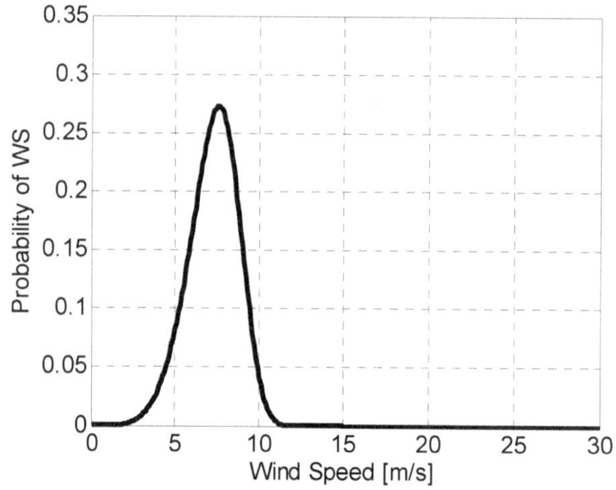

Figure 2. Wind pdf of Roswell, NM

Table 2. Estimated Reynolds at wind speeds

$U(m/s)$	$U_{TAN}(m/s)$	$U_{REL}(m/s)$	Re
5	23.7	24.2	899,224
7.5	23.7	24.9	922,859
10	23.7	25.7	954,966

4. Methodology

In order to calculate the AEP of a given design a set of scripts were written in MATLAB 2014a. It consists on a function which receives a candidate design in the form of a vector containing the control points of the chord, twist, and all airfoils. All lower and upper bounds were mapped to the interval [0, 1] to add simplicity in the coding process. The script receives the vector of the candidate design and a Bezier function will generate the curve distributions for the chord, twist and the three airfoils. Using a DOS shell from Matlab the code executes XFOIL to acquire flow analysis for angles of attack from -8 to 22 at a Re = 925,000 then its results are saved in a text file. A MATLAB script reads the previously written text files and extracts information required for the extrapolation of the coefficients of lift and drag from -180 to 180 degrees by utilizing the AERODAS model as designed by Spera [24]. An example of an extrapolation of the coefficients is shown in Figure 3.

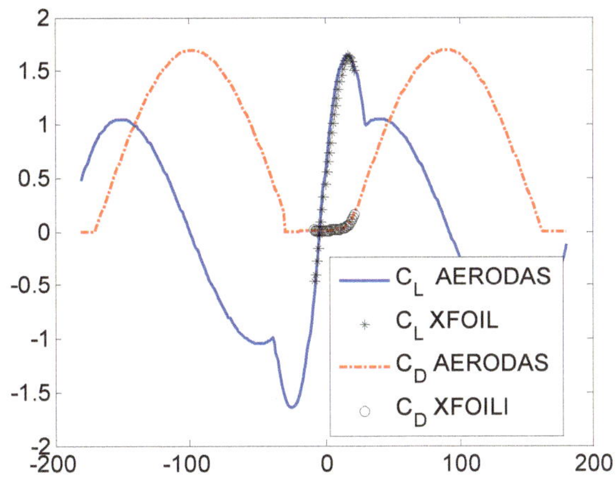

Figure 3. AERODAS extrapolation

A BEM script is executed and returns a Cp curve for the wind speeds 5m/s to 25m/s; finally the formula in Eq. (2) is implemented to calculate the final result of the objective function. A diagram is shown on the process utilized in the calculation of the AEP in Figure 4.

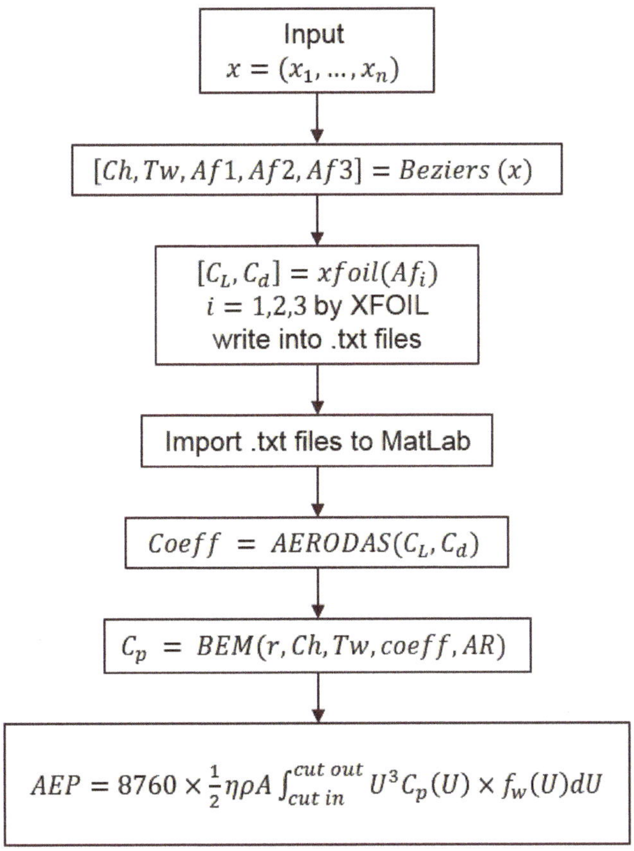

Figure 4. Calculation procedure for the AEP

The SQP algorithm in Matlab with a relative step of 0.15 for the finite difference step was implemented on the AEP calculation for an optimization of the design. The starting point for the search was the design of the Phase VI wind turbine blade in which the airfoils are the S809 for all sections of the blade with a relative thickness of 0.21. The optimization took 18 iterations with a total of 1672 function evaluations.

5. Results

The results of the optimized design increase the AEP by 36%. The initial AEP for the Phase VI blade for the area of Roswell NM yields a value of 4.4x107 W-h. For the designed blade the total energy production is of 6.0×10^7 W-h.

The most important constraint is for the chord to not have a larger profile than that of the Phase VI as this would yield a design with an increased cost. In Figure 5, it is shown that chords for the optimized blade and the Phase VI are almost identical with the optimized being slightly smaller.

Figure 6 presents also twists for both blades where the optimized has a smaller profile for the twist with decreasing values of around 18 to a little over 0 degrees in the tip of the blade.

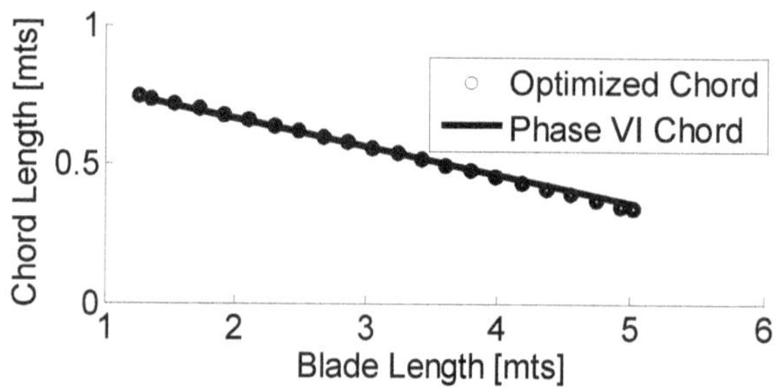

Figure 5. Chords of the optimized and Phase VI

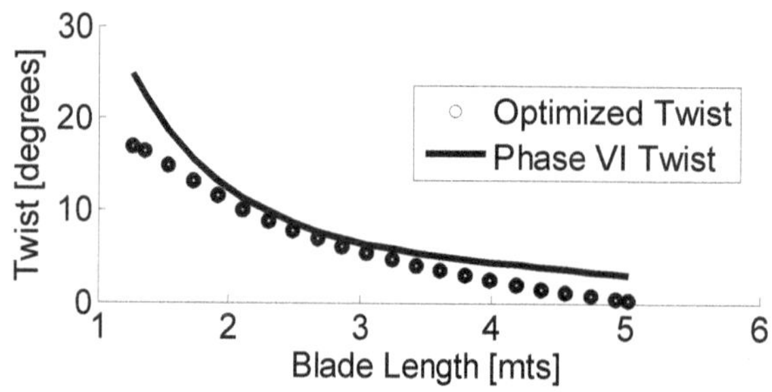

Figure 6. Twists of the optimized and Phase VI

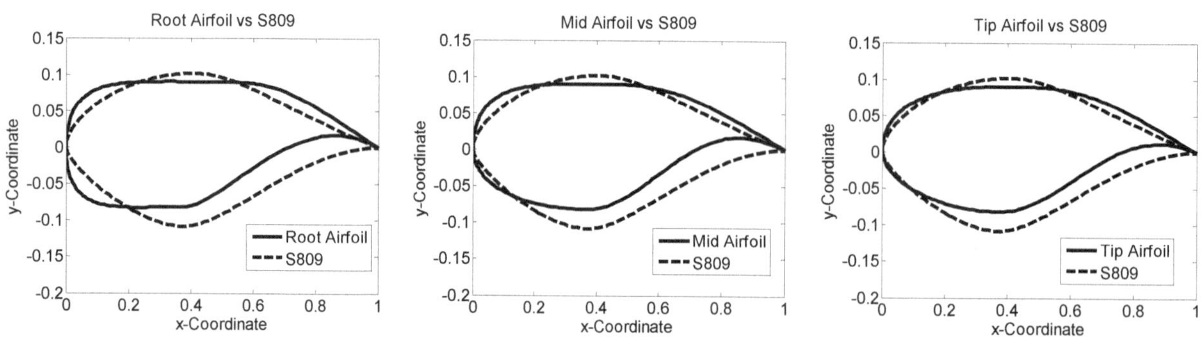

Figure 7. Airfoils for the root, mid-section and tip airfoils compared against the S809

The relative thickness of the three airfoils were identical after the optimization process going from a value of around 0.21 which corresponds to the S809 to a value of exactly 0.1722 for all of them (Figure 7). This indicates as expected that the optimization will tend to shift to smaller thicknesses for better aerodynamic properties, however this may compromise the structural integrity of the blade if the lower bounds of this parameter reaches low levels.

Figure 8 shows the coefficients of lift of the optimized airfoils obtained from XFOIL, with values always above those of the S809 airfoil (Re=$9.25x10^6$), also their *alpha* value at stall are higher than the S809's. The Cp curves of all the designed airfoils for an angle of attack at seven degrees is shown in Figure 9 where they show similar shapes as they share several points in common in account of their identical relative thickness

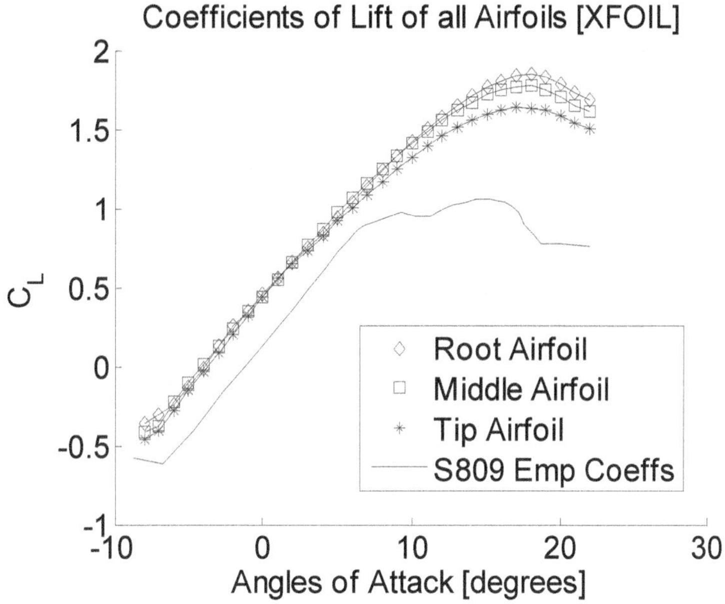

Figure 8. Coefficients of Lift at Re= 925,000 for designed airfoils and S809

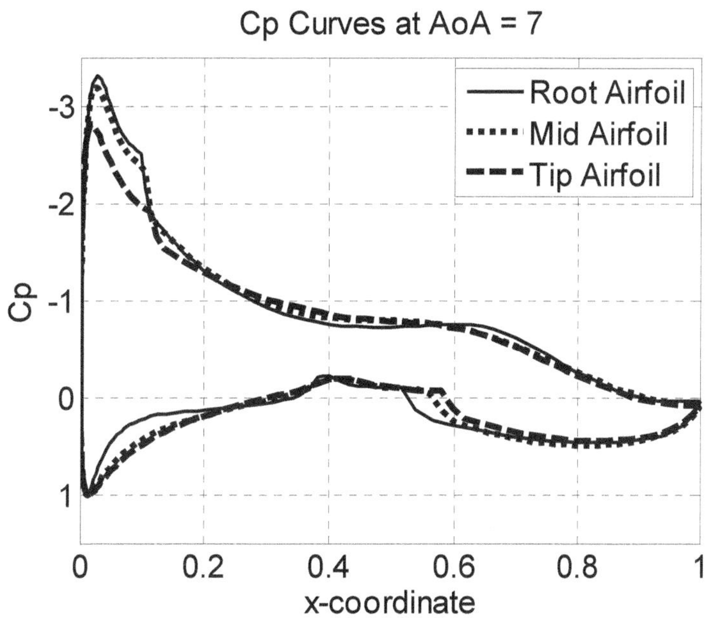

Figure 9. Cp Curves at 7 degrees

The ratio *Cl /Cd* of the designed airfoils have values of 63.7, 67.7, and 69.4 for the root, midsection and the tip respectively at angles of attack 8, 7, and 7 degrees. Figure 10 shows a plot of the C_l/C_d for the designed airfoils and the S809. The plot indicates that the designed airfoils have a maximum aerodynamic efficiency lower than that of the S809's, however they have larger values for a larger range of the angles of attack. This is a possible indication that airfoils suited for a FPFS should have *Cl /Cd* curves which do not necessarily require a high max*(Cl /Cd)* as illustrated by the S809's, but instead a flatter curve with non-steep decrements are of preference. This points to the idea that higher aerodynamic efficiency are required for a large range of *alpha* values since the lack of pitching mechanism and rotor speed variability will expose it to several *alphas* depending on wind speed changes.

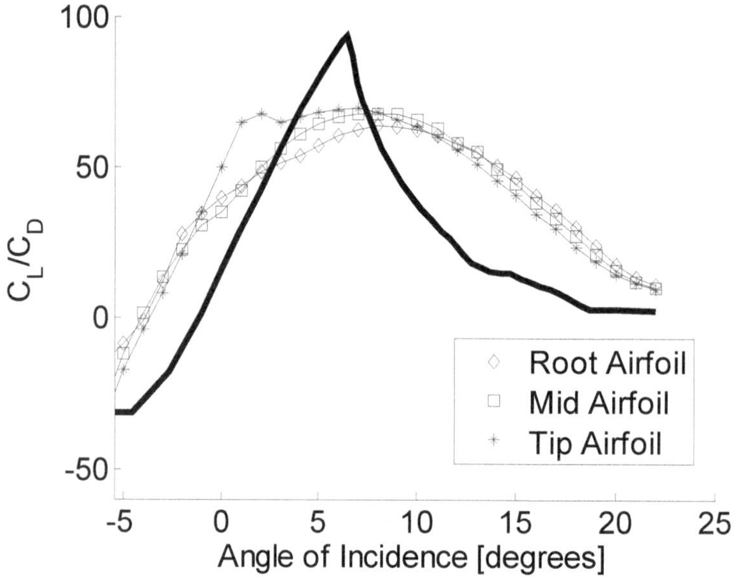

Figure 10. C_l/C_d for the designed airfoils and the S809

A contrast of the energy harvested by each airfoil in the optimized blade is presented in Figure 11 compared against the Phase VI's energy output. Every airfoil curve for outer sections is higher than the inner segments as they can capture more energy on account of their radial position. Figure 11 shows the mid-section of the optimized blade generates almost an identical amount of energy as that of the Phase VI outer portion of the blade which suggests a good optimization level for the design.

A plot of the energy produced by the entire blade for every wind speed is shown in Figure 12, which shows that for wind speeds of 11 m/s and lower most of the energy is harvested.

For a better analysis of this information a plot of the angles of attack in every segment of the blade is shown in Figure 13, which shows that the angles of attack in the airfoils go from around 2 degrees to around 22. This indicates that instead of a maximization of the C_l/C_d of the airfoil for a given *alpha-design*, the optimization should be implemented for a range of angles of attack which would likely hit the blade for the majority wind speeds.

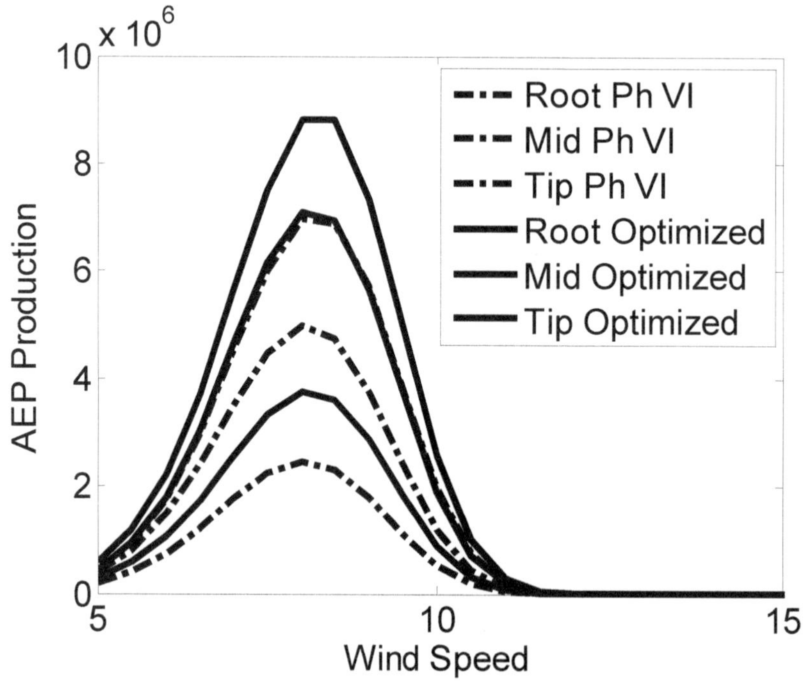

Figure 11. Energy yields by the root, mid-section and tip portion of the optimized and the Phase VI blade depending on wind speeds

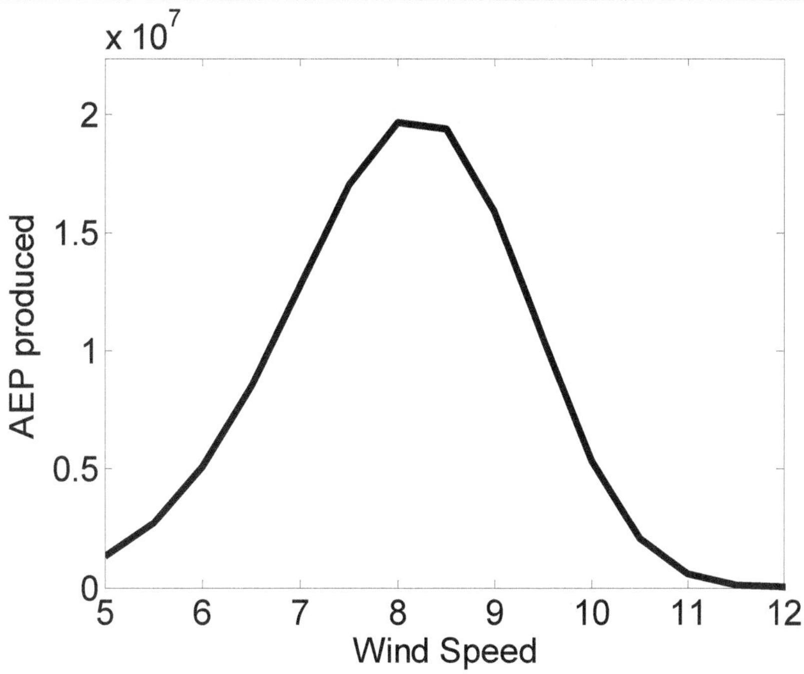

Figure 12 Energy produced by the blade at different wind speeds

For a better analysis of this information a plot of the angles of attack in every segment of the blade is shown in Figure 13, which shows that the angles of attack in the airfoils go from around 2 degrees to around 22. This indicates that instead of a maximization of the C_l/C_d of the airfoil for a given *alpha-design*, the optimization should be implemented for a range of angles of attack which would likely hit the blade for the majority wind speeds.

Finally, in Figure 14 the C_p curve generated by the optimized blade is shown against that of the Phase VI's. The plot shows the optimized curve is above the baseline design for every wind speed, which means it will be more efficient for any variation of wind profiles.

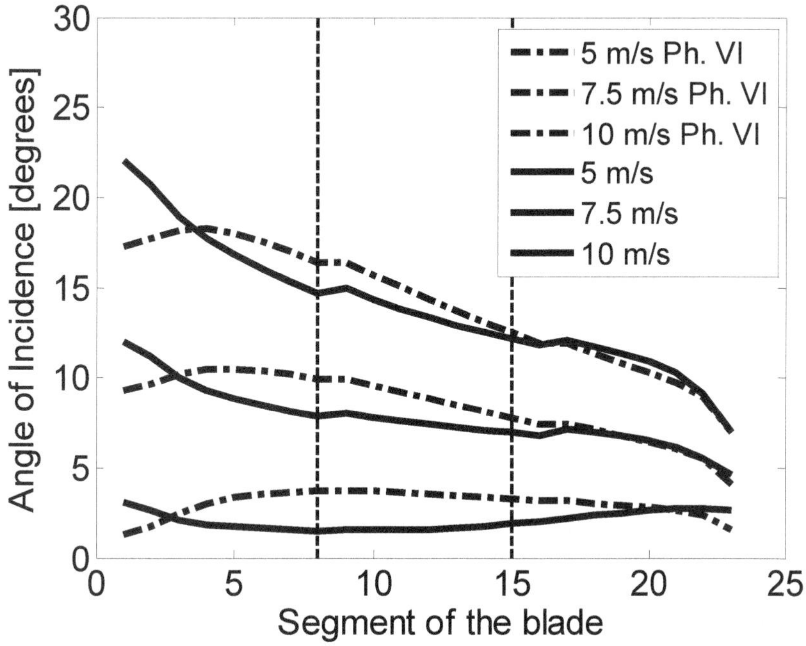

Figure 13. Angle of attacks at 5, 7.5 and 10 m/s at given blade segments

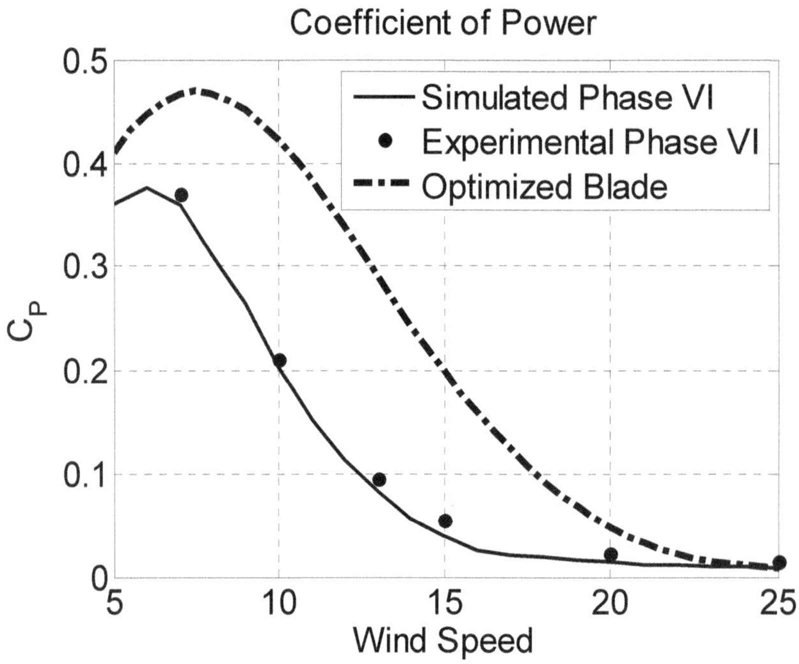

Figure 14. Coefficient of Power for Phase VI and the designed blade

6. Conclusions

A FPFS wind turbine was optimized in this study to maximize the Annual Energy Production using the chord, twist, and three different airfoils with variable relative thickness as design variables which where modeled utilizing Bezier curves. The wind profile used data obtained from the NOAA for a station located in the city of Roswell, NM. A baseline design for the optimization of this blade starts with the Phase VI developed in the NASA-Ames experiment which consists of a variable twist and chord with a length of 5.03m and containing the S809 airfoil with a 0.21 relative thickness. The simulation of the energy production of the rotors was conducted by the use of the Blade Element Momentum Theory and the SQP method was employed to optimize the given design. The results returned three airfoils with identical relative thickness at 0.17, less than the S809's; also a smaller twist and chord distributions were found to be optimal if compared to the Phase VI blade. In total the design blade has an AEP increase of 36%.

The designed airfoils had curves of coefficients of lift above those of the S809 airfoil. Furthermore, the plot of the C_l/C_d indicated that the S809 had a higher max(C_l/C_d) for a given *alpha* than those of the three designed airfoils. However, the aerodynamic efficiency of the S809 quickly goes below those of the designed airfoils. This result is congruent with the fact that a FPFS turbine is always exposed to a wide range of angles of attack as wind speeds change. Taking into account that several airfoils in the past have been designed by maximizing the C_l/C_d --as they are fitted for VP or VS systems-- suggests that a maximization in the aerodynamic efficiency for one *alpha-design* is not be the best technique to implement for a FPFS wind turbine.

Acknowledgement

The authors are grateful to the United States Department of Agriculture (USDA) for their support to this project via the BuildinG a Regional Energy and Educational Network (BGREEN) grant.

References

[1] R. Wiser, 2012 Wind technologies market report. (2014).

[2] R.W. Thresher, D.M. Dodge, Trends in the evolution of wind turbine generator configurations and systems. Wind Energy 1 (1998) 70-86.

[3] F.D. Bianchi, H. De Battista, R.J. Mantz, Wind turbine control systems: principles, modelling and gain scheduling design, Springer, 2006.

[4] D. Saheb-Koussa, M. Haddadi, M. Belhamel, S. Hadji, S. Nouredine, Modeling and simulation of the fixed-speed WECS (wind energy conversion system): application to the Algerian Sahara area. Energy 35 (2010) 4116-4125.

[5] G. Venkatesh, S. Kulkarni, Energy yield of passive stall regulated fixed speed wind turbine with optimal rotor speed. Electric power systems research 76 (2006) 1019-1026.

[6] X. Liu, Y. Chen, Z. Ye, Optimization model for rotor blades of horizontal axis wind turbines. Frontiers of Mechanical Engineering in China 2 (2007) 483-488.

[7] X. Liu, L. Wang, X. Tang, Optimized linearization of chord and twist angle profiles for fixed-pitch fixed-speed wind turbine blades. Renewable Energy 57 (2013) 111-119.

[8] E. Minisci, M. Campobasso, M. Dell'Antonio, A. Montecucco, M. Vasile, Multi-disciplinary robust design of variable speed wind turbines, Conferenza EUROGEN, Settembre, 2011.

[9] P. Fuglsang, C. Bak, M. Gaunaa, I. Antoniou, Design and verification of the Risø-B1 airfoil family for wind turbines. Journal of solar energy engineering 126 (2004) 1002-1010.

[10] C. Bak, P. Fuglsang, Modification of the NACA 632-415 leading edge for better aerodynamic performance. Journal of solar energy engineering 124 (2002) 327-334.

[11] W. Timmer, R. Van Rooij, Summary of the Delft University wind turbine dedicated airfoils. Journal of Solar Energy Engineering 125 (2003) 488-496.

[12] F. Grasso, Development of thick airfoils for wind turbines. Journal of Aircraft 50 (2013) 975-981.

[13] C. Lindenburg, Investigation into rotor blade aerodynamics. Netherlands Society for Energy and the Environment, Paper ECN-C-03-025 (2003).

[14] M. Hansen, Documentation of code and airfoil data used for the NREL 10-m wind turbine. ROTABEM-DTU, November (2000).

[15] J. Manwell, J. McGowan, A. Rogers, Wind energy explained: theory, design and application., 2nd Edition ed., Wiley, New York, NY, USA, 2009.

[16] G.R. Fischer, T. Kipouros, A.M. Savill, Multi-objective optimisation of horizontal axis wind turbine structure and energy production using aerofoil and blade properties as design variables. Renewable Energy 62 (2014) 506-515.

[17] R. Lanzafame, M. Messina, Design and performance of a double-pitch wind turbine with non-twisted blades. Renewable Energy 34 (2009) 1413-1420.

[18] X. Liu, Blade Design Optimisation for Fixed-Pitch Fixed-Speed Wind Turbines. ISRN Renewable Energy 2012 (2012).

[19] M. Drela, XFOIL: An analysis and design system for low Reynolds number airfoils, Low Reynolds number aerodynamics, Springer, 1989, pp. 1-12.

[20] B. Bavanish, K. Thyagarajan, Optimization of power coefficient on a horizontal axis wind turbine using bem theory. Renewable and Sustainable Energy Reviews 26 (2013) 169-182.

[21] T. Kim, S. Lee, H. Kim, S. Lee, Design of low noise airfoil with high aerodynamic performance for use on small wind turbines. Science in China Series E: Technological Sciences 53 (2010) 75-79.

[22] X. Li, K. Yang, J. Bai, J. Xu, A method to evaluate the overall performance of the CAS-W1 airfoils for wind turbines. Journal of Renewable and Sustainable Energy 5 (2013) 063118.

[23] N. Mittal, S. Ahmed, N. Tenguria, EVALUATION OF PERFORMANCE OF HORIZONTAL AXIS WIND TURBINE BLADES BASED ON OPTIMAL ROTOR THEORY. Journal of Urban and Environmental Engineering 5 (2011) 15-23.

[24] D.A. Spera, Models of lift and drag coefficients of stalled and unstalled airfoils in wind turbines and wind tunnels. National Aeronautics and Space Administration, NASA/CR-2008-215434 (2008).

[25] National Oceanic and Atmospheric Administration. (2014, January). Retrieved from http://www.ncdc.noaa.gov

[26] E. Hau, Wind turbines: fundamentals, technologies, application, economics, 2006, in: L. Hansen (Ed.), Springer, 2006.

8

Transient thermal behavior of a homogeneous composite micro-domain: The hyperbolic heat-conduction model

Faisal M. AL- Ghathian

Faculty of Engineering Technology, Al-Balqa' Applied University, P.O. Box 179, Tafila, 66110, Jordan.

Abstract
The transient thermal behavior of a homogeneous composite micro-domain described by the hyperbolic heat-conduction model with neglecting conduction in the fluid domain is investigated semi-analytically. The composite micro-domain consists of a matrix (fluid domain) and inserts (solid domain), each made of different material. The effect of different parameters that affect the local thermal equilibrium assumption under the effect of the hyperbolic heat conduction model is investigated.

Keywords: Composite; Heat conduction; Hyperbolic model; Thermal equilibrium assumption; Porous micro-channel.

1. Introduction

over the past two to three decades, the study of heat transfer in porous media has evolved as result of it is importance in the study of many engineering applications. Nield and Bejan [1] highlighted the new conceptual development and applications of convection in porous media. One of the main issues in the study of porous media is the assumption of local thermal equilibrium (LTE) were it is assumed that both the fluid and solid are in LTE, therefore only one energy equation is considered [1], limiting the results to certain special cases and applications. On other hand [2-7] several studies adopted the two-phase model where there are two energy equations for the solid and fluid domain. It is clear that there is a need to establish the conditions when the LTE can be used in the study of convection in porous media.

Numerous studies [8-12] investigated the validity of LTE assumption in porous media for different flow conditions and geometries. They established a group of dimensionless parameters that control the LTE assumption for different flow conditions in porous media, and derived the criteria necessary for LTE assumption. All the previous studies were described by the parabolic heat conduction models.

Recently Nnanna et al.[13] performed experimental study of non-Fourier thermal response in porous media, in this study a two equation model that uses non-Fourier (dual phase lag) to study the response of a porous medium subjected to a short time thermal disturbance is verified experimentally. Also, they showed that during a rapid transient even when the fluid and solid have the same temperature the Fourier conduction model failed to describe temperature filed.

Rapid transient is encountered in many applications that involves porous medium, such as laser synthesis and processing of thin-film deposition where in this application a heat source such as a laser and/or microwave of extremely short duration or very high frequency is used. In the present study the thermal equilibrium assumption in transient natural convection flow in porous channel as described by a hyperbolic heat-conduction model is investigated.

2. Analysis

Consider the problem of unsteady natural convection fluid flow into a parallel plate channel totally filled with porous media. The unsteadiness in the channel thermal behavior is due to a sudden change in the temperature of the channel wall. Referring to Figure 1,

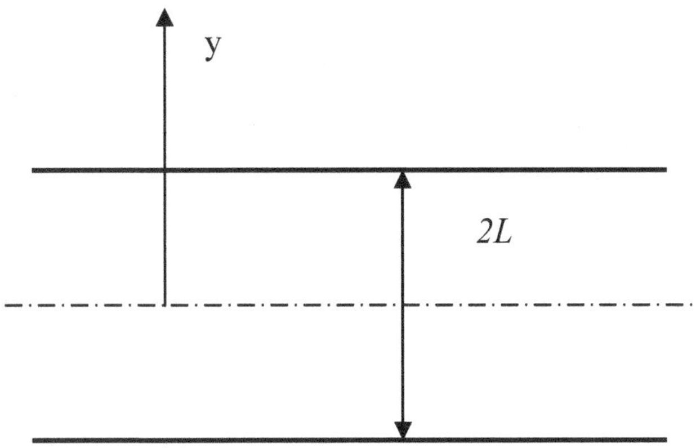

Figure 1. Schematic representation of the domain under consideration

The energy equations with the initial and boundary conditions for both the fluid and solid domains for the hyperbolic heat conduction model are given as:

$$\tau_f \frac{\partial^2 \theta_f}{\partial \eta^2} + \frac{\partial \theta_f}{\partial \eta} = Bi(\theta_s - \theta_f) + \tau_f \frac{\partial}{\partial \eta}(\theta_s - \theta_f) \tag{1}$$

$$\tau_s \frac{\partial^2 \theta_s}{\partial \eta^2} + \frac{\partial \theta_s}{\partial \eta} = \frac{K_R}{C_R}\frac{\partial^2 \theta_s}{\partial Y^2} - Bi(\theta_s - \theta_f) - Bi\tau_s \frac{\partial}{\partial \eta}(\theta_s - \theta_f) \tag{2}$$

where $C_R = \frac{(1-\varepsilon)\rho_s c_s}{\varepsilon \rho_f c_f}$ $Bi = \frac{h_v L}{k_f}$ $K_R = \frac{k_s}{k_f}$

The initial and boundary conditions become:

$$\theta_s(0,Y) = \theta_f(0,Y) = 0 \quad \frac{\partial \theta_s(0,Y)}{\partial \eta} = \frac{\partial \theta_f(0,Y)}{\partial \eta} = 0 \tag{3}$$

$$\theta_s(\eta,1) - 1 = Kn\frac{\Omega}{\Pr}Q(\eta,1) \quad \theta_f(\eta,1) - 1 = Kn\frac{\Omega}{\Pr}Q(\eta,1) \quad \frac{\partial \theta_s(\eta,0)}{\partial Y} = 0$$

Equations (1-3) are solved using Laplace transformation technique. Now with the notation that $L\{\theta_s(\eta,Y)\} = W_s(S,Y)$ and $L\{\theta_f(\eta,Y)\} = W_f(S,Y)$, Laplace transformation of Eqs.(1-3) yields:

$$W_f = \frac{(\tau_f S + Bi)}{(\tau_f S^2 + \tau_f S + S + Bi)} W_s \tag{4}$$

$$\left(\frac{K_R}{C_R}\right)\frac{\partial^2 W_s}{\partial^2 Y} - (\tau_s S^2 + S + Bi + Bi\tau_s)W_s = -(Bi + Bi\tau_s)W_f \tag{5}$$

Also, the Laplace transformation of the boundary conditions is given as:

$$\left.\begin{aligned} & W_s(S,1)-\frac{1}{S}=-Kn\frac{\Omega}{\Pr}W_s'(S,1) \\ & \frac{\partial W_s}{\partial Y}(S,0)=0 \end{aligned}\right\} \tag{6}$$

According to the boundary conditions given in Eq. (6), Eqs. (4-5) are solved to give:

$$W_s=C\left(e^{HY}+e^{-HY}\right) \tag{7}$$

$$W_f=C\frac{\left(\tau_f S+Bi\right)\left(e^{HY}+e^{-HY}\right)}{\left(\tau_f S^2+\tau_f S+S+Bi\right)} \tag{8}$$

where $H^2=\left[\left(\tau_s S^2+S+Bi+Bi\tau_s\right)\left(\frac{C_R}{K_R}\right)-\frac{\left(Bi+Bi\tau_s\right)\left(\frac{C_R}{K_R}\right)}{\left(\tau_f S^2+\tau_f S+S+Bi\right)\left(Bi+Bi\tau_f\right)}\right]$ and $C=\frac{1/S}{\left[\left(e^H+e^{-H}\right)+Kn\frac{\Omega}{\Pr}\left(e^H-e^{-H}\right)\right]}$

Equations (7-8) are inverted using a computer program based on Riemann-sum approximation [14] as:

$$\theta(\eta,Y)\cong\frac{e^{\gamma\eta}}{\eta}\left[\frac{1}{2}W(\gamma,Y)+\mathrm{Re}\sum_{n=1}^{N}W\left(\gamma+\frac{in\pi}{\eta},Y\right)(-1)^n\right] \tag{9}$$

where Re refers to the "real part of" and $i=\sqrt{-1}$ is the imaginary number, N is the number of terms used in Riemann-sum approximation and γ is the real part of the Bromwich contour that is used in inverting Laplace transforms. The Riemann-sum approximation for the Laplace inversion involves a single summation for numerical process. Its accuracy depends on the value of γ and the truncation error dictated by N.

3. Results and discussion

The effect of different parameters on the validity of the thermal equilibrium assumption in transient natural convection flow in porous channel as described by a hyperbolic heat-conduction model is investigated in Figures 2-6 for the case (neglecting conduction in the fluid domain)

Figure 2 shows the transient behavior of the fluid and solid temperatures at different K_R with neglecting the conduction in the fluid domain. As shown, the difference between the fluid and solid temperatures increases as K_R decreases, which implies that as K_R increases the thermal resistance of the solid domain decreases or the thermal resistance of the fluid increases. The effect of total thermal capacity ratio C_R on the transient behavior of the fluid and solid temperatures is shown in Figure 3. It is clear that the difference between the fluid and solid temperatures increases as the value of C_R decreases. The transverse conduction in the fluid domain is neglected which implies that the effect of the thermal disturbance is carried into the channel directly through the solid domain and then the solid domain transfer it to the fluid domain through the volumetric convective heat transfer coefficient.

Figure 4 shows the effect of Biot number on the transient fluid and solid temperatures with neglecting conduction in the fluid domain. It is obvious from these figures that the difference decreases as Biot number increases. This implies that the effect of Bi number on the temperature difference is insignificant at large values of Bi. This is justified, since the time required for both fluid and solid domain to attain the same temperature is inversely proportional to q, where q is the convective heat transfer between the fluid and solid domain. The transient behavior of the difference between the fluid and solid temperatures at different τ_f and τ_s is shown in Figure 5 with neglecting conduction in the fluid domain. It is clear from this figure that the difference increases as τ_f and τ_s decrease. Effect of Knudsen number Kn on the transient fluid and solid temperatures is shown in Figure 6.

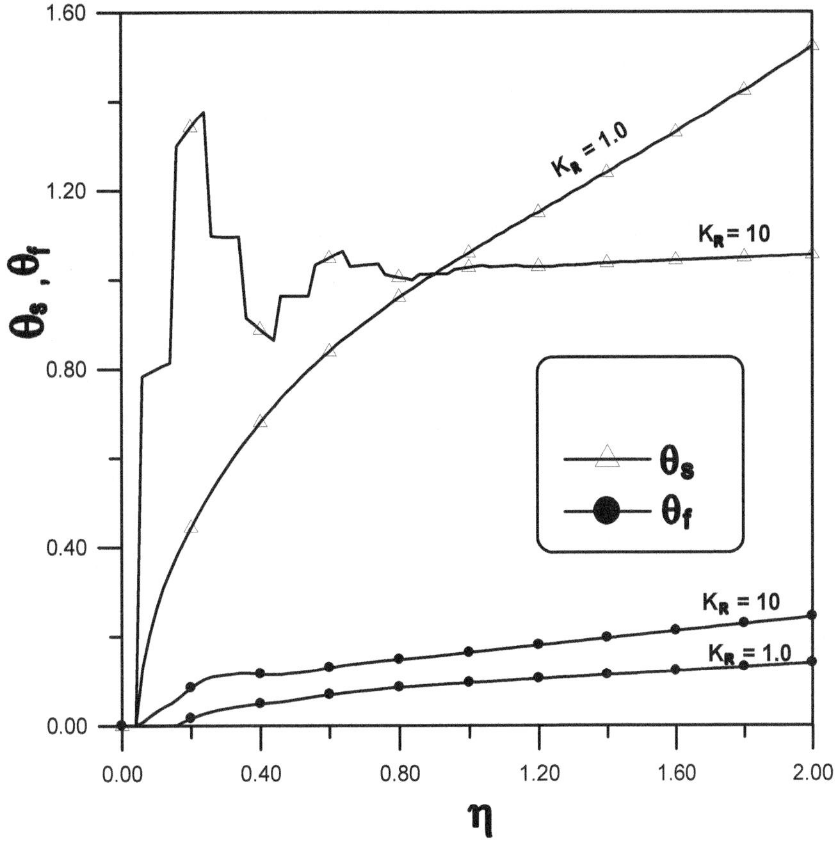

Figure 2. Transient behavior of the fluid and solid temperature at different K_R

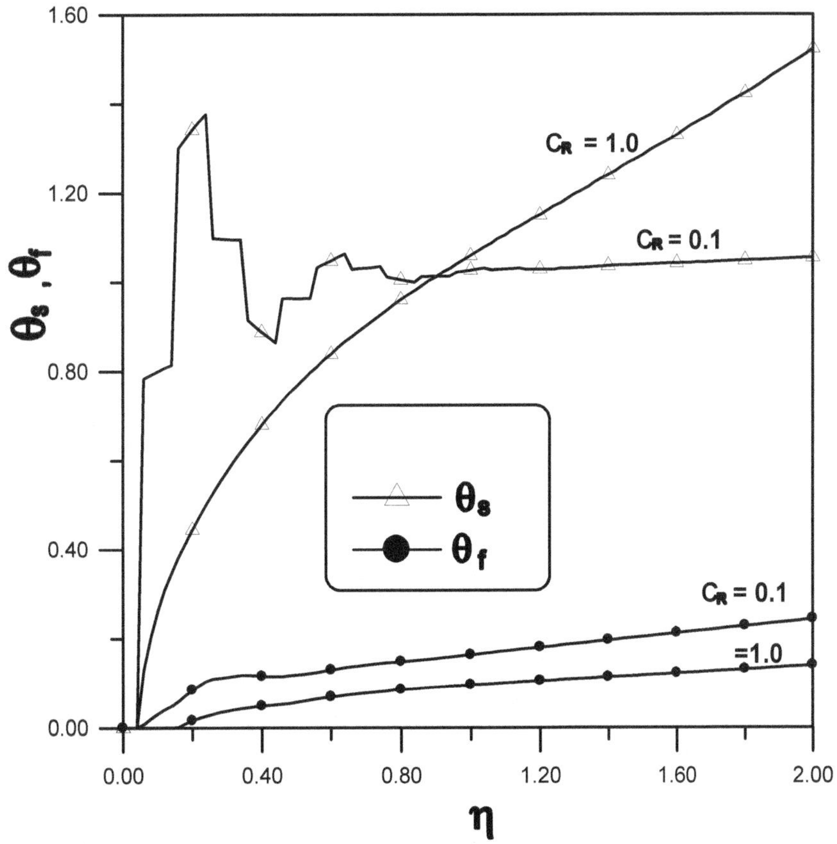

Figure 3. Transient behavior of the fluid and solid temperature at different C_R

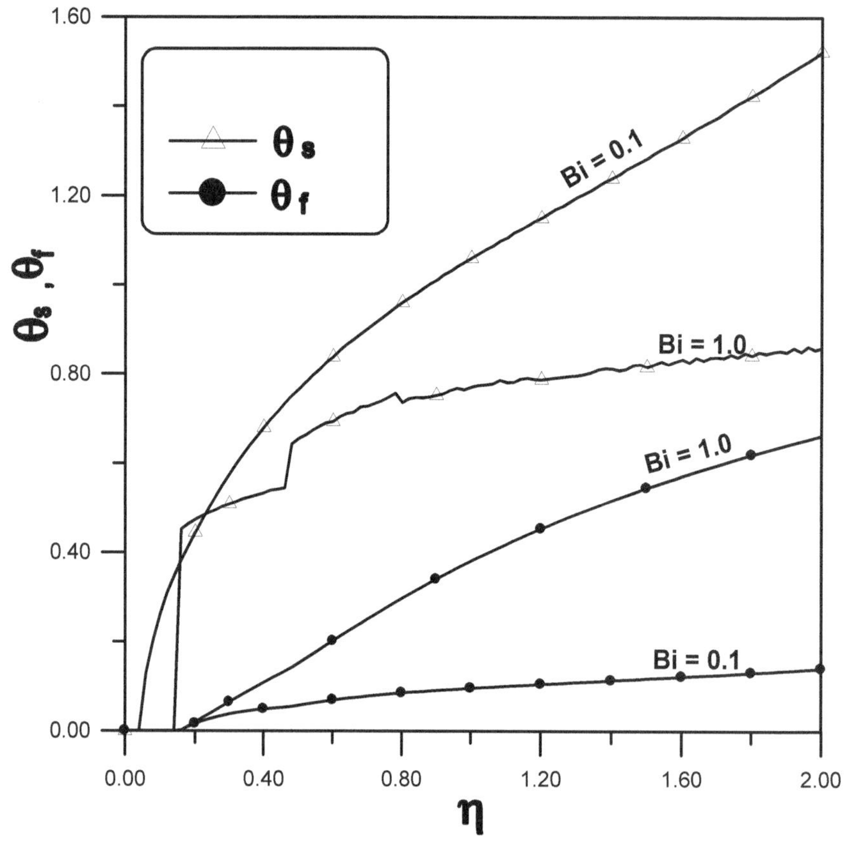

Figure 4. Transient behavior of the fluid and solid temperature at different B_i

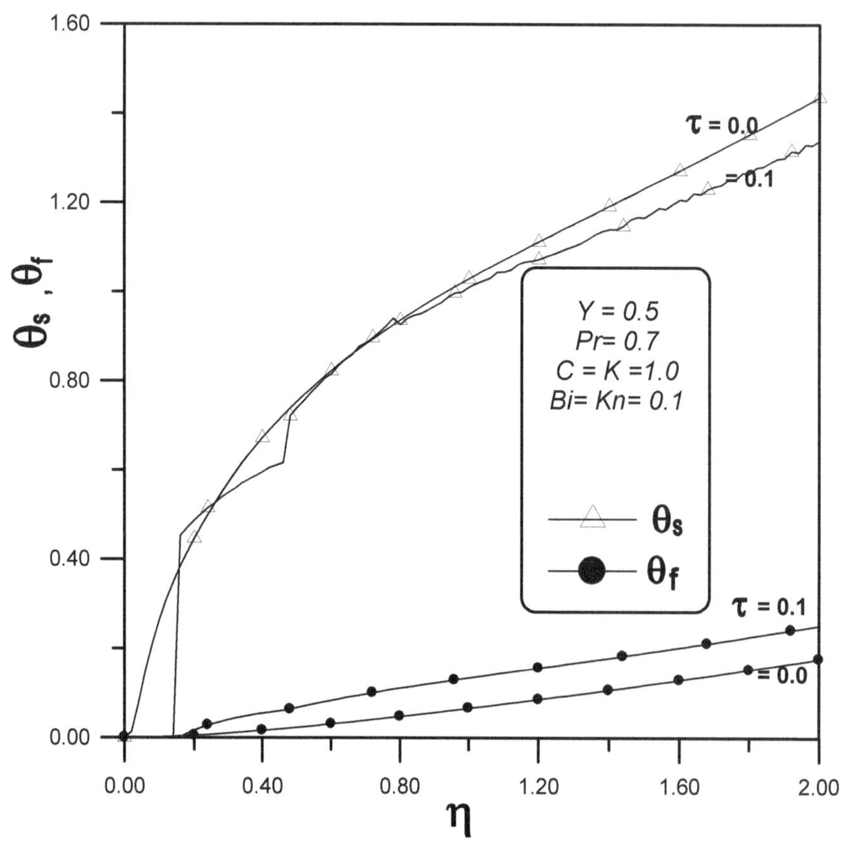

Figure 5. Transient behavior of the fluid and solid temperature at different τ

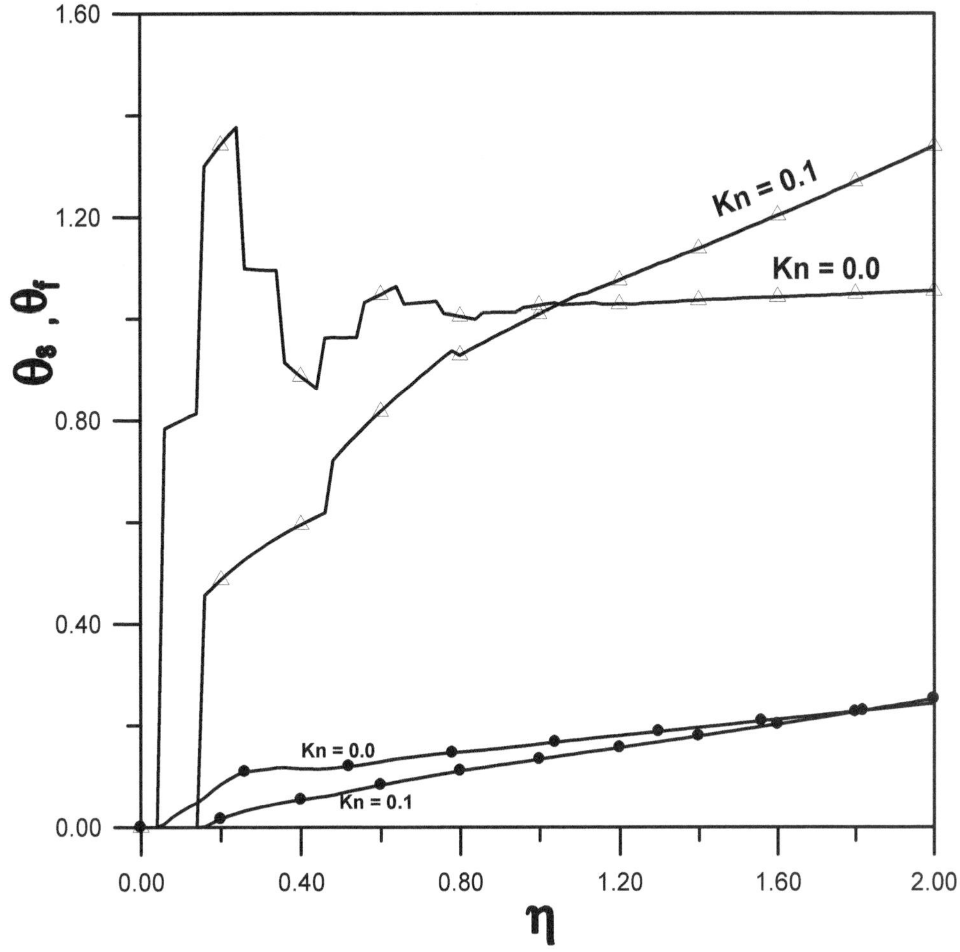

Figure 6. Transient behavior of the fluid and solid temperature at different Kn

4. Conclusions

Thermal equilibrium assumption in transient natural convection flow in porous channel as described by a hyperbolic heat-conduction model is investigated with neglecting the conduction in the fluid domain. It is found that the volumetric Biot number, thermal conductivity ratio, phase lag in heat flux, Knudsen number and total thermal capacity ratio have the most significant effect on the local thermal equilibrium assumption. The local thermal equilibrium assumption is secured for large values of Biot number, Knudsen number and thermal conductivity ratio and small values of total thermal capacity ratio, phase lag in heat flux.

References

[1] D. A. Nield. and A. Bejan "Convection in porous Media", Second edition, Springer 1999.

[2] B. Alazmi and K, Vafai. "Constant wall heat flux boundary conditions in porous media under local thermal non-equilibrium conditions", International Journal of Heat and Mass Transfer 45, 3071-3087, 2002.

[3] Y. X. Tao and D. M., Gray "Validation of local thermal equilibrium in unsaturated porous media with simultaneous-flow and freezing", Int. Comm. Heat Mass Transfer, 20, 323-332, 1993.

[4] W. J. Minkowycz, A. Haji-Sheikh and K.Vafai "On departure from local thermal equilibrium in porous media due to a rapidly changing heat source: the Sparrow number", International Journal of Heat and Mass Transfer 42, 3373-3385, 1999.

[5] D. A. S. Rees, and I. Pop "Vertical free convection boundary-layer flow in a porous medium using a thermal non-equilibrium model", Journal of Porous Media, 3 (1), 31-44, 2000.

[6] D. A. Nield and A. V.Kuznetsov "Local thermal non-equilibrium effects in forced convection in a porous medium channel: a conjugate problem", Int. J. Heat Mass Transfer, 24 (17), 3245-3249, 1999.

[7] M. Spigaand G.L. Morinib "Transient response of non-thermal equilibrium packed beds" International Journal of Engineering Science 37, 179-188, 1999.

[8] M.A. Al-Nimr and S.Kiwan "Examination of the thermal equilibrium assumption in periodic forced convection in a porous channel", J. Porous Medium, 5(1), 35-40 2002.

[9] A. F. Khadrawiand M. A.Al-Nimr "Examination of the thermal equilibrium assumption in transient natural convection flow in porous channel", Transport in Porous Media, 53(3), 317-329, 2003.

[10] O. M. Haddad M. A. Al-Nimrand A. Al-Khateeb "Validity of the local thermal equilibrium assumption in natural convection from a vertical plate embedded in porous medium: Non-Darcian model", Int. J. Heat Mass Transfer, 47, 2037-2042, 2004.

[11] S. A. Khashan, A. M. Al-Amiriand M. A. Al-Nimr "Local thermal non-equilibrium effect on developing forced convection through porous pipes", International Conference on Thermal Engineering Theory and Applications, Beirut, Lebanon, May 31-June 4, 2004.

[12] A. F. Khadrawi M. S. Tahatand M. A.Al-Nimr "Validation of the Thermal Equilibrium Assumption in Periodic Natural Convection in Porous Domains", 26 (5), 1633-1649, 2005.

[13] A. G. Nnanna A. Haji-Sheikh and K. T.Harris "Experimental Study of Non-Fourier Thermal Response in Porous Media", Journal of Porous Media, 8(1) 31-44, 2005.

Nomenclature

Bi	Biot number, $\frac{h_v L}{k_f}$
C	specific heat capacity, $J/kg\,K$
C_R	total thermal capacity ratio, $\frac{(1-\varepsilon)\rho_s c_s}{\varepsilon \rho_f c_f}$
h_v	volumetric heat transfer coefficient, $W/m^2\,K$
k	thermal conductivity, $W/m\,K$
K_R	thermal conductivity ratio, $\frac{k_s}{k_f}$
Kn	Knudsen number (= λ/L)
2L	slab thickness, m
S	Laplacian domain
t	time, s
t_o	reference time, s
T	temperature, K
T_w	wall temperature, K
y	axial coordinate, m
Y	dimensionless axial coordinate, $\frac{y}{L}$

Greek symbols

η	dimensionless time, $\frac{t}{t_o}$
ε	porosity
Ω	$=\frac{2-\sigma_T}{\sigma_T}\left(\frac{2\gamma}{\gamma+1}\right)$
γ	specific heat ratio
λ	mean free path, m
ν	kinematic viscosity, m^2/s
ρ	density, $\frac{kg}{m^3}$
σ_T	thermal accommodation coefficient
σ_v	Tangential-momentum accommodationcoefficient
θ	dimensionless temperature, $\frac{T-T_\infty}{T_\infty}$
$\tilde{\tau}$	phase lag in heat flux, s
τ	dimensionless phase lag in heat flux, $\frac{\tilde{\tau}}{t_o}$

Subscripts

f	fluid domain
s	solid domain
w	wall

Anaerobic digestion of pig manure and glycerol from biodiesel production

Pakamas Chetpattananondh[1], Sumate Chaiprapat[2], Chaisri Suksaroj[2]

[1] Department of Chemical Engineering, Faculty of Engineering, Prince of Songkla University, Hat Yai, Songkhla, 90112, Thailand.
[2] Department of Civil Engineering, Faculty of Engineering, Prince of Songkla University, Hat Yai, Songkhla, 90112, Thailand.

Abstract
Increasing biodiesel production causes a surplus of glycerol. This work aims to investigate the crude glycerol pretreatment method and then apply the glycerol as a co-substrate with pig manure for anaerobic digestion. The optimum crude glycerol pretreatment method was acidification with 6% of H_2SO_4 that highest glycerol recovery was obtained with lowest cost. Co-digestions of glycerol and pig manure enhanced biogas and methane productions compared with mono-digestions. Biogas and methane productions in semi-continuous digestions were highly effected by OLR. The optimum OLR was 3.06 kg SCOD/m^3 that biogas production was maintained at 3 L/d with methane composition of 72% and SCOD removal higher than 80%.

Keywords: Biogas; Crude glycerol; Pig manure; Anaerobic digestion; Biodiesel.

1. Introduction

Anaerobic digestion has become a proven technology for the treatment of organic wastes such as municipal solid waste, industrial organic waste, agricultural residues, and animal manure. The digestion is the decomposition of organic materials by microorganisms in the absence of oxygen to produces biogas which is rich in methane, followed by carbon dioxide, ammonia and traces of other gases, volatile fatty acids, and water [1]. Composition of biogas varies depending upon the raw materials and fermentation conditions [2]. Biogas yield depends on type of substrate. The biogas yield of pig manure is comparable to other feedstock [3]. Pig production is increasing because of increased consumer demands for pork resulting to large volumes of manure to be managed in a sustainable manner by optimizing usage of the nutrients and energy in the manure while at the same time minimizing the negative impact on the external environment, food safety and human health [4]. The anaerobic digestion is a complex sequence composed of four major microbial steps, hydrolysis, acidogenesis, acetogenesis and methanogenesis [5]. Biogas can be used as a vehicle fuel or for co-generation of electricity and heat, and thus, can lead to reductions in greenhouse gas emissions [6].

Biodiesel produced by the transesterification of vegetable oils or animal fats with an alcohol to produce esters has been widely used worldwide. During the biodiesel production process, about 10% (w/w) of glycerol is generated as a primary by-product [7]. The rising of the biodiesel industry causes a surplus of glycerol resulting to a decrease in crude glycerol costs [8]. The crude glycerol is a mixture of glycerol itself, with alcohols, water, salts, heavy metals, free fatty acids, unreacted mono-, di- and triglycerides, and methyl esters. It has few direct uses and possesses a very low value [9]. Crude glycerol purification is high cost and not economic feasibility of the small and medium size plants [10]. Using of co-substrates in anaerobic digestion can enhance the process, because co-substrates can supply nutrients which may be deficient, and at the same time have an overall positive synergistic effect in the digestion medium, leading to stable digestion and enhanced biogas yields [11]. Several successful studies, in batch and/or continuous experiments, have been published with reference to the benefits of the addition of glycerol to enhance the anaerobic digestion of various substrates including animal manure. Co-digestion of animal manure and glycerol shows advantages as (1) the water content in manure acts as solvent for glycerol; (2) the high alkalinity of manure gives a buffering capacity for the temporary accumulation of volatile fatty acids; (3) the wide range of nutrients in manure are essential for bacterial growth; and (4) glycerol supplies rapidly biodegradable matter [12]. Astals et al. [7] studied the co-digestion of the pig manure and glycerol in batch tests. It was found that the mixture of 80% pig manure and 20% glycerol (w/w) produced the highest methane production with 0.215 L/g COD, while the one with 20% pig manure and 80% glycerol was clearly inhibited by the volatile fatty acid due to the low nitrogen concentration of the mixture. Nuchdang and Phalakornkule [13] reported that the anaerobic digestion of acid-treated glycerol with supplement in a synthetic medium was satisfactory at organic loading rates of 1.3-2.6 g COD/L.d with maximum methane yield of 0.32 L/g COD. While anaerobic digestion of acid-treated glycerol with pig manure (80:20 by COD ratio, providing a suitable C:N ratio of 20 g C/g N) was successful at organic loading rates of 1.3-5.0 g COD/L.d with maximum methane yield of 0.24 L/g COD.

In the first part of this study the crude glycerol from biodiesel production was purified by three methods. The treated glycerol from optimum method was then applied as a co-substrate with pig manure for digestion at mesophilic temperature using granular sludge from upflow anaerobic sludge blanket (UASB) as microorganisms in batch and semi-continuous experiments. Effects of chemical oxygen demand: total kjeldahl nitrogen (COD:TKN) ratio, hydraulic retention time (HRT) and corresponding organic loading rate (OLR) on biogas yield were investigated.

2. Materials and methods

2.1 Crude glycerol pretreatment

Crude glycerol was obtained from the Specialized R&D Center for Alternative Energy from Palm Oil and Oil Crops, Faculty of Engineering, Prince of Songkla University, Hat Yai, Thailand from production of biodiesel by transesterification of waste cooking oils and palm oil using potassium hydroxide catalyst. The crude glycerol was a brown viscous liquid at room temperature. Three crude glycerol purification methods, (a) acidification with 6% H_2SO_4, (b) acidification with 30% H_2SO_4, and (c) acidification with HCl and coagulation with 6% cationic polyamine (PA) blending with 94% poly-aluminium chloride (PACl) [14], were performed as follows:

(a) Acidification with 6% H_2SO_4, 6% of H_2SO_4 was added to 500 ml of crude glycerol and pH of the mixture was adjusted from 9.8 to 2. The mixture was left over night to separate into three layers. The top layer was methyl ester and free fatty acid, the middle layer was glycerol, water and methanol. Potassium sulfate and sodium sulfate were found in the bottom layer. The glycerol layer was obtained by using separatory funnel.

(b) Acidification with 30% H_2SO_4, The procedure was the same as method (a), but using 30% H_2SO_4 instead of 6% H_2SO_4

(c) Acidification with HCl and coagulation with 6% PA blending with 94% PACl, 500 ml of crude glycerol was pH adjusted from 9.8 to 5 with 2% HCl and left for one night. The mixture was separated into two layers. The top one was methyl ester and free fatty acid and the bottom one contained glycerol. The bottom layer was obtained to adjust pH to 8 with 30% NaOH and then 6% PA blending with 94% PACl with 25%v/v was added. The mixture was left for separation and the glycerol layer at the bottom was separated.

2.2 Substrate preparation

Fresh pig manure was collected from Faculty of Natural Resource, Prince of Songkla University, Hat Yai, Thailand. The pig manure was from the fattening pigs as it possesses high nutrient content. The pig manure was blended and kept at 4 °C before used. Glycerol was from the optimum pretreatment method. Granular sludge was obtained from the UASB unit at Chotiwat Manufacturing Co., Ltd. Hat Yai, Thailand.

2.3 Anaerobic batch digestion

Anaerobic batch digestion was performed in mono-digestion and co-digestion with a substrate composition prepared as shown in Table 1. In co-digestion the amount of pig manure was kept constant while varied amount of glycerol was added in order to investigate optimum COD:TKN ratio. The inoculums of 37,500 mg volatile suspended solids/L (VSS/L) granular sludge and deionized (DI) water were filled up to a volume of 1 L in each digester. The mixture was mixed with stirrer to obtain uniform properties, purged with nitrogen gas for five minutes to create an anaerobic environment and then kept at mesophilic temperatures (33±3 °C) with initial pH of 7.2 for 1 week. Each digester was shake-mixed manually once a day [15]. The biogas produced was sampled from digester headspace for quality analysis and collected in gas collection bottles for volume determination using liquid displacement. All experiments were performed in duplicate and average results are reported.

Table 1. Anaerobic batch digestion

Pig manure [g]	Glycerol [g]	COD [mg/L]	TKN [mg/L]	COD:TKN ratio
20	45	26,962	392	70:1
20	40	24,335	375	65:1
20	35	22,105	354	60:1
20	30	18,482	341	55:1
20	25	14,216	285	50:1

2.4 Anaerobic semi-continuous digestion

Digestion was operated in 3-L continuous stirred tank reactors (CSTR) with operational volume of 2.5L. The experiments were performed in semi-continuous mode with daily feeding. In the start-up, the inoculums of 37,500 mg VSS/L granular sludge was added followed by the mixture of pig manure and glycerol and DI water with COD:TKN ratio of 40:1 and HRT of 50 days corresponding to OLR of 0.16 kg SCOD/m^3.d into three identical reactors. Other three identical reactors were operated with COD:TKN ratio of 50:1. The digestion was maintain at pH 6.8-7.2 and mesophilic temperatures (33±3 °C) until reached steady state conditions. The HRT was reduced stepwise from 10 to 5 and 2.5 days corresponding to OLRs as shown in Tables 2 and 3. The biogas produced was sampled from reactor headspace for quality analysis and collected in gas collection bottles for volume determination using liquid displacement. The digesters were operated for 96 days.

2.5 Analytical methods

Crude and treated glycerol were analyzed for glycerol content by standard method (ASTM D7637) [16], water content by Karl Fischer titration method and ash and matter organic non-glycerol (MONG) by standard method developed by the International Union of Pure and Applied Chemistry [17]. Performance of anaerobic digestion was evaluated by biogas yield. The biogas production was collected by liquid displacement method and biogas composition was analyzed by a gas chromatography (HP6890N) equipped with a silcosteel packed column (Restek Corporation, USA) and a thermal conductivity detector (TCD). Determination of COD, soluble COD (SCOD), TKN, pH, alkalinity, volatile fatty acid (VFA), and VSS were carried out according to the Standard Methods for the examination of Water and Wastewater [18].

Table 2. Anaerobic semi-continuous digestion with COD:TKN ratio of 40:1

Reactor	SCOD [mg/L]	HRT [d]	Flow rate [L/d]	OLR [kg SCOD/m^3.d]
1	7,646	10	0.25	0.76
		5	0.50	1.53
		2.5	1.00	3.06
2	3,852	10	0.25	0.39
		5	0.50	0.77
		2.5	1.00	1.54
3	1,917	10	0.25	0.19
		5	0.50	0.38
		2.5	1.00	0.77

Table 3. Anaerobic semi-continuous digestion with COD:TKN ratio of 50:1

Reactor	SCOD [mg/L]	HRT [d]	Flow rate [L/d]	OLR [kg SCOD/m^3.d]
1	10,679	10	0.25	1.07
		5	0.50	2.14
		2.5	1.00	4.27
2	5,339	10	0.25	0.53
		5	0.50	1.07
		2.5	1.00	2.14
3	2,645	10	0.25	0.26
		5	0.50	0.53
		2.5	1.00	1.06

3. Results and discussions

3.1 The optimum crude glycerol pretreatment method

The crude glycerol consisted of 36-50% glycerol, 4-12% water, 3-6% ash and 35-47% MONG with pH 9.6-10.2. Highest glycerol recovery (30%) was obtained by acidification with HCl and coagulation with 6% PA blending with 94% PACl followed by acidification with 6% H_2SO_4 (26%) and acidification with 30% H_2SO_4 (12%). However, cost of acidification with HCl and coagulation with 6% PA blending with 94% PACl was extremely higher than acidification with 6% of H_2SO_4. Therefore, acidification with 6% of H_2SO_4 was selected as the optimum crude glycerol pretreatment method.

3.2 Anaerobic batch digestion

It was distinctively noticed that the biogas and methane productions were all enhanced in co-digestions of pig manure and glycerol compared with mono-digestions. The highest biogas production (446 mL/d) was obtained by fermentation with 50:1 COD:TKN ratio at day 2 (Figure 1) and the highest accumulative biogas production (1,062 mL) was also obtained by this COD:TKN ratio (Figure 2). The biogas compositions were analyzed and it was found that methane increased markedly during the initial 2 days while CO_2 reduced. The CH_4 composition was about 64% at day 3 for all COD:TKN ratios. Therefore, 50:1 COD:TKN ratio was chosen for anaerobic semi-continuous digestion. In addition, 40:1 COD:TKN ratio was also investigated.

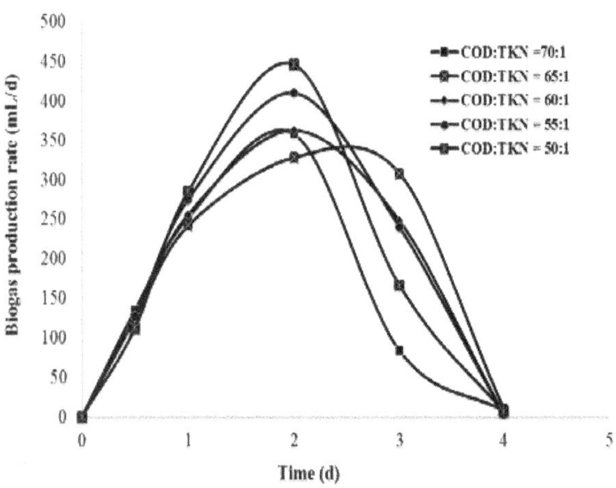

Figure 1. Biogas production rate from anaerobic batch digestion

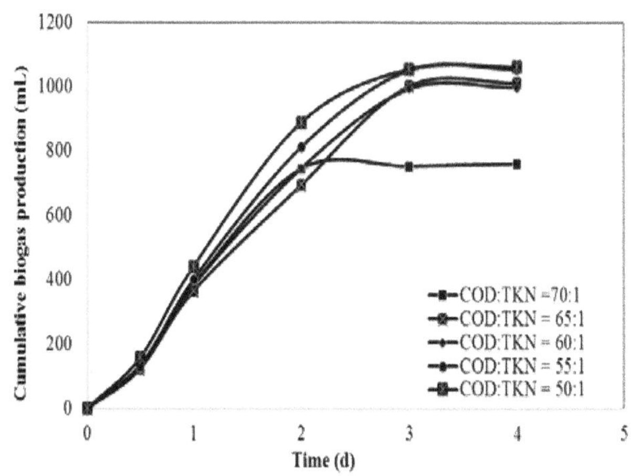

Figure 2. Cumulative biogas production from anaerobic batch digestion

3.3 The optimum conditions for anaerobic semi-continuous digestion

The start-up periods with HRT of 50 days for both 40:1 and 50:1 COD:TKN ratios were carried out for 16 days. After that HRT was decreased stepwise from 10 to 5 and 2.5 days, respectively. The pH of each digester decreased when HRT was changed from 50 to 10 days due to increasing of volatile fatty acids. Volatile fatty acids are the main intermediates products formed during the anaerobic breakdown of organic matter to methane and carbon dioxide. Sufficient buffering capacity (alkalinity) is required to solve pH decreasing problem. Sodium bicarbonate was then added to adjust the pH to maintain within the optimum values, 6.8-7.2 [19]. Daily biogas production increased with reducing of HRT for both COD:TKN ratios (Figures 3 and 4). At lower organic loadings biogas production rates were more stable. With the highest OLR (3.06 kg SCOD/m^3.d) of 40:1 COD:TKN ratio the biogas production immediately increased and fluctuated at the beginning, then remained approximately at 3 L/d. With the highest OLR (4.27 kg SCOD/m^3.d) of 50:1 COD:TKN ratio the biogas production rates were initially high (4-4.3 L/d) and then reduced sharply to maintain approximately at 0.9 L/d because of high amount of volatile fatty acids (700-1290 mg/L CH_3COOH) and high VFA/Alkalinity ratios (1.2-4.4). The optimum VFA concentrations were reported to be 50-500 mg/L CH_3COOH [20] and VFA/Alkalinity ratio > 0.8 inhibited methane production [21]. Methane composition in biogas was about 66-72% excepted at OLR 4.27 kg SCOD/m^3.d that methane composition was about 46%. The theoretical methane composition is 72% for glycerol substrate [13].

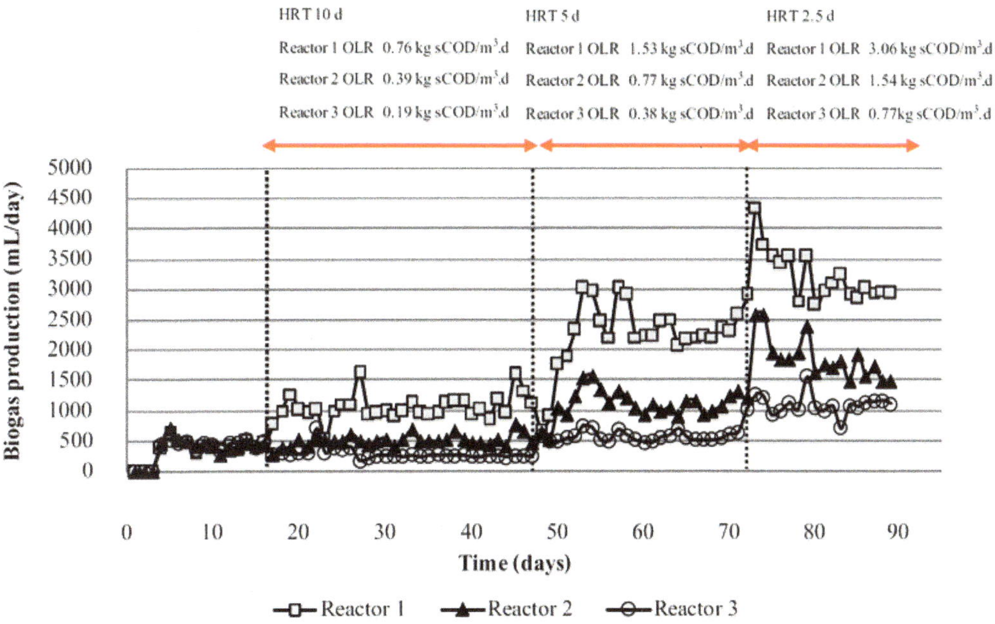

Figure 3. Biogas production from anaerobic semi-continuous digestion with COD:TKN ratio of 40:1

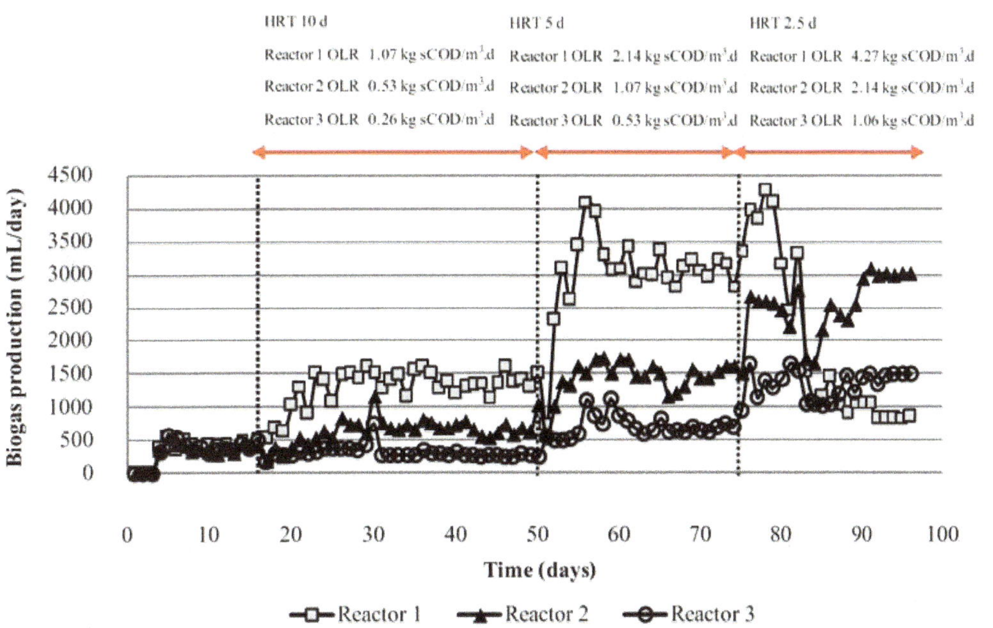

Figure 4. Biogas production from anaerobic semi-continuous digestion with COD:TKN ratio of 50:1

SCOD removal efficiencies were maintained higher than 80% during OLR of 0.16-3.06 kg SCOD/m^3.d. With OLR of 4.27 kg SCOD/m^3.d average SCOD removal efficiencies were 75.76 ± 6.73%. Microbial concentration in the digesters was monitored in term of mixed liquor volatile suspended solids (MLVSS). At the end of digestion MLVSS in all reactors increased from 37,500 to 40,900-41,564 mg/L. It can be seen that biogas and methane yields increased with increasing of OLR for both COD:TKN ratios (Tables 4 and 5) excepted at OLR of 4.27 kg SCOD/m^3.d that biogas and methane yield were lowest since pH could not be maintained within the optimum values. Therefore, 3.06 kg SCOD/m^3.d was the optimum OLR. The biogas production with COD:TKN ratios of 40:1 and 50:1 were not significantly different as OLR was more predominant.

Table 4. Biogas yield and methane yield from anaerobic semi-continuous digestion with COD:TKN ratio of 40:1

Reactor	HRT [d]	OLR [kg SCOD/m^3.d]	Biogas yield [L/g $SCOD_{removed}$]	Methane yield [L/g $SCOD_{removed}$]	pH
1	10	0.76	0.75 ± 0.143	0.51 ± 0.093	6.97 ± 0.22
	5	1.53	0.70 ± 0.061	0.47 ± 0.036	6.82 ± 0.10
	2.5	3.06	0.57 ± 0.029	0.36 ± 0.023	6.82 ± 0.42
2	10	0.39	0.69 ± 0.199	0.48 ± 0.136	6.96 ± 0.15
	5	0.77	0.70 ± 0.170	0.49 ± 0.116	6.80 ± 0.08
	2.5	1.54	0.72 ± 0.031	0.49 ± 0.025	6.80 ± 0.57
3	10	0.19	0.60 ± 0.016	0.41 ± 0.017	6.96 ± 0.15
	5	0.38	0.75 ± 0.203	0.53 ± 0.139	6.84 ± 0.21
	2.5	0.77	0.63 ± 0.150	0.46 ± 0.093	6.80 ± 0.66

Table 5. Biogas yield and methane yield from anaerobic semi-continuous digestion with COD:TKN ratio of 50:1

Reactor	HRT [d]	OLR [kg SCOD/m^3.d]	Biogas yield [L/g $SCOD_{removed}$]	Methane yield [L/g $SCOD_{removed}$]	pH
1	10	1.07	0.56 ± 0.004	0.37 ± 0.006	6.83 ± 0.16
	5	2.14	0.61 ± 0.034	0.40 ± 0.027	6.89 ± 0.13
	2.5	4.27	0.11 ± 0.004	0.05 ± 0.009	5.39 ± 0.06
2	10	0.53	0.48 ± 0.020	0.33 ± 0.014	6.86 ± 0.17
	5	1.07	0.64 ± 0.047	0.44 ± 0.028	6.87 ± 0.16
	2.5	2.14	0.70 ± 0.017	0.46 ± 0.016	6.80 ± 0.11
3	10	0.26	0.43 ± 0.005	0.30 ± 0.009	6.84 ± 0.29
	5	0.53	0.59 ± 0.050	0.42 ± 0.032	6.81 ± 0.22
	2.5	1.07	0.64 ± 0.028	0.47 ± 0.032	6.80 ± 0.10

4. Conclusions

The crude glycerol from biodiesel production consisted of 36-50% glycerol, 4-12% water, 3-6% ash and 35-47% MONG. The optimum crude glycerol pretreatment method was acidification with 6% of H_2SO_4. The glycerol obtained after treatment could be applied as the co-substrate with pig manure in biogas production successfully. Biogas and methane productions in semi-continuous digestions were primarily effected by OLR. The optimum OLR was 3.06 kg SCOD/m^3 that biogas production was maintained at 3 L/d with methane composition of 72% and SCOD removal higher than 80%.

Acknowledgements

The authors gratefully acknowledged the financial supports from Prince of Songkla University via Grant no. ENG520108S. Department of Chemical Engineering and Department of Civil Engineering, Faculty of Engineering, Prince of Songkla University were very appreciated.

References

[1] Adeoti, O., Ayelegun, T.A., Osho, S.O. Nigeria biogas potential from livestock manure and its estimated climate value. Renewable and Sustainable Energy Reviews. 2014, 37, 243-248.

[2] Ismail, Z.Z., Talib, A.R. Assessment of anaerobic co-digestion of agro wastes for biogas recovery: A bench scale application to date palm wastes. International Journal of Energy and Environment. 2014, 5(5), 591-600.

[3] Arthur, R., Brew-Hammond, A. Potential biogas production from sewage sludge: A case study of the sewage treatment plant at Kwame Nkrumah university of science and technology, Ghana. International Journal of Energy and Environment. 2010, 1(6), 1009-1016.

[4] Huong, L.Q., Madsen, H., Anh, L.X., Ngoc, P.T., Dalsgaard, A. Hygienic aspects of livestock manure management and biogas systems operated by small-scale pig farmers in Vietnam. Science of the Total Environment. 2014, 470-471, 53-57.

[5] Khanal, S.K. Anaerobic biotechnology for energy production: Principles and applications. Wiley-Blackwell, 2008.

[6] Ye, J., Li, D., Sun, Y., Wang, G., Yuan, Z., Zhen, F., Wang, Y. Improved biogas production from rice straw by co-digestion with kitchen waste and pig manure. Waste Management. 2013, 33, 2653-2658.

[7] Astals, S., Ariso, M., Gali, A., Mata-Alvarez, J. Co-digestion of pig manure and glycerine: experimental and modeling study. Journal of Environmental Management. 2011, 92, 1091-1096.

[8] Yazdani, S.S., Gonzalez, R. Anaerobic fermentation of glycerol: a path to economic viability for the biofuels industry. Current Opinion in Biotechnology. 2007, 18, 213–219.

[9] Pagliaro, M., Rossi, M. The future of glycerol: New uses of a versatile raw material. The Royal Society of Chemistry, 2008.

[10] Pachauri, N., He, BH. Value-added utilization of crude glycerol from biodiesel production: A survey of current research activities. American Society of Agricultural and Biological Engineers Annual International Meeting, Portland, 9-12 of July 2006.

[11] Athanasoulia, E., Melidis, P., Aivasidis, A. Co-digestion of sewage sludge and crude glycerol from biodiesel production. Renewable Energy. 2014, 62, 73-78.

[12] Mata-Alvarez, J., Dosta, J., Mace, S., Astals, S. Codigestion of solid wastes: a review of its uses and perspectives including modelling. Critical Reviews in Biotechnology. 2011, 31(2), 99-111.

[13] Nuchdang, S., Phalakornkule, C. Anaerobic digestion of glycerol and co-digestion of glycerol and pig manure. Journal of Environmental Management. 2012, 101, 164-172.

[14] Xie, Q-G, Taweeprada, W., Musikavong, C., Suksaroj, C. Separation of oily sludge and glycerol from biodiesel processing waste by coagulation. Songklanakarin Journal of Science and Technology. 2011, 33, 699-703.

[15] Zhang, R.H., El-Mashad, H.M., Hartman, K., Wang, F., Liu, G., Choate, C., Gamble, P. Characterization of food waste as feedstock for anaerobic digestion. Bioresource Technology. 2007, 98, 929–935.

[16] ASTM D7637. Standard test method for determination of glycerin assay by titration. 2010.

[17] Hautfenne, A. International union of pure and applied chemistry. Standard methods for the analysis of oils, fats and derivatives. Pure and Applied Chemistry. 1980, 52, 1939-1954.

[18] APHA (Ed.). Standard methods for the examination of water and wastewater. American Public Health Associations, 2005.

[19] Cioabla, A.E., Ionel, I., Dumitrel, G-A, Popescu, F. Comparative study on factors affecting anaerobic digestion of agricultural vegetal residues. Biotechnology for Biofuels. 2012, 5(39), 2-9.

[20] Metcalf and Eddy, Inc. Wastewater Engineering: Treatment and Reuse. McGraw-Hill, 2003.

[21] Zhao, H.W., Viraraghavan, T. Analysis of the performance of an anaerobic digestion system at the Rigna wastewater treatment plant. Bioresource Technology, 2004, 95(3), 301-307.

10

Performance analysis of an endoreversible rectangular cycle with heat transfer loss and variable specific heats of working fluid

Chao Wang [1,2,3], Lingen Chen [1,2,3], Yanlin Ge [1,2,3], Fengrui Sun [1,2,3]

[1] Institute of Thermal Science and Power Engineering, Naval University of Engineering, Wuhan 430033, China.
[2] Military Key Laboratory for Naval Ship Power Engineering, Naval University of Engineering, Wuhan 430033, China.
[3] College of Power Engineering, Naval University of Engineering, Wuhan 430033, China.

Abstract
The performance of an air-standard rectangular cycle with heat transfer loss and variable specific heats of working fluid is analyzed by using finite-time thermodynamics. The relations between the work output and the compression ratio, between the efficiency and the compression ratio, and the optimal relation between work output and the efficiency of the cycle are derived by detailed numerical examples. Moreover, the effects of heat transfer loss and variable specific heats of working fluid on the cycle performance are analyzed. The results show that the effects of heat transfer loss and variable specific heats of working fluid on the cycle performance are obvious. The results may provide some guidelines for the application of the rectangular cycle.

Keywords: Finite-time thermodynamics; Endoreversible rectangular cycle; Working fluid with variable specific heats; Performance analysis.

1. Introduction

The application of Finite-time Thermodynamics [1-7] in performance analysis and optimization of thermal engine has achieved series of results. Rubin [8] defined the endoreversible cycle model earliest. Mozurkewich et al. [9] and Hoffman et al. [10] derived the optimal motion of the piston by using finite time thermodynamics and optimal control theory. Chen et al. [11] modeled the Diesel cycle with friction loss and studied the effect of friction loss on cycle performance. Klein et al. [12] and Chen et al. [13, 14] studied the performance of Diesel cycle and Otto cycle with heat transfer loss, and analyzed the effect of heat transfer loss on the performance. Al-Hinti et al. [15] studied the performance of Diesel cycle by using different heat transfer model. Qin et al. [16] and Ge et al. [17] derived the performance characteristics of Diesel cycle with friction loss and heat transfer loss. The works mentioned above were performed without considering the variable specific heats of the working fluid. Ghatak and Chakraborty [18] analyzed the performance of Dual cycle by considering the effect of heat transfer loss and variable specific heats of working fluid. Chen et al. [19] studied the performance characteristics of an irreversible Dual cycle with friction loss and linear variable specific heats of working fluid. Ge et al. [20-22] studied the performance of endoreversible and irreversible Otto cycle and Diesel cycle with variable specific heats of the working fluid. Chen et al. [23] modeled a class of universal heat engine cycle with friction loss and heat transfer loss by considering the effect of variable specific heats of working fluid.

Rectangular cycle consists of four processes: an isochoric and an isobaric heat addition process, an isochoric and an isobaric heat rejection process. Ferreira Da Silva [24] derived the power output and the efficiency of rectangular cycle by using classical thermodynamics. Liu et al. [25] modeled endoreversible rectangular cycle with heat transfer loss and studied the performance characteristics of the cycle. Liu et al. [26] modeled irreversible rectangular cycle with friction loss and heat transfer loss by using finite time thermodynamics, and analyzed the effect of friction loss and heat transfer loss on cycle performance. Based on Refs. [24-26], this paper will study the performance characteristics of endoreversible rectangular cycle with heat transfer loss and variable specific heats of working fluid.

2. Cycle model

An air standard rectangular cycle is shown in Figure 1. The heat additions are an isochoric process 1-2 and an isobaric process 2-3; the heat rejections are an isochoric process 3-4 and an isobaric process 4-1.

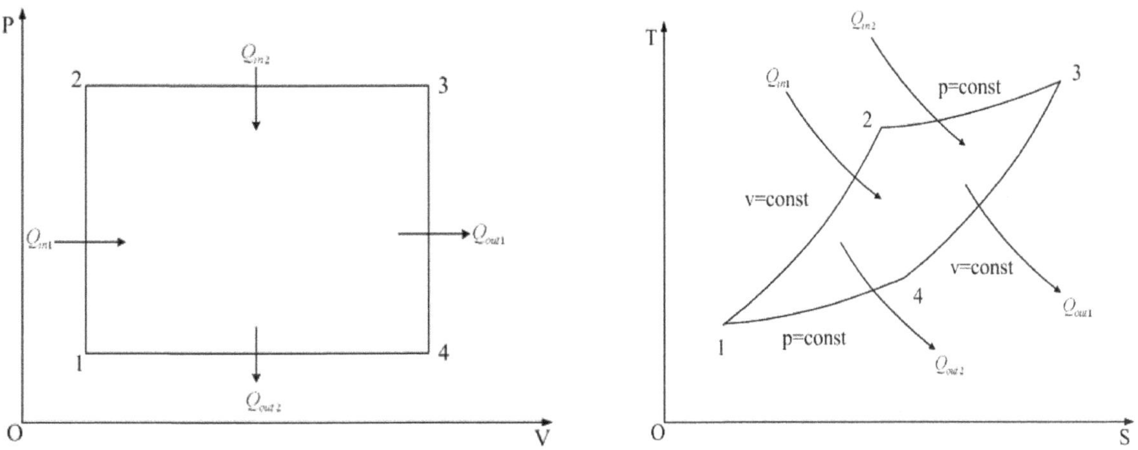

(a) P-V diagram of the cycle model (b) T-S diagram of the cycle model

Figure 1. Endoreversible rectangular cycle model

The specific heats of working fluid are variable in practical cycle, and the performance of the cycle is affected greatly by the variation. According to Refs. [20, 24], it can be supposed that the specific heats of the working fluid are only related to its temperature, and over the temperature ranges generally encountered for gases in heat engines (300-2200K), the specific heats show a linear relationship with the temperature, which may be closely approximated in the following forms:

$$C_{pm} = a_p + kT \tag{1}$$

$$C_{vm} = b_v + kT \tag{2}$$

where a_p, b_v and k are constants. According to the relation between C_{pm} and C_{vm}

$$R = C_{pm} - C_{vm} = a_p - b_v \tag{3}$$

where R is the molar gas constant of the working fluid.

The heat added to unit mass of working fluid per cycle may be written as

$$\begin{aligned} Q_{in} &= Q_{in1} + Q_{in2} = Q_{12} + Q_{23} = \int_{T_1}^{T_2} C_{vm} dT + \int_{T_2}^{T_3} C_{pm} dT \\ &= b_v(T_2 - T_1) + 0.5k(T_2^2 - T_1^2) + a_p(T_3 - T_2) + 0.5k(T_3^2 - T_2^2) \\ &= b_v(T_2 - T_1) + a_p(T_3 - T_2) + 0.5k(T_3^2 - T_1^2) \end{aligned} \tag{4}$$

The heat rejected from unit mass of working fluid per cycle may be written as

$$\begin{aligned} Q_{out} &= Q_{out1} + Q_{out2} = Q_{34} + Q_{41} = \int_{T_4}^{T_3} C_{vm} dT + \int_{T_1}^{T_4} C_{pm} dT \\ &= b_v(T_3 - T_4) + 0.5k(T_3^2 - T_4^2) + a_p(T_4 - T_1) + 0.5k(T_4^2 - T_1^2) \\ &= b_v(T_3 - T_4) + a_p(T_4 - T_1) + 0.5k(T_3^2 - T_1^2) \end{aligned} \tag{5}$$

The work output of the cycle is

$$\begin{aligned} W &= Q_{in} - Q_{out} = b_v(T_2 + T_4 - T_3 - T_1) + a_p(T_3 + T_1 - T_2 - T_4) \\ &= (a_p - b_v)(T_3 + T_1 - T_2 - T_4) = R(T_3 + T_1 - T_2 - T_4) \end{aligned} \tag{6}$$

According to the ideal gas equation $pV = nRT$, one has

$$V_3 / V_2 = T_3 / T_2 \tag{7}$$

$$V_4 / V_1 = T_4 / T_1 \tag{8}$$

The compression ratio is defined as $\gamma = V_3 / V_2$, therefore

$$T_3 = \gamma T_2 \tag{9}$$

$$T_4 = \gamma T_1 \tag{10}$$

$$Q_{in} = b_v(T_2 - T_1) + a_p T_2(\gamma - 1) + 0.5k(\gamma^2 T_2^2 - T_1^2) \tag{11}$$

$$Q_{out} = \gamma b_v(T_2 - T_1) + a_p T_1(\gamma - 1) + 0.5k(\gamma^2 T_2^2 - T_1^2) \tag{12}$$

$$W = R(\gamma - 1)(T_2 - T_1) \tag{13}$$

There are no losses in an ideal rectangular cycle, but the heat transfer loss can not be ignored in an endoreversible rectangular cycle. One can assume that the heat transfer loss through the cylinder wall is proportional to the temperature difference between the working fluid and the atmosphere, and that the wall temperature is a constant at T_0. One has the heat added to unit mass of working fluid by combustion as the following relation [14-16].

$$Q_{in} = \alpha - \beta(T_1 + T_3) = \alpha - \beta(\gamma T_2 + T_1) \tag{14}$$

where α and β are two constants related to the combustion and heat transfer.
The efficiency of the cycle is

$$\eta = \frac{W}{Q_{in}} = \frac{R(\gamma - 1)(T_2 - T_1)}{b_v(T_2 - T_1) + a_p T_2(\gamma - 1) + 0.5k(\gamma^2 T_2^2 - T_1^2)} \tag{15}$$

When γ and T_1 are given, T_2 can be obtained from Eq. (11) and Eq. (14).

$$T_2 = \frac{-[b_v + a_p(\gamma - 1) + \beta\gamma] + \sqrt{[b_v + a_p(\gamma - 1) + \beta\gamma]^2 + 2k\gamma^2(b_v T_1 + 0.5kT_1^2 + \alpha - \beta T_1)}}{k\gamma^2} \tag{16}$$

Defining

$$A=b_v+a_p(\gamma-1)+\beta\gamma \tag{17}$$

$$B=b_vT_1+0.5kT_1^2+\alpha-\beta T_1 \tag{18}$$

The work output and the efficiency of the cycle are as follows:

$$W=R(\gamma-1)(\frac{-A+\sqrt{A^2+2k\gamma^2B}}{k\gamma^2}-T_1) \tag{19}$$

$$\eta=\frac{R(\gamma-1)(\sqrt{A^2+2k\gamma^2B}-A)}{[b_v+a_p(\gamma-1)](\sqrt{A^2+2k\gamma^2B}-A)+A^2+k\gamma^2B-A\sqrt{A^2+2k\gamma^2B}-b_vk\gamma^2T_1-0.5k^2\gamma^2T_1^2} \tag{20}$$

3. Numerical examples and discussion

According to Refs. [14, 20], ranges of parameters are as follows: $\gamma=1.0-10.0$, $\alpha=60000-70000J/mol$, $\beta=20-30J/mol\cdot K$, $k=0.003844-0.009844J/mol\cdot K^2$, $b_v=19.868-23.868J/mol\cdot K$, $T_1=300-400K$. Using the above ranges of parameters, the characteristics curves of $W-\gamma$, $\eta-\gamma$, and $W-\eta$ are plotted as in Figures 2-14.

Figures 2-13 show the effects of different parameters on cycle performance when T_1=300K. One can see that the work output versus compression ratio characteristics and the efficiency versus compression ratio characteristics are parabolic curves, and the work output versus efficiency is loop shaped.

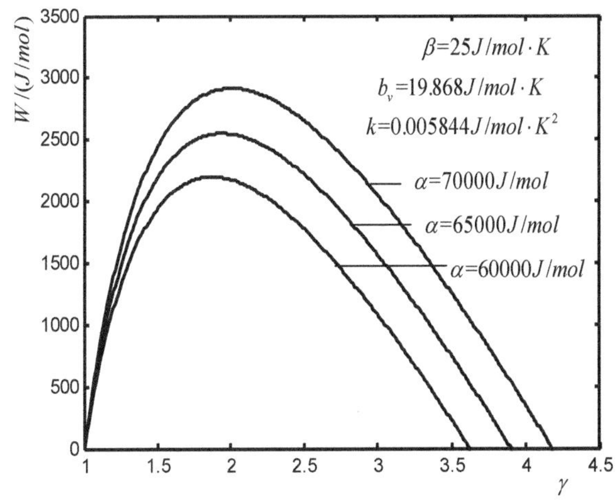

Figure 2. The influence of α on cycle work output

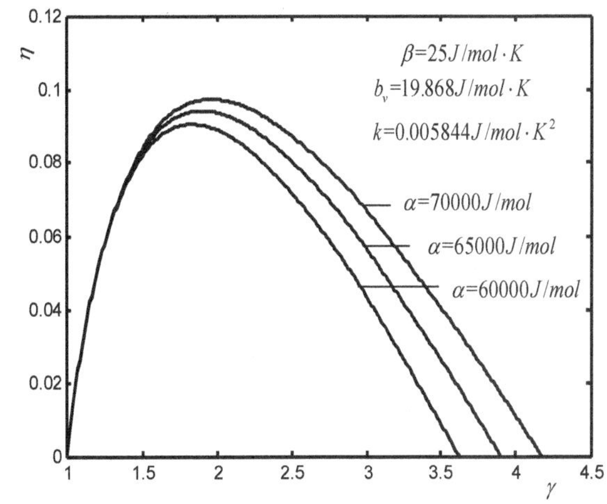

Figure 3. The influence of α on cycle efficiency

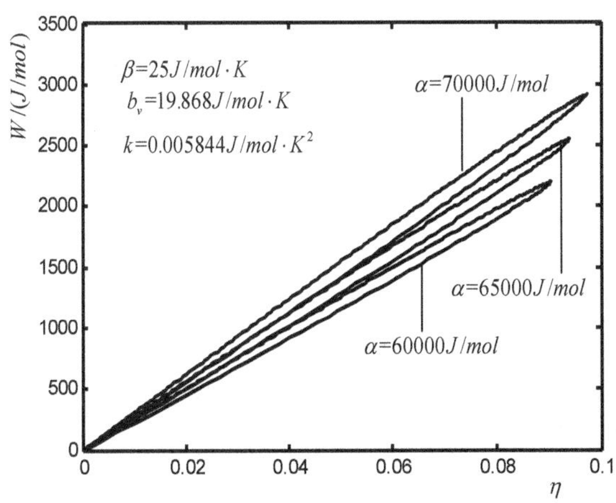

Figure 4. The influence of α on cycle work output versus efficiency

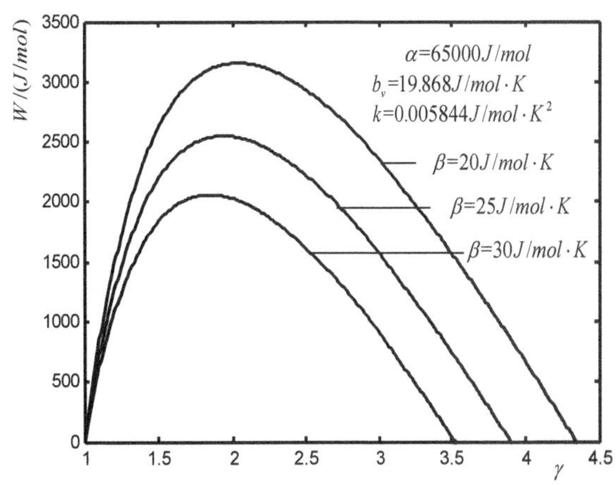

Figure 5. The influence of β on cycle work output

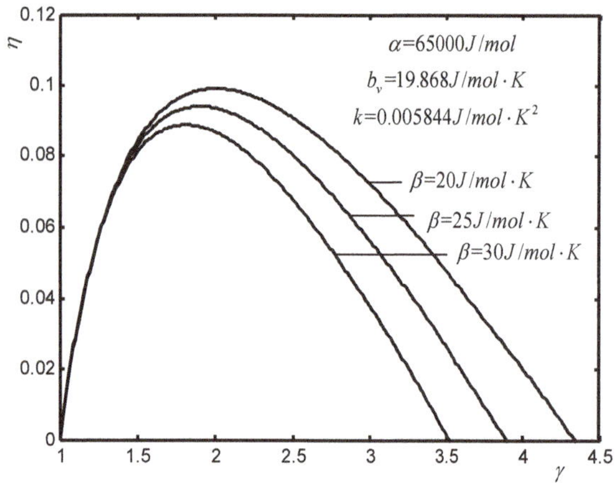

Figure 6. The influence of β on cycle efficiency

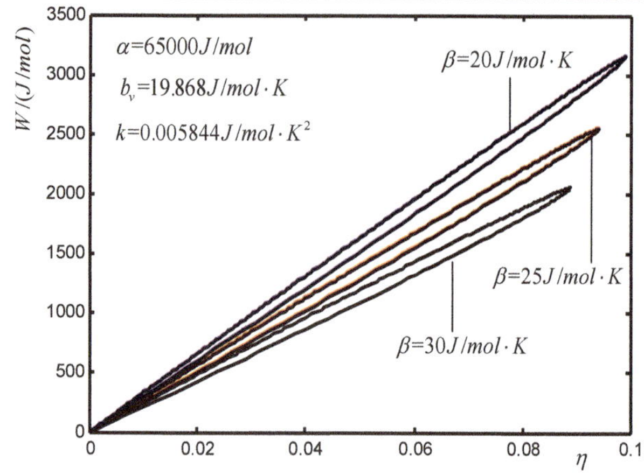

Figure 7. The influence of β on cycle work output versus efficiency

Figures 2-7 show the effects of combustion and heat transfer. α is a parameter related to the combustion and it reflects the heating value of the fuel. β is a parameter related to heat transfer loss. In this case, when α increases about 16.75%, the maximum work output increases about 32.73%, the efficiency at maximum work output increases about 7.52%, the compression ratio at maximum work output increases about 6.95%; the maximum efficiency increases about 7.51%, the work output at maximum efficiency increases about 32.75%, the compression ratio at maximum efficiency increases about 7.10%. And when β decreases about 33.3%, the maximum work output increases about 53.37%, the efficiency at maximum work output increases about 11.60%, the compression ratio at maximum work output increases about 10.27%; the maximum efficiency increases about 11.47%, the work output at maximum efficiency increases about 53.43%, the compression ratio at maximum efficiency increases about 10.50%.

Figures 8-10 show the effects of a_p and b_v on cycle performance. From Eqs. (1) and (2), one can see that when k=0, $C_{pm} = a_p$ and $C_{vm} = b_v$, a_p and b_v are constant specific heats of working fluid. Because $R = C_{pm} - C_{vm} = a_p - b_v = R$= constant, a_p and b_v must change synchronously. The results show that the maximum work output and the maximum efficiency of the cycle will decrease with the increases of a_p and b_v, and the values of compression ratios at maximum work output point and maximum efficiency point will decrease too. Furthermore, when b_v increases about 20.13%, the maximum work output decreases about 9.57%, the efficiency at maximum work output decreases about 14.26%, the compression ratio at maximum work output decreases about 2.06%; the maximum efficiency decreases about 14.24%, the work output at maximum efficiency decreases about 9.58%, the compression ratio at maximum efficiency increases about 2.10%.

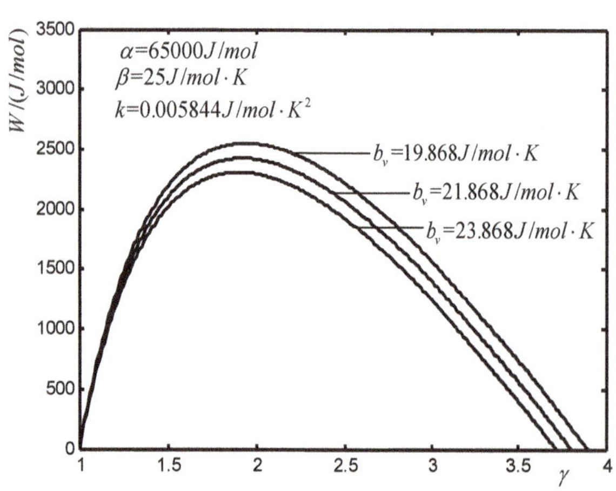

Figure 8. The influence of b_v on cycle work output

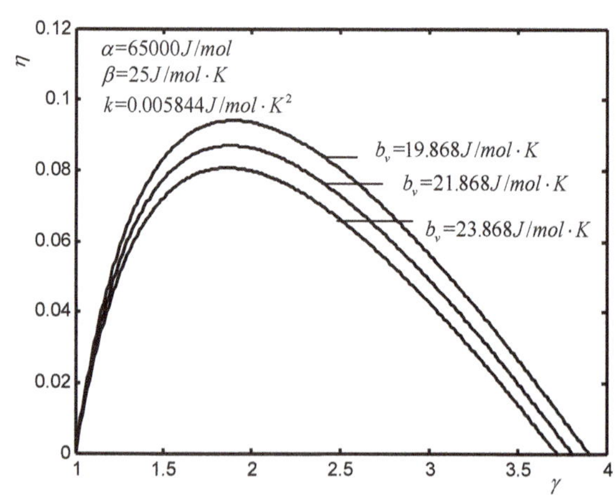

Figure 9. The influence of b_v on cycle efficiency

Figures11-13 show the effect of k on cycle performance. The degree of variation of specific heats with temperature will be acute when k increases. One can see that the maximum work output and the maximum efficiency of the cycle will decrease with the increase of k, and the values of compression ratios at maximum work output point and maximum efficiency point will decrease too. Here, when k increases about 156%, the maximum work output decreases about 10.81%, the efficiency at maximum work output decreases about 16.37%, the compression ratio at maximum work output decreases about 2.05%; the maximum efficiency decreases about 16.35%, the work output at maximum efficiency decreases about 10.78%, the compression ratio at maximum efficiency decreases about 1.57%.

Figure 14 show the relationship between the work output and the efficiency of the cycle in different initial temperature T_1. One can see that the maximum work output and the maximum efficiency of the cycle will decrease with the increase of T_1. Furthermore, when T_1 increases about 33.33%, the maximum work output decreases about 27.14%, the efficiency at maximum work output decreases about 19.04%, the compression ratio at maximum work output decreases about 12.89%; the maximum efficiency decreases about 19.12%, the work output at maximum efficiency decreases about 28.12%, and the compression ratio at maximum efficiency decreases about 12.10%.

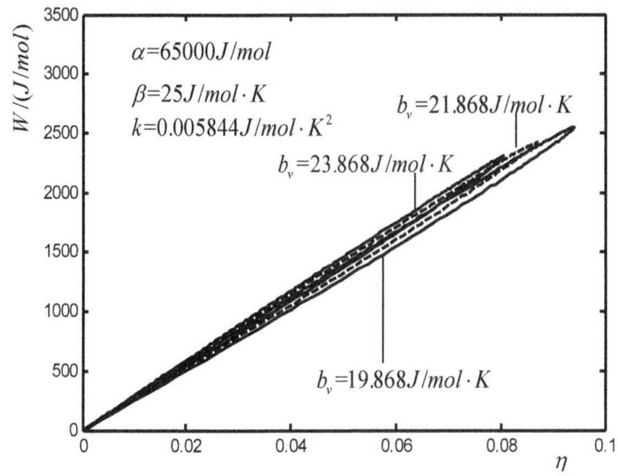

Figure 10. The influence of b_v on cycle work output versus efficiency

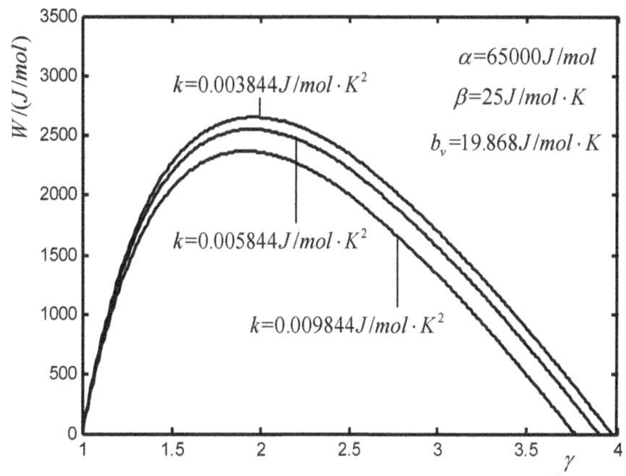

Figure 11. The influence of k on cycle work output

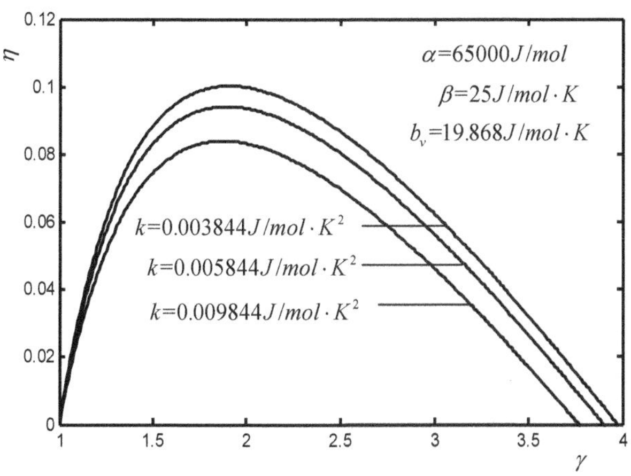

Figure 12. The influence of k on cycle efficiency

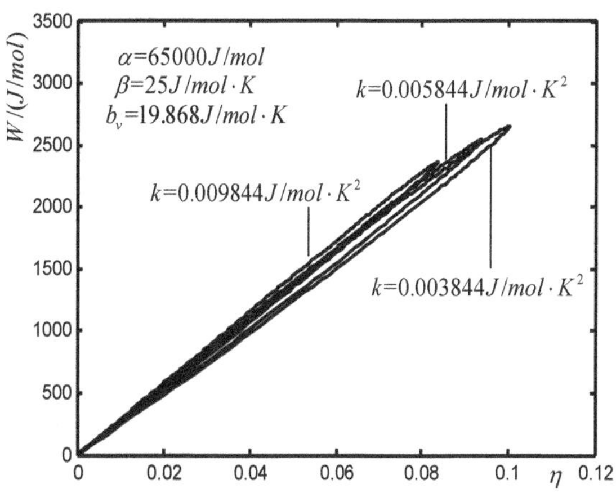

Figure 13. The influence of k on cycle work output versus efficiency

4. Conclusion

In this paper, an air-standard rectangular cycle with heat transfer loss and variable specific heats of working fluid is analyzed by using finite-time thermodynamics. The analytical functions of the work output and the efficiency are derived, and the performance characteristics of the cycle are obtained by detailed numerical examples. The results show that the effects of heat transfer loss and variable specific heats of working fluid on the cycle performance are obvious. The results may provide some guidelines for the application of the rectangular cycle.

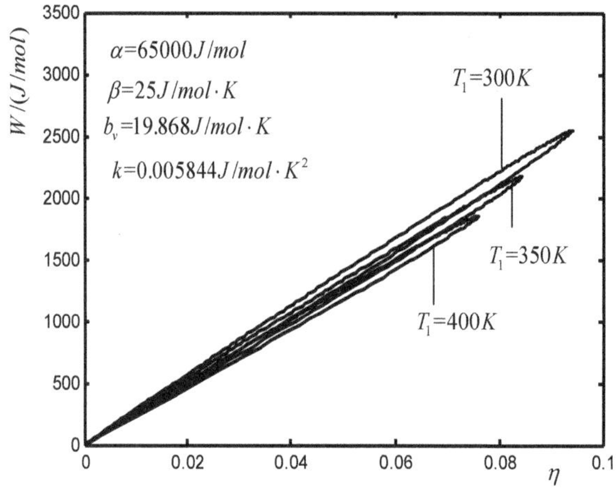

Figure 14. The influence of T_1 on cycle work output versus efficiency

Acknowledgments

This paper is supported by the National Natural Science Foundation of P. R. China (Project No. 10905093).

Reference

[1] Curzon F L, Ahlborn B. Efficiency of a Carnot engine at maximum power output. Am. J. Phys., 1975, 43(1): 22-24.

[2] Bejan A. Entropy generation minimization: The new thermodynamics of finite-size devices and finite-time processes. J. Appl. Phys., 1996, 79(3): 1191-1218.

[3] Berry R S, Kazakov V A, Sieniutycz S, Szwast Z, Tsirlin A M. Thermodynamic Optimization of Finite Time Processes. Chichester: Wiley, 1999.

[4] Chen L, Wu C, Sun F. Finite time thermodynamic optimization or entropy generation minimization of energy systems. J. Non-Equilib. Thermodyn., 1999, 24(4): 327-359.

[5] Chen L, Sun F. Advances in Finite Time Thermodynamics: Analysis and Optimization. New York: Nova Science Publishers, 2004.

[6] Chen L. Finite Time Thermodynamic Analysis of Irreversible Process and Cycles. Beijing: Higher Education Press, 2005(in Chinese).

[7] Andresen B. Current trends in finite-time thermodynamics. Angewandte Chemie International Edition, 2011, 50(12) : 2690-2704.

[8] Rubin M H. Optimal configuration of a class of irreversible heat engines. Phys. Rev. A, 1979, 19(3): 1272-1287.

[9] Mozurkewich M, Berry R S. Optimal paths for thermodynamic systems: the ideal Otto cycle. J. Appl. Phys., 1982, 53(1): 34-42.

[10] Hoffman K H, Watowich S J, Berry R S. Optimal paths for thermodynamic systems: The ideal Diesel cycle. J Appl. Phys., 1982, 85(6): 2125-2134.

[11] Chen L, Lin J, Luo J, Sun F, Wu C. Friction effect on the characteristic performance of Diesel cycles. Int. J. Energy Research, 2002, 26(11): 965-971.

[12] Klein S A. An explanation for observed compression ratios in internal combustion engines. Trans. ASME J. Engng. Gas Turbine Pow. 1991, 113(4): 511-513.

[13] Chen L, Zen F, Sun F. Heat transfer effects on the net work output and power as function of efficiency for air standard Diesel cycle. Energy, 1996, 21(12): 1201-1205.

[14] Chen L, Zen F, Sun F. Heat transfer effects on the net work output and efficiency characteristics for an air standard Otto cycle. Energy Convers. Magmt. 1998, 39(7): 643-648.

[15] Al-Hinti I, Akash B, Abu-Nada E. Performance analysis of air-standard Diesel cycle using an alternative irreversible heat transfer approach. Energy Convers. Manage, 2008, 43(15): 2019-2031.

[16] Qin X, Chen L, Sun F, Wu C. The universal power and efficiency characteristics for irreversible reciprocating heat engine cycles. Eur. J. Phys., 2003, 24(4): 359-366.

[17] Ge Y, Chen L, Sun F, Wu C. Reciprocating heat-engine cycles. Ap. Energy, 2005, 81(4): 397-408.

[18] Ghatak A, Chakraborty S. Effect of external irreversibilities and variable thermal properties of working fluid on thermal performance of a Dual internal combustion engine cycle. Strojnicky Casopsis (J. Meh. Energy), 2007, 58(1): 1-12.

[19] Chen L, Ge Y, Sun F, Wu C. Effects of heat transfer, friction and variable specific heats of working fluid on performance of an irreversible Dual cycle. Energy Convers. Mgnt. 2006, 47(18/19):3224-3234.

[20] Ge Y, Chen L, Sun F, Wu C. Thermodynamic simulation of performance of an Otto cycle with heat transfer and variable specific heats of working fluid. Int. J. Therm. Sci., 2005, 44(5):506-511.

[21] Ge Y, Chen L, Sun F, Wu C. The effects of variable specific heats of working fluid on the performance of an irreversible Otto cycle. Int. J. Exergy, 2005, 2(3):274-283.

[22] Ge Y, Chen L, Sun F, Wu C. Performance of an endoreversible Diesel cycle with variable specific heats of working fluid. Int. J. Ambient Energy, 2008, 29(3):127-136.

[23] Chen L, Ge Y, Sun F. Unified thermodynamic description and optimization for a class of irreversible reciprocating heat engine cycles. Proc. IMechE, Part D: J. Automob. Eng., 2008, 222(D8): 1489-1500.

[24] Ferreira da Silva M F. Some considerations about thermodynamic cycles. Eur. J. Phys., 2012, 33(1): 13-42.

[25] Liu X, Chen L, Qin X, Ge Y, Sun F. Finite-time thermodynamic analysis for an endoreversible rectangular cycle. Energy Conservation, 2013, 32(1): 19-21(in Chinese).

[26] Liu C, Chen L, Ge Y, Sun F. The power and efficiency characteristics for an irreversible rectangular cycle. Power and Energy, 2013, 34(2): 113-117(in Chinese).

11

Assessment of wind resources in Braşov region (Romania)

E. Eftimie[1], N. Eftimie[2]

[1] Department of Product Design, Mechatronics and Environment, Transilvania University of Braşov, Eroilor 29, 500036, Romania.
[2] Department of Manufacturing Engineering, Transilvania University of Braşov, Eroilor 29, 500036, Romania.

Abstract
The accurate assessment of wind potential for an area requires both the knowledge of probability density function and the power density of wind. For this purpose, for the area of interest are required databases that to contain direct measurements of wind parameters recorded during at least one year. In order to model the wind speed, Weibull distribution is used. However, the use of Weibull distribution is most often difficult due to the need of knowing the Weibull distribution parameters (the shape parameter and the scale parameter). Considering this aspect, this paper proposes a method for estimating the Weibull parameters using their dependence on the Gamma function. The exemplification is made through a case study, by processing of data recorded by two weather stations located in two different areas of Braşov. The proposed algorithm is a method easily to apply to any location that has a secure database.

Keywords: Maximum energetic wind speed; Mean wind speeds; Power density; Weibull parameters; Weibull probability density function.

1. Introduction

At present, the use of wind energy is widespread due to some benefits such as the significant reducing of toxic gases produced to power generation by burning fossil fuels; diversification of sources for energy production and the development of new technologies to capitalize renewable energy; the capitalization of renewable resources energy of isolate areas for their introduction into the economic circuit.
Nowadays, the use of wind energy refers primarily to non-polluting electricity produced at a significant scale with wind turbines.
The full knowledge of the conversion technology of wind energy into electricity involves knowledge in different areas: aerodynamics, electrical engineering, mechanical engineering, civil engineering, etc., but also in the field of meteorology. It is envisaged that the formulation of a correct decision regarding the investments opportunity for implementation of wind energy conversion systems requires information on wind energy resources of the site where it will be built a possible wind power plant.
At present though the technical literature proposes wind maps, this does not publish the evaluation methods used for their development [1-3].
Regarding the expressions of shape parameter we mention, commonly technical literature [4, 5] offers a power expression of this depending on the turbulence intensity.
In addition, although there are a number of specialized software (for instance WAsP, UPMORO, UPMPARK, Wind Power, WindPro programs) for calculating wind climatology statistics (maps, graphs,

images characteristic to common wind atlases but also numerical information in tabular form regarding to the wind speed and power density, estimations on wind energy potential) the estimation models used are not known [6, 7].

It should also be noted that the results provided – even complex – can be applied only to the station site where the measurements were carried out. It is envisaged that wind speed can vary both in time, depending on weather conditions and also over short distances; therefore the assessment procedures of wind turbines location must consider all regional parameters that are likely to influence the wind conditions, namely: obstacles in close proximity, description of land roughness, orography (such as hills, these could cause acceleration or deceleration effects of air flow), climate, environmentally protected areas.

Considering this aspect, the paper proposes a method to estimate the Weibull parameters (and finally the wind power density) and that to be easily to apply if a decision on the opportunity of placement of wind turbines in certain geographical areas must be taken.

2. Material and methods

2.1 Weibull probability distribution function

The widely used distribution for statistical modelling of wind speed variation for a given site is represented by the Weibull probability distribution with two parameters.

The probability density function of wind speed is given by Weibull distribution, as follows [7-9]

$$f(v) = k / A(v / A)^{k-1} exp\left(-(v / A)^k\right) \tag{1}$$

where A and k represent the parameters of Weibull probability distribution, A – the scale parameter (m/s) and k – the shape parameter.

Equation (1) can be written depending on Gamma function [10]:

$$f(v) = k / \bar{v}\Gamma(1+1/k)(v / \bar{v}\Gamma(1+1/k))^{k-1} exp\left(-(v / \bar{v}\Gamma(1+1/k))^k\right) \tag{2}$$

where $\Gamma(1+1/k)$ represents the Gamma function for argument $(1 +1/k)$.

The scale parameter is calculated as below

$$A = \bar{v} / \Gamma(1+1/k) \tag{3}$$

where $\bar{v}$ represents the mean of wind speed observations and it is defined as:

$$\bar{v} = 1/n \sum_{i=1}^{n} v_i \tag{4}$$

where n is the number of wind speed observations, v_i – the ith wind speed data value.

The cumulative distribution function represents the integral of Weibull probability density function and is calculated with the following equation [10]:

$$F(v) = 1 - exp\left(-(v / \bar{v}\Gamma(1+1/k))^k\right) \tag{5}$$

This function is often used to compute probabilities.

One of the phenomena that negatively influence the turbine rotor is turbulence; this leads to increase mechanical stress caused by short gusts of wind, the propeller material wearies and could destroy.

The indicator that characterizes the wind turbulence is the turbulence intensity defined as the ratio of standard deviation and average speed over a specified period of time, respectively

$$\sigma / \bar{v} = \sqrt{\Gamma(1+2/k)/\Gamma^2(1+1/k) - 1} \tag{6}$$

The wind turbulence refers to fluctuations in wind speed on a short period of time. The turbulence is caused by two phenomena: the friction between the airflow and the surface of the earth (often enhanced by topographical features of the relief, namely the presence of valleys, hills and mountains); the second phenomenon relates to thermal effects which cause the vertical movements of air masses [7].
Concomitant with increasing height, wind turbulence decreases.
It must be mentioned that the most frequent wind speed, v_{mp}, and the maximum energetic wind speed, v_{maxE}, can be also described using Weibull distribution parameters [8, 9, 11-13]:

$$v_{mp} = A(1-1/k)^{1/k} \tag{7}$$

$$v_{max\,E} = A(1+2/k)^{2/k} \tag{8}$$

2.2 Wind power density and wind energy
The mean wind power represents the product of mass flow rate that crosses the wind turbine blades, ρvS, and the kinetic energy of wind per unit mass, $v^2/2$ [7]:

$$\overline{P(v)} = 0.5\,\rho S \bar{v}^3 \tag{9}$$

where ρ = 1.225 kg/m^3 represents the air density considering normal air pressure and the air temperature of 15°C), $P(v)$ (W) – the wind power and S (m^2) – the swept area of rotor blades.
The wind power density (wind power per unit area) of the selected site, $p(v)$ (W/m^2), can be expressed depending on Weibull parameters as [4, 5, 7]:

$$p(v) = P(v)/S = 0.5\,\rho A^3 \Gamma(1+3/k) \tag{10}$$

3. Calculations
3.1 Relations for Gamma function and the shape parameter of Weibull distribution
The determination of probability density function, the wind turbulence (the ratio between the standard deviation and average wind speed), the value of the most frequent wind speed, the value of the maximum energetic wind speed and even the wind power density can be achieved using the Gamma function. For situations where it is desired the development of own procedures for estimating the parameters mentioned above and the software used for this purpose do not have functions for statistical analysis, this paper proposes the estimation of Gamma function using equation (11) or equation (12) depending on the argument value:

$$\Gamma(x) = 0.7282\,x^4 - 4.1991\,x^3 + 9.3776\,x^2 - 9.6245\,x + 4.7116 \tag{11}$$

$$\Gamma(x) = 0.0416\,x^6 - 0.67\,x^5 + 4.6327\,x^4 - 17.192\,x^3 + 36.15\,x^2 - 40.544x + 19.673 \tag{12}$$

As it was seen in previous chapters, a complete analysis of wind statistic requires the knowledge of shape parameter. It is obvious the fact that the use of a more accurate expression of shape parameter leads to more precise estimations of Weibull probability density function, Weibull cumulative distribution function, wind power density and maximum energetic wind speed.
This paper proposes the determination of shape parameter using equation (6). Knowing the values of Gamma function, the right term of equation (6) can be graphically represented for different values of shape parameter (Figure 1); thus, the expressions of shape parameter, k, that best approximates the curve plotted on the graph, can be determined.
We propose for the shape parameter, k, the use of equation (13),

$$k = 1.0275(\sigma/\bar{v})^{-1.073} \tag{13}$$

Or for a more accurate determination, depending on the intensity turbulence value, the use of one of the equations (14), (15) or (16)

$$k = 1.0618(\sigma / \bar{v})^{-1.056} \ for \ \sigma / \bar{v} \leq 0.2 \tag{14}$$

$$k = 0.9736(\sigma / \bar{v})^{-1.111} \ for \ 0.2 < \sigma / \bar{v} \leq 0.8 \tag{15}$$

$$k = 1.013(\sigma / \bar{v})^{-0.922} \ for \ 0.8 < \sigma / \bar{v} \tag{16}$$

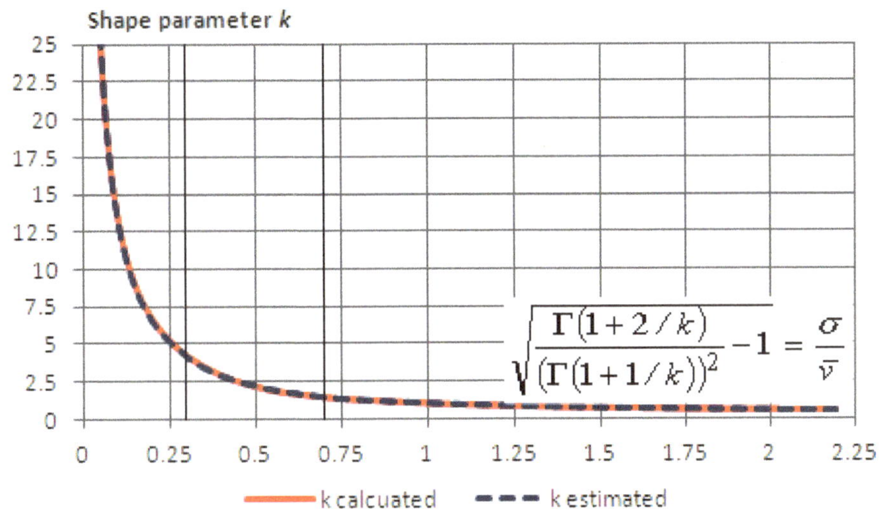

Figure 1. Determination of the shape parameter, *k*

The comparison test of proposed relations for determination of shape parameter, *k*, with respect to its actual values was assessed using three statistical indicators: Mean Percentage Error (MPE), Mean Bias Error (MBE) and Root Mean Square Error (RMSE). These indicators are calculated with the following equations [14]:

$$MBE = \left(\sum_{i=1}^{N} \left(k_{estimated\,i} - k_i \right) \right) / N \tag{17}$$

$$MPE = \left(\sum_{i=1}^{N} \left(\left(k_i - k_{estimated\,i} \right) / k_i \right) 100 \right) / N \tag{18}$$

$$RMSE = \sqrt{\left(\sum_{i=1}^{N} \left(k_{estimated\,i} - k_i \right)^2 \right) / N} \tag{19}$$

where i is the index, $k_{estimated\ i}$ – the ith estimated value of shape parameter and k_i – the actual value of shape parameter calculated from equation (6).

The statistical parameters were calculated for two variation ranges of turbulence intensity values (Table 1), namely:

- between 0.05 and 2.23 that corresponds to values of shape parameter in the limits of range of 25 and 0.5;
- between 0.12 and 2.23 respectively for values of shape parameter between 10 and 0.5.

Table 1. Statistical indicators values (MBE, MPE, RMSE)

Range limits for *k*	*k* estimated using technical literature [4, 5, 11]			*k* estimated using equation (13)			*k* estimated using equations (14)-(16)		
	MPE (%)	MBE	RMSE	MPE (%)	MBE	RMSE	MPE (%)	MBE	RMSE
0.5÷25	-0.3762	0.1722	0.3313	-0.8372	*0.1175*	*0.2158*	**0.0495**	**-0.0037**	**0.0550**
0.5÷10	0.9932	-0.0137	0.1073	*-0.5826*	*0.0124*	*0.0906*	**0.0041**	**-0.0077**	**0.0466**

The comparison between the statistical test results for the three estimations proposed emphasizes very good results for the estimation of shape parameter using one of the equations (14), (15) or (16).

The MPE and MBE variation, for shape parameter values between 0 and 10 are presented by Figure 2 and Figure 3.

According to the values of MPE and MBE, calculated for *k* values between 0.5 and 10, the best performance is found at the estimation made using one of the equations (14), (15) or (16). We also mention the estimation achieved using equation (13) leads, for *k* values between 4 and 10, to a better approximation compared with the approximation made with equation proposed by literature.

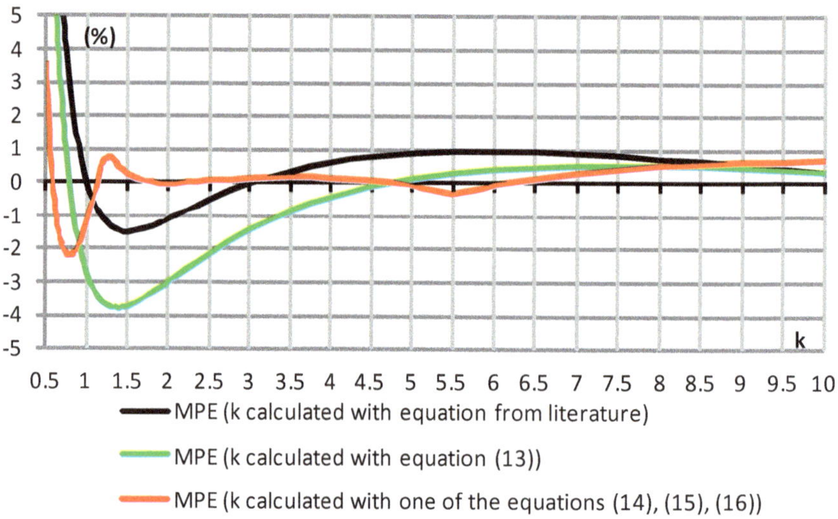

Figure 2. Mean Percentage Error variation for shape parameter *k* calculated with different equations

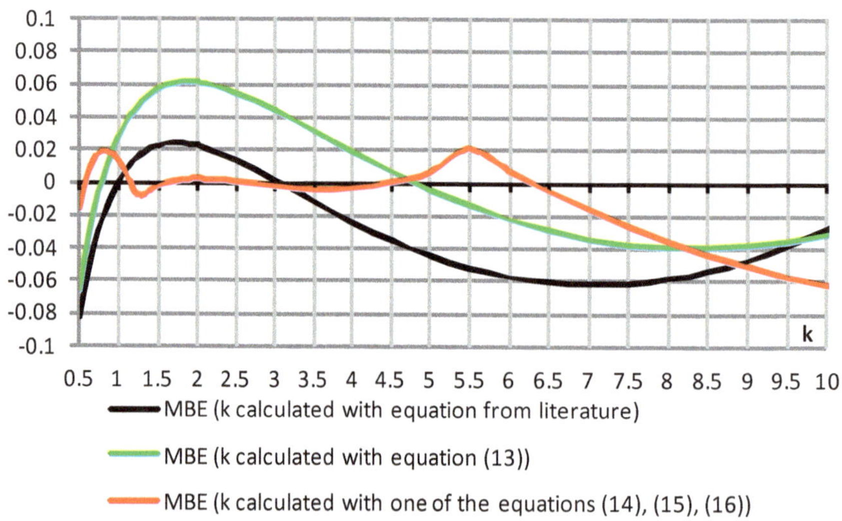

Figure 3. Mean Bias Error variation for shape parameter *k* calculated with different equations

Taking into consideration those above mentioned the present paper proposes the following algorithm:

- considering the available wind database the following values are determined: the average speed, standard deviation and the turbulence intensity;
- the shape parameter is determined using one of the equations (13)-(16);
- using equation (3), the scale parameter A is determined;
- the following parameters can be calculated: the most frequent wind speed (equation (7)), the maximum energetic wind speed (equation (8));
- the wind power density can be estimated using equation (10);
- the Weibull probability density function can be plotted.

3.2 Experimental and computational methods

Braşov is located in the centre of Romania, in Braşov Depression, located at an altitude of 625 m.

As a consequence of the geographical position of Braşov (45.65°N, 25.59°E), this benefits by a temperate continental climate with four distinct seasons: spring, summer, autumn and winter. Local climatic differences are determined mainly by relief and latitude, and less by the oceanic influences from the West, that South-Western Mediterranean and those Eastern continental.

The wind statistics climatology was carried out for two locations in Braşov: one in the town area and the other one in the suburbs area. Therefore, a case study with experimental validation was done, the presented algorithm being applied for the two proposed places of Braşov.

The meteorological data were monitored using two Weather Stations (Delta T), including AN1 anemometers for wind speed and wind direction (resolution and error: 1% ±0.1m.s-1; sensitivity: 0.8Hz per m.s-1).

The first weather station is located in town area, on the roof terrace of the Department of Renewable Energy Systems and Recycling, from Transilvania University of Braşov; this weather station records meteorological data (solar radiation, temperature, the amount of precipitations, wind speed and wind direction) from October 2005.

It is mentioned, the Romanian National Meteorological Administration has installed in the same location a weather station (that works in parallel with Delta-T weather station) for data comparing and validation.

As it can be noticed, the large amount of recorded data makes possible a complete and reliable energy analysis.

The second weather station is installed on the roof terrace of the Research & Development Institute located in the suburbs area. The meteorological database comprises data recorded from January 2013.

At the two weather stations the measurements of both wind parameters, the direction and speed, are carried out, according to recommendations of World Meteorological Organization, at 10 m height above the ground.

Meteorological data management was performed using the Visual FoxPro programming environment. To the choice of software environment for meteorological data processing, the following issue was considered: the large amount of data to be stored and processed makes necessary the use of a specialized database management system, that to make possible the obtaining of specialized results. In this way, there were possible: data management, defining the rules for databases, creating queries using visual design tools, and finally building applications.

4. Results and discussion

Therefore, for the case study there were considered two areas of Braşov, namely: town area of Braşov, for that a weather database corresponding to a period of 8 years is available, and the suburbs area of Braşov, which has weather data for a period about a year and a half.

The results presented further were obtained after processing the entire database available for each weather station (from January 2006 to April 2014 for town area and from January 2013 to April 2014 for suburbs area).

A summary of the monthly values of Weibull parameters (shape and scale parameters) calculated for the measured wind speeds at 10 m, for the two locations of Braşov area (town and suburbs areas) are presented in Table 2.

For the town area, it was found that the monthly values of Weibull shape parameter, k, has values in the range of 0.75-1.16. The monthly values of Weibull scale parameter, A, vary in the range of 0.77-1.51 m/s, these corresponding to monthly mean wind speeds values between 0.9-1.5 m/s.

Table 2. Monthly Weibull parameters for Braşov site

Month	Town area		Suburbs area	
	Shape parameter k	Scale parameter A (m/s)	Shape parameter k	Scale parameter A (m/s)
January	0.815	0.924	1.003	1.784
February	0.928	1.021	1.054	1.760
March	0.988	1.520	1.172	2.940
April	0.993	1.293	1.313	2.549
May	1.090	1.348	1.290	2.428
June	1.159	1.409	1.419	1.981
July	1.117	1.433	1.308	2.530
August	1.099	1.151	1.344	1.854
September	0.988	1.135	1.290	3.077
October	0.849	0.776	1.000	1.429
November	0.855	0.895	1.105	1.881
December	0.750	0.805	0.766	1.216

In the case of suburbs area, the monthly values of Weibull shape parameter, *k*, vary between 0.76-1.42, the monthly values of Weibull scale parameters, A, being contained in the range of 1.21-3.05 m/s, thus corresponding to the monthly values of mean wind speeds in the range of 1.5-2.9 m/s.

The monthly probability density functions and the cumulative distribution functions of wind speeds for the two analyzed sites are presented in Figures 4 and 5.

For town area it can be seen a tendency of obtaining the wind speeds around 1.2 m/s, for all the months for the entire year. The monthly peak frequencies range between 47% (in March) and 60% (in August and October), Figure 4.

In the case of suburbs area the monthly peak frequencies are recorded for winds speeds of 1-2 m/s and the range limits are between 20% (September) and 47% (December), Figure 5.

Table 3 presents the monthly values of maximum energetic wind speed calculated using equation (8) and the monthly values of wind power density.

Comparative analysis of the calculated values (based on measured data) and the estimated wind power density shows that the percentage differences between these varies within the limits of -5.4% (September) and 2.24% (January); so it is recorded a slight overestimation of the wind power density.

Comparing the calculated and estimated values of wind power density obtained for suburbs area, the percentage differences are between -4.3% (July) and 4.7% (October).

It is obvious that wind power density values recorded in suburbs area are higher than those recorded for the town area, the highest monthly differences being obtained for the month of September (wind power density for suburbs area is about eight times higher than for town area); the lowest difference was obtained for June.

The annual value of wind power density for suburbs area is about 3.7 times higher than its value obtained for town area.

Figures 6 and 7 show the monthly variation of wind power density depending on the wind speed, for the two areas subjected to analysis.

Considering the town area, the maximum wind power density was obtained during March, the diagram for this month validating the value of maximum energetic wind speed presented by Table 4.

In the case of suburbs area, the maximum wind power density was also obtained for March, values higher than 5.25 W/m^2 being obtained for wind speeds values between 6 m/s and 12 m/s.

It must also be noted the October month with high values of wind power density (higher than 6 W/m^2) for wind speeds between 5 m/s and 9 m/s.

However it should be mentioned that much more important is the study of distribution of wind power depending on wind direction. For this purpose Figure 8 presents the wind power density depending on wind direction (the energetic rose).

For town area, the maximum power wind density is given by the winds from North-East, West, and North-West.

In the case of suburbs area, the rose of wind power density shows as predominant direction, North-West direction. Compared to the town area the wind power density has a more significant value.

The monthly values of wind power density obtained on each sector are presented by Table 4 for town area and by Table 5 for suburbs area.

For the town area, the monthly distribution of wind power density shows a predominant direction from North-East during January, February, May, July and November. The highest value of wind power density is obtained during January, for wind from North-East. A significant value of wind power density is recorded during December, for winds from South-West direction.

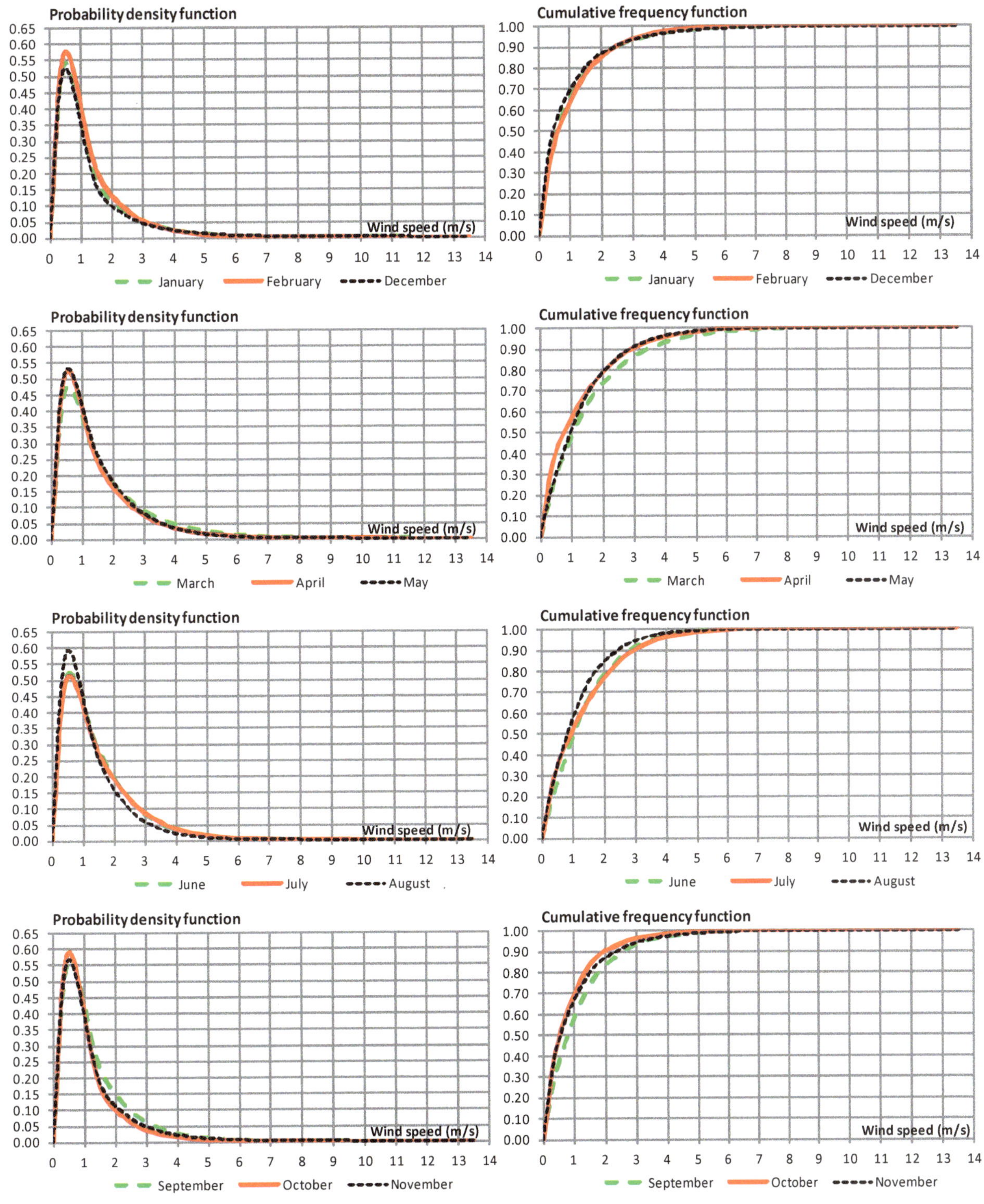

Figure 4. Monthly wind speed probability density and cumulative distribution functions for town area

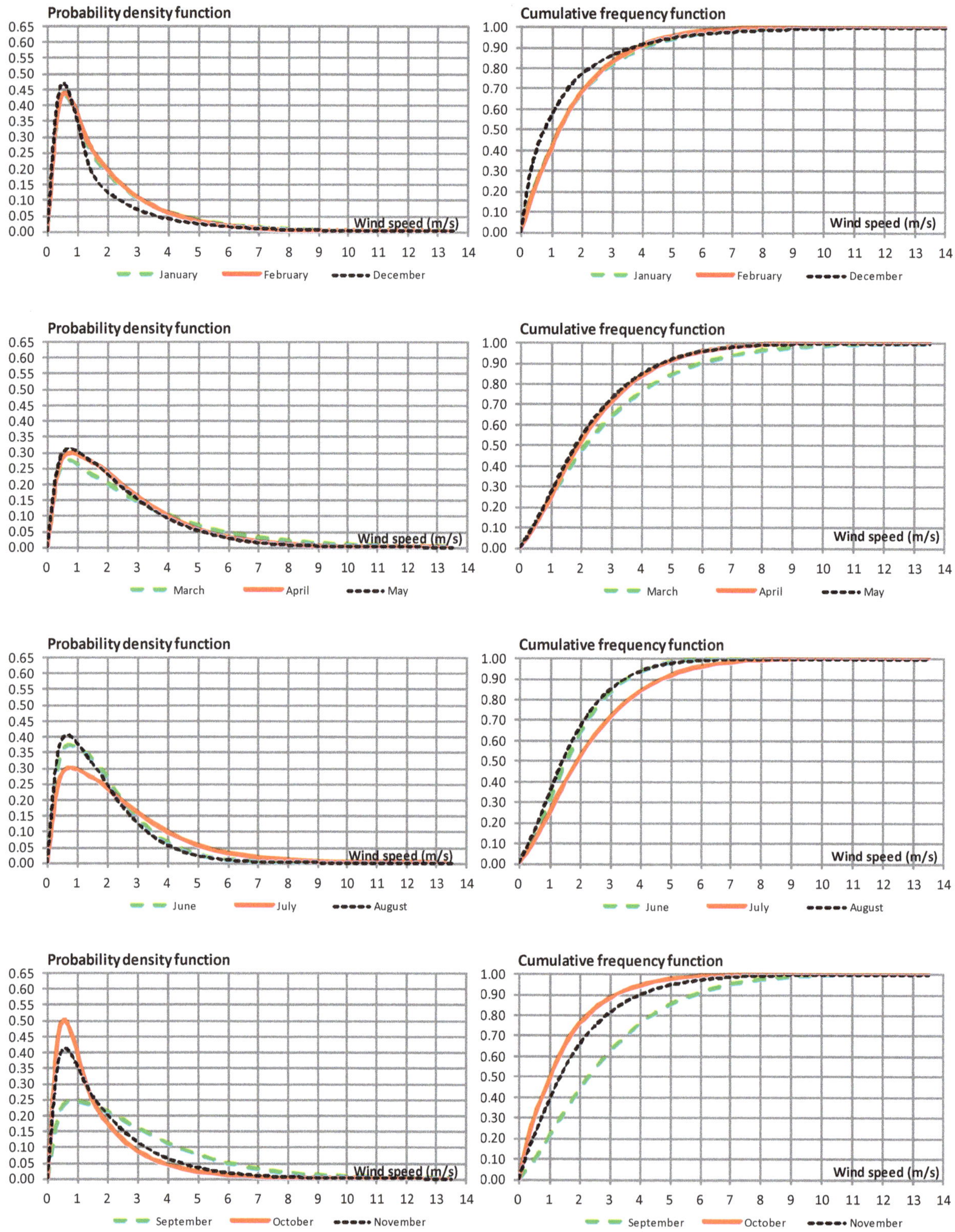

Figure 5. Monthly wind speed probability density and cumulative distribution functions for suburbs area

In the case of suburbs area, for six months (April, June, July, August, September and October) the highest values of wind power density are obtained due to the winds from North-West; five months (January, February, March, November, December) show as predominant direction for the wind power density, the West direction.

The highest value of wind power density is obtained due to winds from West, during March. We mention, during March, important values of wind power density are also obtained due to winds from West and North-West.

Tables 4 and 5 present also the annual values of wind power density depending on wind direction; it is ascertained, the highest values of wind power density are obtained for winds from North-East and North-West if town area is considered; for suburbs area the highest values of wind power density are obtained for winds from West and North-West.

Table 3. Monthly maximum energetic wind speed and monthly calculated and estimated wind power

Month	Maximum energetic wind speed (m/s) – equation (7)		Calculated wind power density (W/m^2)		Estimated wind power density (W/m^2) – equation (9)	
	Town area	Suburbs area	Town area	Suburbs area	Town area	Suburbs area
January	4.23	5.32	7.100	21.425	7.260	20.629
February	3.52	4.83	5.013	19.064	5.200	18.877
March	**4.66**	**6.87**	**12.491**	**56.786**	**12.987**	55.227
April	3.93	5.16	7.807	26.245	8.183	26.798
May	3.51	5.02	6.403	24.049	6.665	24.160
June	3.35	3.68	5.997	10.496	6.283	10.598
July	3.59	5.14	7.052	25.337	7.388	26.444
August	2.96	3.65	3.833	9.921	4.036	9.779
September	3.48	6.36	5.270	**45.894**	5.558	46.164
October	3.23	4.28	3.385	11.323	3.480	10.781
November	3.66	4.79	4.742	17.902	4.925	17.315
December	4.55	6.51	7.187	24.045	7.560	23.402

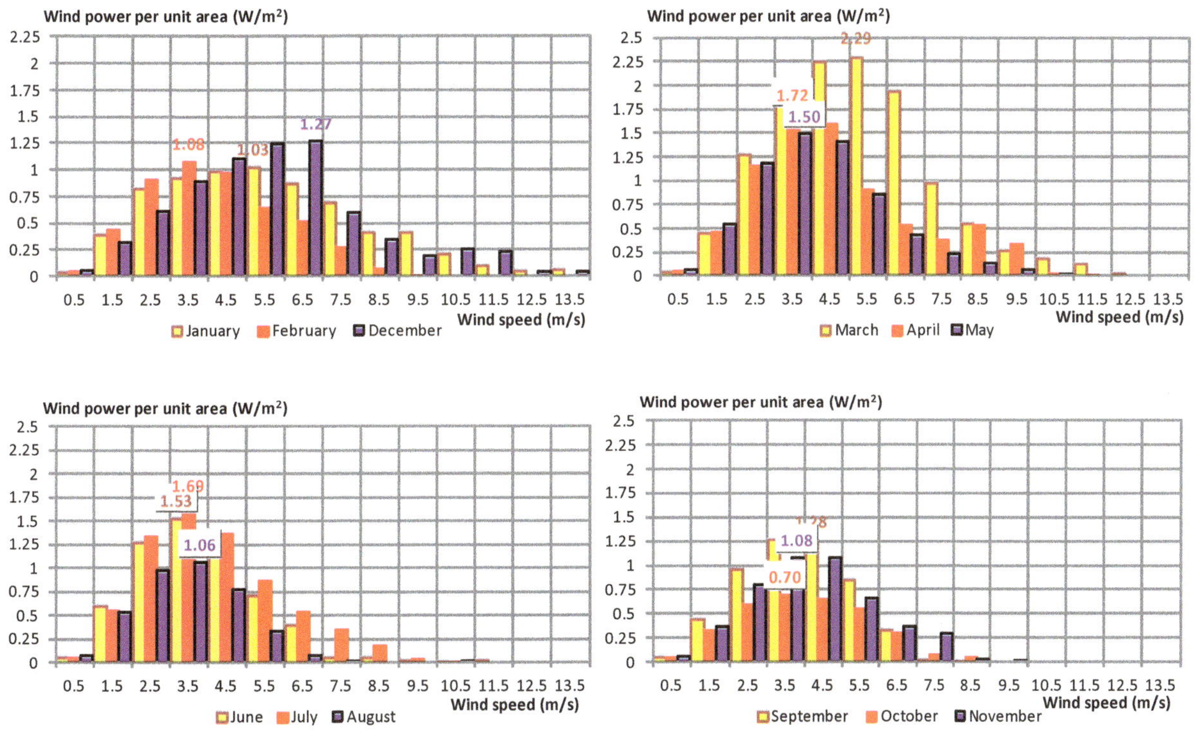

Figure 6. Monthly wind power density for town area

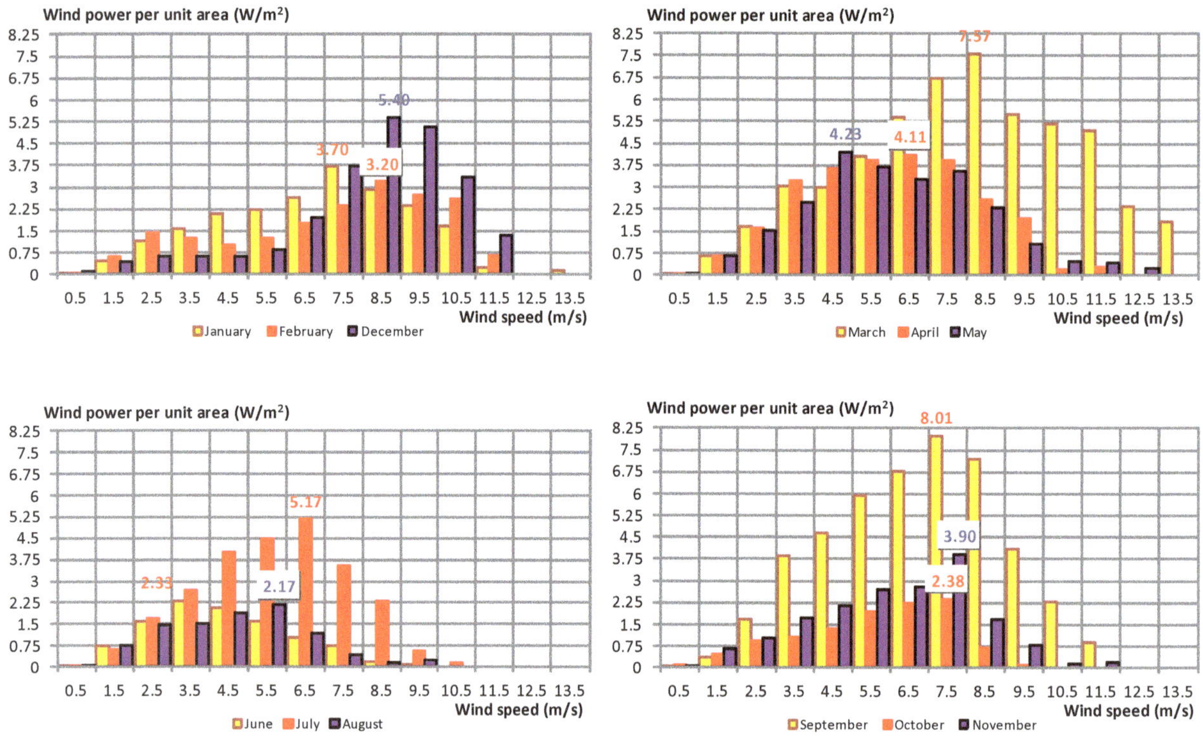

Figure 7. Monthly wind power density for suburbs area

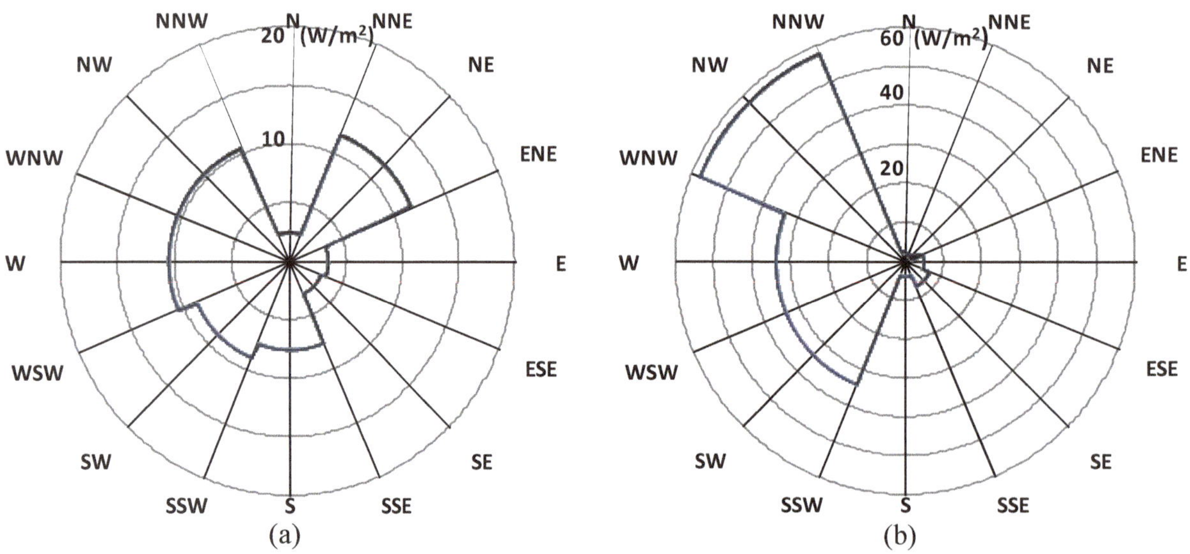

Figure 8. Wind power density rose: (a) Town area; (b) Suburbs area

Table 4. Monthly values of wind power density depending on wind direction for town area

Month	Wind power density (W/m²)							
	N	NE	E	SE	S	SW	W	NW
January	0.51	**28.12**	3.55	2.15	0.91	4.93	8.15	5.32
February	2.67	**10.19**	4.11	3.17	2.56	5.17	5.40	7.13
March	6.97	14.90	4.98	3.43	4.61	17.36	**19.25**	**20.88**
April	3.22	13.21	5.04	4.34	2.51	5.63	10.77	**19.71**
May	3.06	**11.39**	5.09	4.50	3.78	7.51	5.80	9.83
June	3.79	10.68	3.98	3.29	1.98	5.99	6.38	**12.03**
July	1.53	**17.28**	3.69	3.39	1.92	4.26	7.51	12.32
August	2.82	3.61	1.26	1.88	2.51	3.32	4.83	**10.00**
September	1.53	7.48	1.82	1.71	2.71	3.11	**9.72**	**9.88**
October	0.32	2.76	1.53	1.21	1.17	6.75	5.39	6.59
November	0.26	**7.59**	2.28	2.80	1.97	7.22	**7.86**	5.18
December	1.10	2.43	3.02	3.18	9.23	**19.37**	10.38	5.84
Annual values	2.54	**11.67**	3.35	2.91	2.90	7.48	8.80	**10.54**

Table 5. Monthly values of wind power density depending on wind direction for suburbs area

Month	Wind power density (W/m²)							
	N	NE	E	SE	S	SW	W	NW
January	1.14	1.00	4.81	7.19	1.35	29.69	**49.50**	46.30
February	1.80	1.61	4.94	4.46	1.19	44.43	**44.65**	23.26
March	2.04	4.66	7.92	8.75	11.14	83.82	**133.44**	104.96
April	9.29	1.96	8.16	13.67	3.88	44.39	40.92	**50.91**
May	2.75	5.28	5.95	17.16	9.12	**42.84**	36.27	33.11
June	6.25	1.53	2.58	2.49	2.20	14.30	16.00	**19.56**
July	8.38	1.48	3.34	6.56	3.06	9.65	35.99	**69.92**
August	0.21	1.94	2.36	3.87	4.24	10.92	19.66	**33.18**
September	4.54	0.74	2.98	2.11	3.13	24.00	76.65	**82.07**
October	0.23	0.66	1.62	1.89	0.90	19.51	19.78	**43.93**
November	1.18	0.99	4.49	3.70	2.71	38.00	**43.48**	25.11
December	1.46	0.42	1.03	1.58	0.94	2.70	**75.55**	40.62
Annual values	2.70	1.72	4.86	6.75	3.73	33.81	**57.70**	**53.62**

5. Conclusions

The wind variability also implies the variability of generated electricity. This is the difference compared to the most sources of conventional energy where the fuel is usually kept constant. The primary source of energy in producing wind energy does not have a constant flow.

The site climate describes this variability statistically. Different sites have different wind climates; the wind energy dependence of the cube of wind speed leads to variation of annual average power depending on site.

The geographical position of Braşov, in the internal curvature of the Carpathian, at the junction of the Southern Carpathians, the Eastern Carpathians and the Transylvanian Plateau, induces a high variability for the spatial distribution of wind parameters.

Braşov area is characterized by the monthly values of mean wind speeds ranging between 0.9 m/s (October) and 1.55 m/s (March). In order for an area to be viable for wind power turbines installation, this must have a minimum mean wind speed of 4 m/s at a height of 10 m above the ground.

Considering this aspect, the Braşov town area does not fit in the category of areas with significant values of wind potential.

Regarding the suburbs area, this is characterized by the monthly values of mean wind speed between 1.5 m/s (October) and 2.9 m/s (September); therefore the suburbs area does not fall into the minimum speed limit for that this could be taken into account.

However, given the significant values of wind power density obtained for the suburbs area and taking into account the following aspects:

- the influence of different local factors that can lead to the intensification of wind speed on small areas (for example the effect of speed-up due to the hills),
- the anemometric data measure the wind speed at a height of 10 m above the ground, but if it is taken into account the increasing of wind speed with the height above ground, its values can increase significantly (it is mentioned, the towers of modern turbine have heights of at least 60 m [6, 7]; so high towers are needed to raise the wind turbine above the turbulence generated by obstacles on the ground and trees),

then the studies development may lead to the identification of locations that may be exploited in view of electricity production.

Therefore, an investment in wind field and the concrete indication of the placement location of wind turbines are made as a result of a detailed study of available wind potential.

Acknowledgements

The meteorological data used in this study were provided by two weather stations; the database recorded by the weather station from town area is provided by Department of Renewable Energy Systems and Recycling, and the datasets for suburbs area, by Research & Development Institute, from Transilvania University of Braşov, Romania.

References

[1] Ramachandra T.V., Shruthi B.V. Wind energy potential mapping in Karnataka, India, using GIS. Energy Conversion and Management 2005, 46, 1561–1578.

[2] Jewer P., Iqbal M.T., Khan M.J. Wind energy resource map of Labrador. Renewable Energy 2005, 30, 989-1004.

[3] Acker T.L., Williams S.K., Duque E.P.N., Brummels G., Buechler J. Wind resource assessment in the state of Arizona: Inventory, capacity factor, and cost. Renewable Energy 2007, 32, 1453–1466.

[4] Gökçek M., Bayülken A., Bekdemir S. Investigation of wind characteristics and wind energy potential in Kirklareli, Turkey. Renewable Energy 2007, 32, 1739-1752.

[5] Ohunakin O.S. Wind resource evaluation in six selected high altitude locations in Nigeria. Renewable Energy 2011, 36, 3273-3281.

[6] Migoya E., Crespo A., Jiménez A., Garcia J., Manuel F. Wind energy resource assessment in Madrid region. Renewable Energy 2007, 32, 1467-1483.

[7] Bostan I., Dulgheru I.V., Sobor I., Bostan V., Sochirean A. Conversion Systems of Renewable Energies. Tehnica-Info Publishing, 2007.

[8] Waewsak J., Chancham C., Landry M., Gagnon Y. An analysis of wind speed distribution at Thasala, Nakhon Si Thammarat, Thailand. Journal of Sustainable Energy & Environment 2011, 2, 51-55.

[9] Arslan O. Technoeconomic analysis of electricity generation from wind energy in Kutahya, Turkey. Energy 2010, 35, 120-131.

[10] Montgomery D.C., Runger G.C. Applied statistics and probability for engineers. Fifth Edition, Hoboken, NJ: John Wiley & Sons, Inc. 2011.

[11] Justus C.G., Hargraves R, Mikhail A, Graber D. Methods for estimating wind speed frequency distributions. Journal of Applied Meteorology 1977, 17, 350–353.

[12] Elamouri M., Amar F.B. Wind energy potential in Tunisia. Renewable Energy, 2008, 33, 758–768.

[13] Karamanis D., Tsabaris C., Stamoulis K., Georgopoulos D. Wind energy resources in the Ionian Sea. Renewable Energy 2011, 36, 815-822.

[14] Sayegh A.S., Munir S., Habeebullah T.M. Comparing the performance of statistical models for predicting PM10 concentrations. Aerosol and Air Quality Research 2014, 14, 653-665.

12

Performance of single and multiple pressure wind catchers in terms of air flow changes

M. R. Abouseba[1], J. Khodakarami[2]

[1] Sepehr Institute of Isfahan, Isfahan, Iran.
[2] Engineering Department, Ilam University, Ilam, Iran.

Abstract
Although air current is the major cause of flow inside the wind catcher networks, the effect of wind direction changes on the performance of different types of wind catchers is unknown yet. Therefore, present paper aims at creating desirable conditions for comparing single-pressure and multiple-pressure types of wind catchers in order to discover the effect of wind direction changes on the performance of each type of wind catchers. To this end, The CFD model is devised for both types exposed to wind blowing at the speed of 3 m/s and at the angles of 0, 45, 90, 135, and 180 degrees. The obtained results indicated that the type of high-pressure area against the cage inflow determined the direction of wind inside their networks. Despite of this fact, during every blow towards the single-pressure type only one flow was generated whereas in multi-pressure type both downward and upward flows were generated. Moreover, while in the single-pressure types only the high-pressure area in front of the inflow was been used, in the multiple-pressure-type the entire high-pressure area around the cage was been practically in used.

Keywords: Wind catcher; Natural ventilation; Thermal simulation; CFD.

1. Introduction

Wind catchers are building elements which have been used in traditional buildings for many years in the Middle East [1-3]. These elements have been providing natural ventilation and passive cooling in hot and arid as well as hot and humid areas [4]. Moreover, they are ornamental elements which have greatly influenced the urban architecture and helped present impressive views of the urban areas where they have been employed.

Different classifications for wind catchers based on local names [5], function [6], typology [7], and wind-receiving surfaces [8] have been proposed by several authors. In view of the subject under the study and wind catchers function analysis, the classification based on the pressure inside the wind catcher network was used in this paper [9].

1.1 Single pressure wind catchers

Single-pressure is used to refer to a type of wind catcher which constantly has positive or negative pressure inside its columns. Therefore, depending on wind direction towards the wind catcher's inflow, the wind catcher may operate a Badkhor (wind scope) or Badkhan (wind tower) at any moment. This type of wind catcher doesn't have any shared blade inside its column and the external frame determines the wind catcher's network connections (Figure 1). The inflow is located only on one cage front and the outflow often leads to a space. According to some classifications this type of wind catcher is called one way or Ardakani wind catcher.

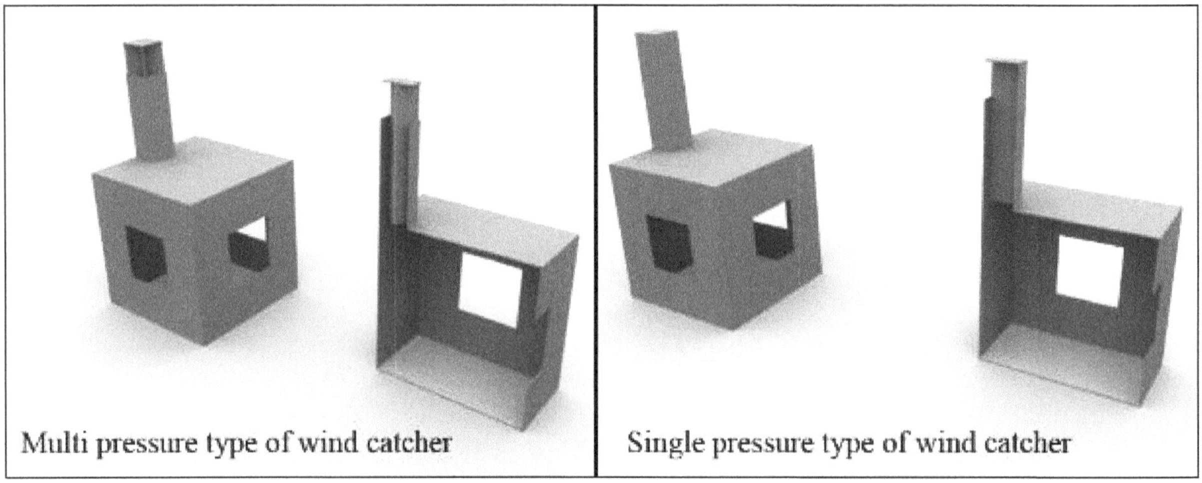

Figure 1. Traditional wind catcher' types

This type of wind catcher is used in some countries such as: Egypt, Iraq, Iran, Pakistan, and Afghanistan [10] (Figure 2). The direction of single-pressure wind catchers' inflow is determined by the quality of the prevailing wind. Therefore, when the prevailing wind is free of pollution, and dust, or carries lower temperature, and more desirable humidity, the inflow of the wind catcher will have the same direction as the prevailing wind. Otherwise, the direction of the wind catcher will take a direction opposite to that of the prevailing wind. The height of the wind catcher is determined by the kind of technology employed for their construction and the height of the prevailing wind [11].

1.2 Multiple-pressure wind catchers

In wind catchers which have internal dividing walls inside their column, different gusts of air generate diverse pressures around the wind catcher cage which directly affect the internal pressure of the internal networks of the wind catcher [2]. Therefore, several positive and negative pressures are simultaneously generated [12]. Inside the wind catcher, dividing walls are of various types and the external form no longer determines the relationship of the wind catcher's internal networks (Figures 1, 2). The inflow of the wind catcher is located on all fronts of the cage and the outflow leads to one or two floor (first floor and basement) [13] (Figure 2). In some classifications they are called two, four, six, and eight sided as well as Yazdi and Kermani in some other classifications [14]. These types of wind catchers are used in Iran and some other countries [15] (Figure 2).

Due to consuming clean and complimentary energies, the wind catcher is considered as one of the most sustainable and economical energy production systems, but because of some functional difficulties their use has been restricted, and therefore a lot of efforts have gone into optimizing the performance of traditional wind catchers. A handful of such efforts are mentioned below.

Later, Karakatsanis et al [16] defined the coefficients of wind pressure at different angles in proportion to each other. Within the same year, Cunningham and Thompson [17] offered a model as a combination of single pressure wind catcher and solar chimney (Figure 3). Givoni [18] used their data during a semi-empirical investigation and stated that the performance of these wind catchers is subject to temperature difference between humid and dry climate, specifications of wind catcher, and thermal behavior of the lateral space of wind catcher.

Formerly, in EXPO92 exhibition, a model of single-pressure wind catcher for rural areas was displayed within which full evaporation was carried out by means of suitable sprays and a pleasant environment for walking was created under the wind catcher [19].

In 2002, Bahadori Nejad & Pakzad [20] analyzed the performance of traditional multi-pressure wind catchers and investigated the mass flow and outgoing pressure of these types of wind catchers in different directions and rates . Also Badran [4] studied the efficiency of single-pressure wind catchers in different climates of Jordan and concluded that in all these climates the height of these kinds of wind catcher's column should be about 9 meters. In 2008, Azizian & Montazeri [21] particularly conducted an empirical study on the hydraulic efficiency of traditional single-pressure wind catchers which a blow at 0° angle provided the best efficiency. Within the same year, taking into account the wide variety of multi-pressure wind catchers in plan as well as sculpting and insertion style, Mahmudi & Mofidi [14] conducted a typology of wind catchers of Yazd.

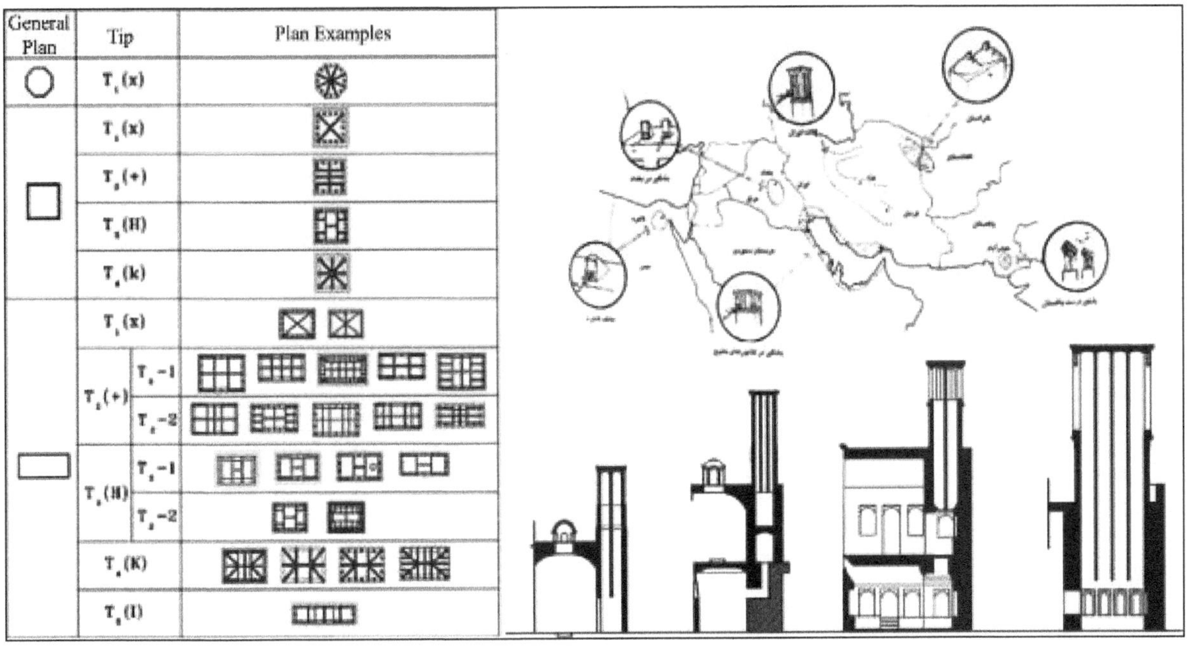

Figure 2. wind catchers' dispersal [10] and cross-section and plan typology of Multi pressure type [20]

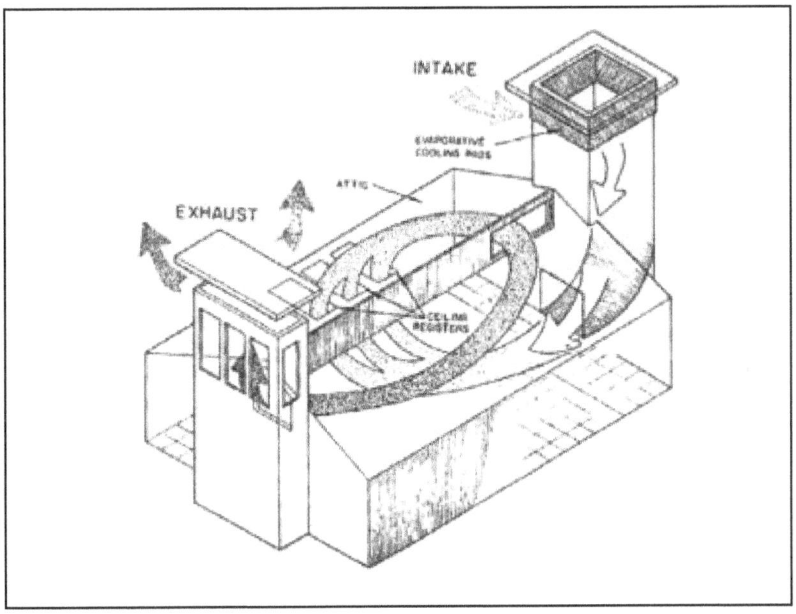

Figure 3. Experimental model of Cunningham and Thompson

Huges et al [22] introduced the background of conducted studies on complementary trend of wind catcher's commercial type. Within the same year, Abousaba [23] conducted studies on recognizing the behavior of the performance of traditional wind catchers and determined the preferred type by using simulation method. in these studies, the performance of traditional multiple – pressure wind catchers' were investigated in multi – floor buildings and, ultimately, pressure separating two – floor cage was used as a strategy to establish pressure stability in wind catcher networks.

Taking into account the above studies, it is certain that no study has so far been conducted on the effect of wind direction on the single-pressure and multiple-pressure wind catchers. Therefore, we aim to study the performance of single-pressure and multiple-pressure wind catchers exposed to different wind directions during which the external air current with a rate of 3m/s hits the both types of wind catchers at 0°, 45°, 90°, 135°, 180° angles. The air current angle in proportion to the wind catcher's cage is shown in Figure 4.

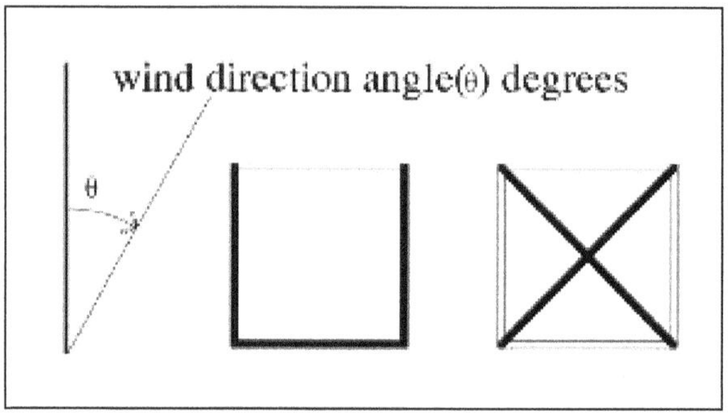

Figure 4. Air current angle in proportion to the wind catcher's cage

2. CFD models properties

The dynamic performance of wind catchers can be simulated by means of CFD model. Before using the commercial package of CFD offered in the Fluent Software, a model of the intended volume has to be designed and networked in other software, and for this purpose we used Gambit software. Therefore, in the first stage two volumes with the same conditions were constructed in the Gambit Software which is shown in Figure 5 in detail.

According to this figure the dimensions, the height of the channel, and the height of the cage are respectively 0/5 * 0/5, 1/5, and 0/5.The right image shows a single-pressure wind catcher whose channel doesn't contain any internal blade and the left image shows a multiple-pressure wind catcher which has the internal space of its column divided by two crossing blades into four parts. It should be mentioned here that there are many different plans for traditional multiple-pressure wind catchers, but in order to obtain the kind of results which allow comparison between the two types of wind catcher, a simple model with crossing blades has been used. Moreover, the height of the wind catcher's column is determined in proportion with the dimensions of the plan; however, basically the height of the wind catcher's column is determined by the level of favorable winds and as at this stage of the study the fluid is single-phase and no moisture is injected into it, the temperature fall resulting from momentum transfer between water and air phases doesn't occur. Therefore, the height of the channel is not increased excessively. In most spaces leading to the wind catcher, the internal blades of the wind catcher stretch down to a height equal to the height of an average man to foster surface evaporation by increasing pressure in the outlet as well as to cause a pleasant flow in a lower level. Therefore, since more complex performance is created in the bottom side of the wind catcher the internal blades of the wind catcher are stretched down up to the height of the ceiling. On the other hand, in most of similar studies the same technique is employed. Another important point which should be mentioned here is the construction of three openings in the eastern, southern, and western fronts of the model. Mainly in traditional buildings will be opening in along the yard and the wind catcher but in order to generalize to different spaces and determine wind catcher's disadvantages in different situations, we used two additional openings in the eastern and western fronts.

After devising the model volume in a range close to 320,000 faces, Volumes were meshed by using Tet / Hybrid networks and over 150,000 cells were created and to increase model accuracy, the number of cells was increased by approaching the important areas. Meanwhile, according to the simple geometry of the wind catcher channel and lower channel space and also having highly important the flow treatment in this area; Therefore, in order to regulate the flow treatment and the reduction cell model, this area was meshed by using Hex / Wedge networks. After building and meshing the model, boundary conditions were determined according to Figure 6.

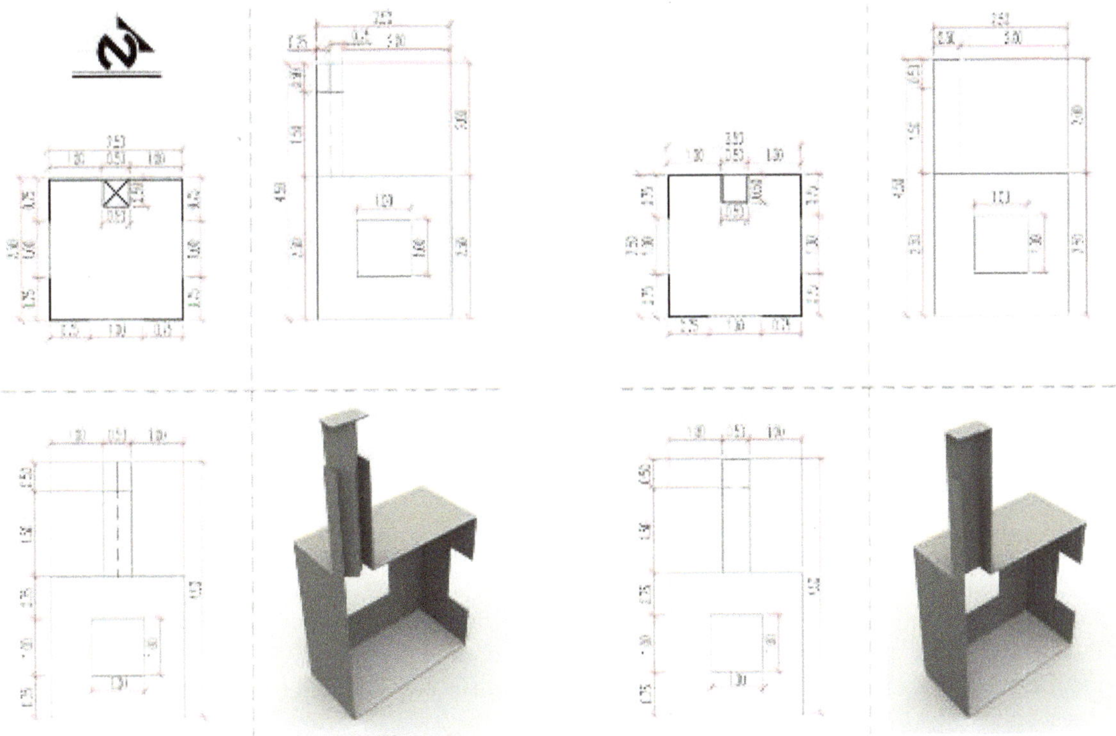

Figure 5. Details of CFD model traditional wind catchers

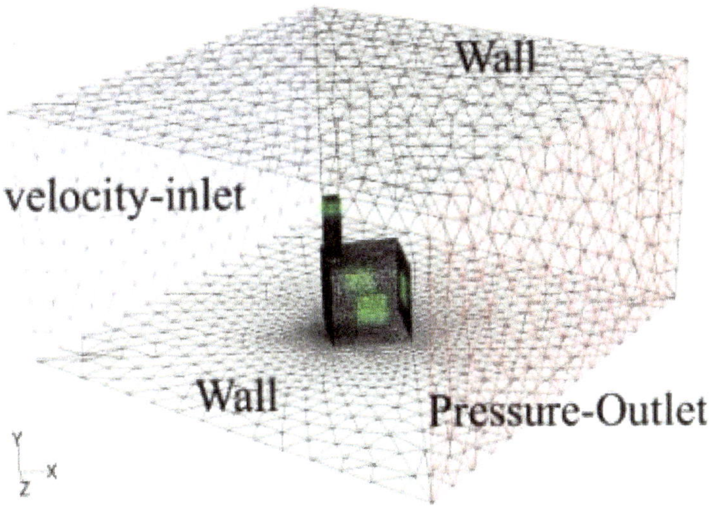

Figure 6. CFD model boundary conditions

Secondly, devised model was read in FLUENT software (Fluent 6.3) and solution conditions were defined as follows:

1- Solver: Pressure Based
2- Space: 3D
3- Formulation: Implicit
4- Time: Steady
5- Operating Pressure: 101325 Pa
6- Pressure-Velocity Coupling: SIMPLE

Because at this stage of investigation, flow velocity and pressures will be enough to compare species of wind catcher and air is the fluid used (and not the ideal gas), so energy equation was considered inactive. Considering that the 10-year average wind speed in the warm months in the Yazd (origin of wind catcher in Iran) was 3.17 m/s; Velocity inlet was considered 3.00 m/s (Table 1).

Table 1. 10-year average wind speed in the warm months in Yazd (m/s)

Year/Month	JUNE	JULY	AUG	SEP	Annual
1996	2.47	2.72	2.52	2.72	2.61
1997	3.80	3.55	3.39	2.98	3.43
1998	3.34	3.96	3.55	3.24	3.52
1999	3.14	3.86	3.03	2.67	3.17
2000	3.39	3.75	3.03	2.52	3.17
2001	3.34	3.86	3.39	2.78	3.34
2002	3.50	3.65	2.98	2.83	3.24
2003	3.34	3.39	3.34	2.83	3.23
2004	3.29	3.44	3.03	2.67	3.11
2005	2.93	3.03	3.08	2.52	2.89
Annual	3.25	3.52	3.14	2.78	**3.17**

According to previous studies and simulations of natural flow, Viscose Model was used as the standard k-epsilon model. Continuity, Momentum, k and epsilon equations were governing on the flow. Following two criteria were measured to determine the validity of data.

Wall Y plus: Because turbulent flows are strongly affected by the wall, therefore it is necessary to evaluate the sensitivity of mesh near the wall through measurements criteria y+. Usually when the k-epsilon model is used, the mesh near the wall must be arranged so that it should be provided y +> 30 or y +<5.

In Figure 7, red line shows the rate of Y^+ in the center of the inner blade of wind catcher, which shows the range $45<Y^+<251$ in height 2.5-4.5 m. Due to the lack of wall at a height less than 2.5 m, the rate of Y^+ is considered as zero. The black line shows Y^+ rate on the vertical wall along the north and wind catcher, which is zero due to absence of cage wall in the range 4.0-4.5 m. At other altitudes, the observed range is $33<Y^+<216$.

Flux Reports: In this model the mass flow rate is reported as follows:

Mass Flow Rate (kg / s)

Flow Outlet -440.99994

Flow Inlet 441.00001

Difference 0.00007

Given the above report, the balance difference the flow rate is very low. Thus the model seems to be necessary conditions for the flow simulation.

2.1 Analyzing the behavior of a single pressure type at different degrees wind blowing

Figure 8-a-S shows the flow behavior at zero-degree angle to the cage of wind catcher. According to this picture, it will be determined that the air flow is moving toward the model with speed of three meters per second(3m/s) that due to collisions with the walls of the wind catcher, the speed of flow has increased in the cage of wind catcher. After the collision outflow with the cage, part of it entered into against the wind channel and the remaining passed from the cage which caused the creation of a negative pressure zone behind it. Afterwards, the outflow crosses the volume and created a weaker negative pressure zone behind the south opening.

Analysis of the flow behavior inside model shows that entering the flow into channel and collision with the walls has increased air velocity in the wall facing the wind, but it has decreased in the wall behind the wind. Then, the flow out of the end of the channel is about two meters per second (2m/s) wind speed and has created a vortex flow within the model. More analysis is needed to examine velocity vectors shown in the Figure 8-a-P for vertical view. This picture indicates that speed of airflow is intensified after hitting the northern wall and through it which has caused a negative pressure zone at the back of all openings. At the same time, the negative pressure in the outer west and east openings and outflow of the wind catcher has led internal flow drawn out. In order to balance the volume of incoming and outgoing flow, outflow is drawn through the southern opening. It was already predicted that the flow enters through the western and eastern openings and flow out of the wind catcher should be exited from the southern opening while suction flow out of the east and west openings are much larger than suction in south opening.

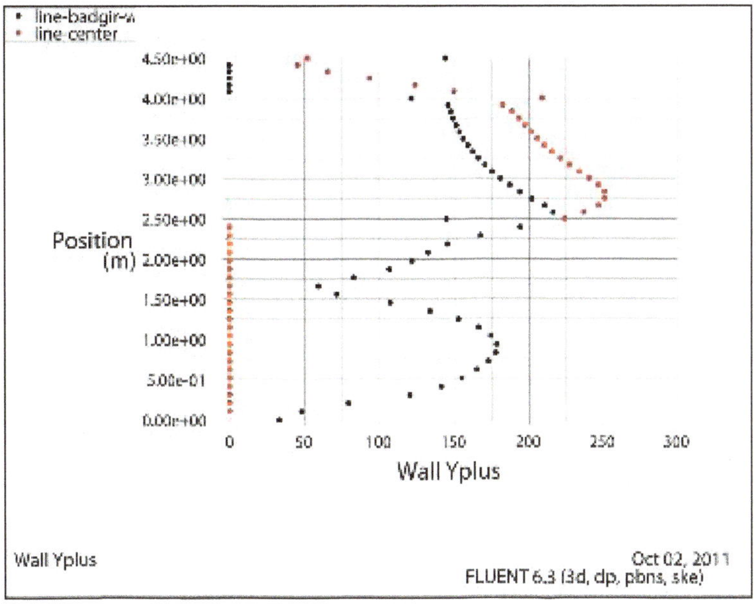

Figure 7. Y^+ chart near the outer wind catcher wall and the inner blade center

Figure 8-b-S was obtained when the wind tunnel was rotated 45 degrees. In the 45 degrees blowing, same situation with the zero degree has occurred because the airflow through the channel of wind catcher into the model. But here the entrance to the channel flow rate has decreased. Figure 8-b-P also makes clear that due to flow entry from the East opening and exit from the South and West openings, only one vortex flow is established within the South West model.

When airflow dealing with the model at 90 degrees (Figure 8-c-S), the flow path inside the channel is contrary and moves slowly upward. The reason of such trend can be seen in creating negative pressure area around the cage and moving the flow from east to west opening. It is clear that airflow is denser after colliding with the east wall and crossing from it. Next, this event is repeated around the wind catcher. For this reason, there have been rapider and denser vectors. Horizontal velocity vectors in 90 degree angle blowing are shown in Figure 8-c-P. This picture shows off the air flow after the collision with the east wall; Considerable portion of it is moved through the opening and the rest of it has passed around the model.

As is shown in Figures 8-d-S&P, air flow deals the model with angle of 135 degrees and passes on the southern and eastern walls. Simultaneously, the air flow has created a negative pressure zone in the north and west which has increased the air velocity inside the channel of wind catcher. In other words, at this angle to 90 degrees, not only the negative pressure is increased around the cage, but the eastern and southern openings increased air velocity upward within the channel.

Figure 8-e-S shows the details of the wind angle of 180 degrees. According to this picture, it will be clear that the main cause of airflow movement is the entrance of external airflow into the model through the southern opening. Also, like the blowing 135 degrees, upward flow within the channel is increased. It is necessary to mention that the two-vortex flow is in the upper and lower of internal space of model which is obtained from a combination of inflow of southern openings and outflow of wind catcher and other openings. The most important thing which should be noted is that as shown in Figure 8-e-P, the airflow enters from an opening and exits from two other openings. Unlike blowing 135 degrees, the airflow enters from two openings and exits from an opening.

Now, it can be stated that the performance of single pressure type of wind catcher strongly depends on the wind flow direction as well as the number and location of openings of model (Figure 9).

In order to conduct a quantitative study on passing volume flow rate in openings and cage, Table 2 is presented which confirms the above results and also shows that a more accurate comparison is possible.

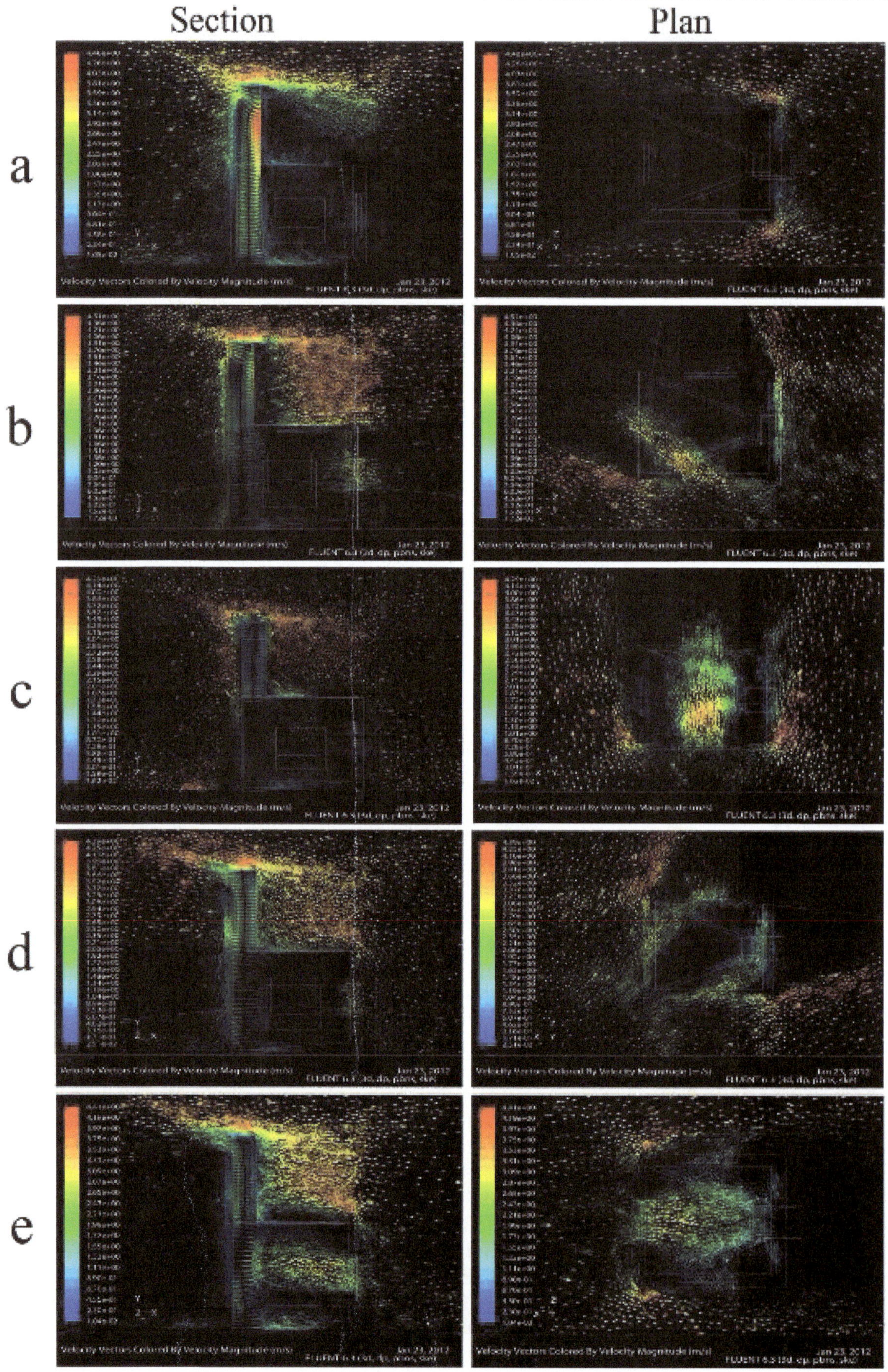

Figure 8. Sections and plans of vectors of velocity magnitude of single pressure wind catchers in different directions blowing

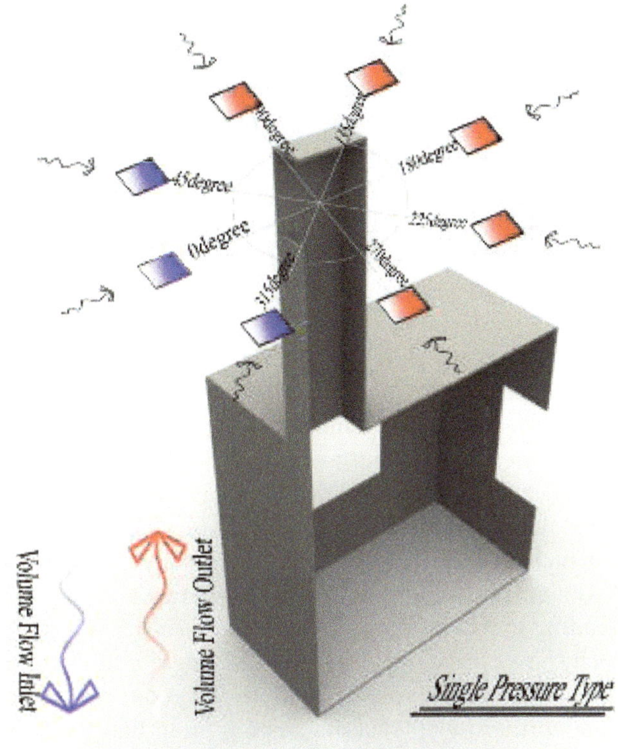

Figure 9. Direction of the airflow inside the single pressure type of wind catchers in different angles blowing

Table 2. Volumetric flow rate in openings and cage

		Volumetric Flow Rate (m3/s)				
Single pressure type	East Opening	-0.4456	1.7306	2.3984	1.1173	-0.9102
	West Opening	-0.4335	-0.4486	-1.6027	-1.7604	-0.9498
	South Opening	0.3591	-1.6274	-0.5939	1.0275	2.1732
	Cage	0.5201	0.3454	-0.2018	-0.3844	-0.3132

2.2 Analyzing the behavior of a multi pressure type at different degrees wind blowing

Figure 10-a-S shows the flow behavior at zero-degree angle to the cage of this type of wind catcher. As in this figure shown, the flow direction and velocity in terms of its magnitude, is shown. This picture indicates that the airflow is denser near the model and its speed is reduced in places where escape is not found. Also, the edges that were created under pressure are released so that its velocity is increased. This is well visible in the roof of the wind catcher. In contrast, the southern opening, the low negative pressure area is created so that it cannot stop being drawn into the external flow. After the airflow passes through the cage of wind catcher, a set of regions of positive and negative pressure is created that has led in the direction of airflow above and below the wind catcher move through the channel. Flow downward into the cage is affected by the positive pressure and then compressed with a speed close to four meters per second (4m/s) is downward. In other channel, under the influence of negative pressure around the cage area, about one meter per second (1m/s) flow velocity rises. That can be one of the causes of relatively high difference between upward and downward flow velocity of the suction openings of the flow from the east and west along said. Within the model, the vortex flow is created has occurred under the influence of the end of the outflow channel and outflow of both Eastern and Western openings and inflow of Southern opening. This theory is obtained by studying the Figures 10-a-S&P.

2.3 Analyzing the behavior of a multi pressure type at 45 degree blowing

According to Figure 10-b-S which shows this type of wind catchers at 45 degree blowing wind, can be express that airflow is dense in the eastern and northern walls and then passed on model of nearly four meters per second (4m/s) speed. There is a range of high speed airflow created under the influence of airflow entering from the East opening and exit from the south opening. Comparing Figures 10-b-S&10-

a-S shows that in the 45 degrees blowing, the flow velocity downward is reduced and upward flow velocity is increased. Another point is that at the end of the wind catcher, withdrawal is occurred easier than zero blowing conditions so it, may operate by vortex flow inside the model.

As shown in Figure 10-b-P is how the airflow has passed around the model and the negative pressure area is being created behind it. The interestingly point is how to create the vortex flow within the interior space occurred by the movement of the air flow from east opening to south opening.

Figure 10-c-S shows the displacement quality of airflow around the model under 90° blowing. Due to its wind catcher, flow velocity has decreased in border wall that seems to be caused by the collision with the wall of the east wind catcher. Noteworthy, same direction of flow in both North and South channels which may be due to the suction flow is derived from the cage. Given the experience that we gained in zero-degree blowing, we forecast that the flow moves down through the Eastern Channel, but this picture is not clear. In the area south of the volume; flow is drawn through the opening which may be due to the entry from eastern opening and exit from western opening (Figure 10-c-P).

It is depicted in Figure 10-c-P that the air has continued to move in three directions after hitting the model. It goes in both directions around the model that has been associated with increased speed and the other that passes through the model; its speed has been increased slightly. We are witnessing a four-vortex flow inside the model which seems that movement mainstream of the building is the most important factor. Likewise, the southern opening, the flow of air to take out and also due to wind catcher down and lift flow same time, leading to a compressive force in the vertical direction is on flow inside.

According to Figure 10-d-S which shows the wind catcher at 135 degree blowing, it can be stated that the airflow movement is due to the positive pressure created in the southern wall of wind catcher and negative pressures occurred in the northern wall. The interesting thing that we have not to deal with it is that wind catcher output flow is rapider than input flow thrust because one can see the flow entering from the two opening and flow out of an opening (Figure 10-d-P). So we are inside flow congestion control that leads to increased negative pressure in outlet channels.

According to Figure 10-d-P, flow path is changed to the north and west after dealing with the model and then is entered into the wall openings. Flow into the model is dense behind the western wall and out of the opening. The vortex created in this situation is happened behind the northern wall model which affected the movement of the northeast region and ceiling model is occurred.

Based on Figure 10-e-S in which flow blows with a 180° angle relative to the model, it is obvious that the collision with the southern wall, condenses it and moves it on the roof. Then get to the wind catcher channels, speed has been increased along the vertical and has reached the cage. Part of the airflow around the cage is entered into the South Channel. The rest of airflow passes the cage and causes a negative pressure area behind it so the flow is moves upward.

In comparison with the wind blowing at zero degree, it is indicated that flow velocity downward is reduced but upward flow velocity is increased. It can be seen as the most important factor in air flow getting into and out of the openings. The important thing in the Figure 10-e-P is a strong vortex that is the northern side and the air flow entering through the southern opening which causes the current compression and wake vortex formation in the interior space.

2.4 Brief analysis of the behavior of multi pressure type of wind catcher at different angles blowing

At the end of this discussion, it can be concluded that in all angles of the wind blowing, multi pressure type of wind catcher, simultaneously provides entry and exit of airflow due to the use of multiple internal channels. In general, we can say where the wind blowing was perpendicular to the wind catcher (zero, 90 and 180 degrees) only the channel against the wind is directed to flow downward and other channels are directed to flow upward. But in other wind blowing (45 and 135 degrees), to flow downward through the two channels (Channels against the wind) and upwards is drawn through two other channels (Figure 11). Finally, the location and number of openings are considered as the most important factors in the wind catcher flow velocity inside the channels.

Finally, Table 3 is represented which confirms above results and makes it possible to compare them more precisely.

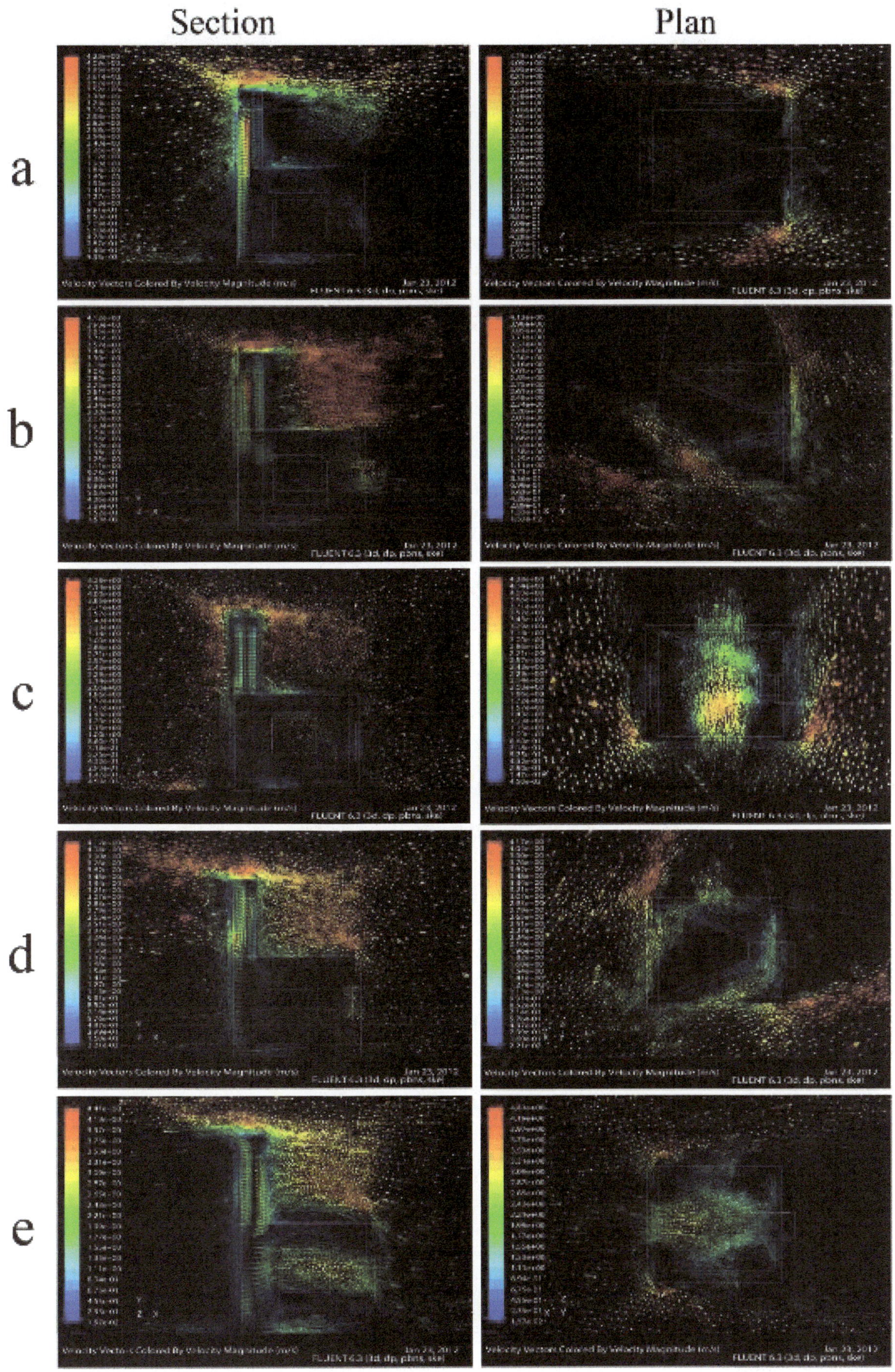

Figure 10. Sections and plans of vectors of velocity magnitude of multi pressure wind catchers in different directions blowing

Table 3. Volumetric flow rate in openings and channels

			Volumetric Flow Rate (m3/s)				
Multi pressure type	Openings	East	-0.2930	1.7577	2.3882	1.0065	-0.9498
		West	-0.2888	-0.1883	-1.6339	-1.7815	-0.9803
		South	0.6368	-1.6275	-0.6604	0.8815	2.1543
	The canal of wind catcher	East	-0.0697	0.1160	-0.1280	0.0625	-0.1069
		West	-0.0507	-0.0867	0.0758	0.0736	0.1208
		South	-0.0703	-0.0870	0.0565	-0.1211	-0.1072
		North	0.1358	0.1159	0.0897	-0.1214	-0.1308

3. Conclusion -Comparing the behavior of Single and Multi pressure types of traditional wind catcher under the influence of changing the angle blowing

At the end of this paper and due to the same conditions for Multi pressure type and Single pressure type of wind catcher, we can compare them together. The results show that in Single pressure type, the entire space inside the channel is used for motion flow in one direction. But in Multi pressure type the entire space inside the channel is used as part of the internal space channel for motion flow downward and the remaining space to create an upward flow. That is how to move the internal flow channel in a function of wind direction (Figures 9, 11 and 12). The wind flow is caused by a series of low compression and congested areas around the cage of both species that Single pressure type uses pressure area in front of the inlet of cage (only one side) but Multi pressure type uses all areas around the cage (Figure 13).

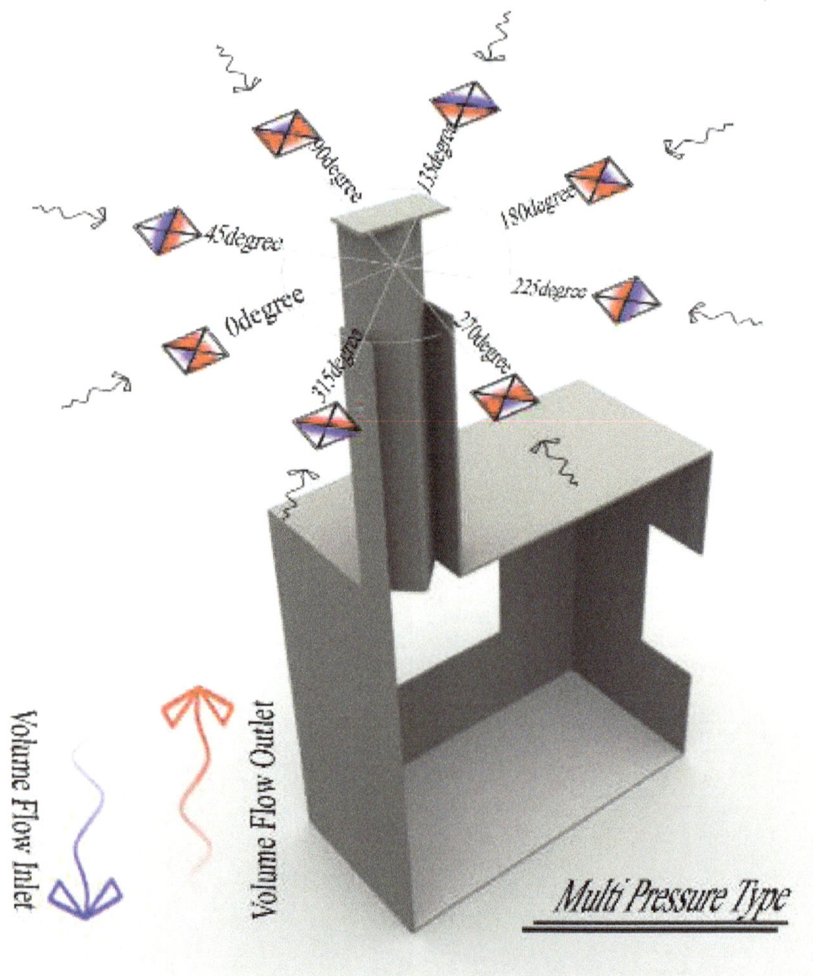

Figure 11. Direction of the airflow inside the multi pressure type of wind catcher in different angles blowing

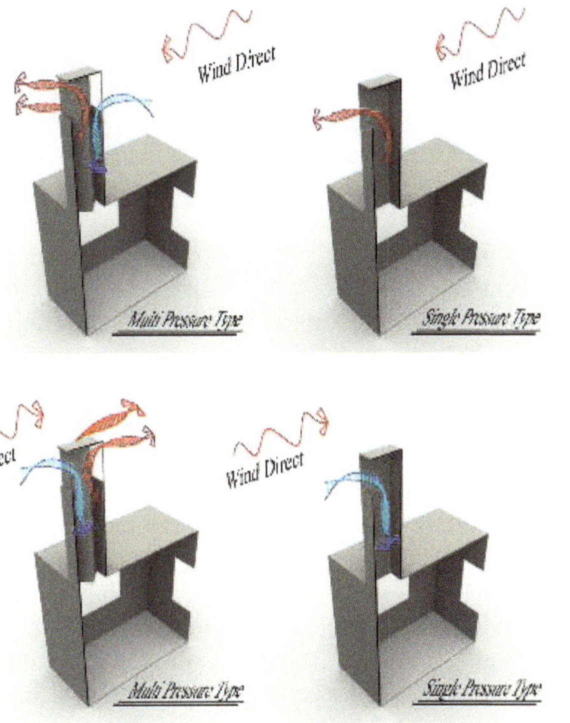

Figure 12. Behaviors of Multi pressure type and Single pressure type of traditional wind catcher in different angles blowing

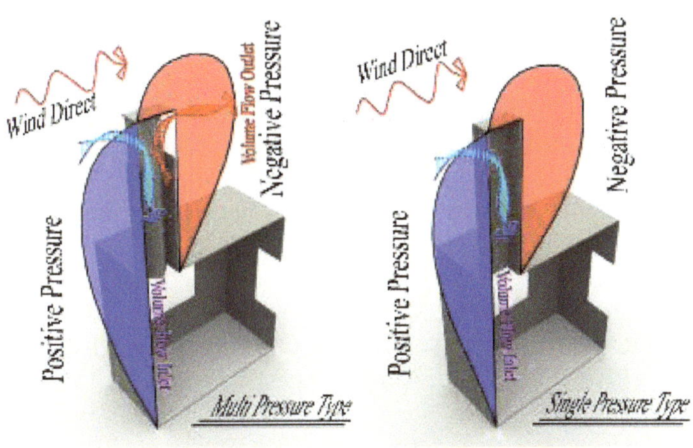

Figure 13. The utilization of Multi pressure type and Single pressure type of traditional wind catcher of high pressure areas around cage

References

[1] Bahadori M.N., 1994, Viability of wind towers in achieving summer comfort in the hot arid regions of the Middle-East, Renewable Energy 5(879–892).

[2] Elmualim A.A., 2006, Effect of damper and heat source on wind catcher natural ventilation performance, Energy and Buildings 38 (939–948).

[3] Sue, R., 2007, Ecohouse, Architectural Press, Third edition.

[4] Badran, A., 2003, Performance of cool towers under various climates in Jordan, Energy and Buildings 35, 1031–1035.

[5] Shariat Zadeh, A.A., 1992, The wind catchers of Yazd and their constructional methods, Jodaganeh, Tehran, 345-358, (in Farsi).

[6] Memarian, Gh. R., 1996, Introducing the Iranian residential architecture, Iran University of science and technology, Tehran, (in Farsi).

[7] Mahmodi, M., 2009, Wind catcher symbol of Iranian architecture, Yazda, Tehran, (in Farsi).

[8] Roaf. S; 1981; Badgir; Living with the desert; oxford press; p-p57-71.
[9] Abouseba M. and Khodakarami J., 2011, Physic analysis of traditional wind catchers and their disadvantages with an approach to practical performance, proceeding of ETEC, Tehran, (in Farsi).http://www.civilica.com/Paper-ETEC01-ETEC01_086.html
[10] Ahmad Nejad, M., 2001, Badkhan, Nashre Khak, Isfahan, (in Farsi)Tavasoli, M., 1982, the construction of city and architecture in hot and dry climate of Iran, Payam and Payvandeno, Tehran, (in Farsi).
[11] Tavasoli, M., 1982, the construction of city and architecture in hot and dry climate of Iran, Payam and Payvandeno, Tehran, (in Farsi).
[12] Bahadori N.M. and yaghobi, M., 2006, Natural ventilation and cooling in traditional buildings, Markaze Nashre Daneshgahi, Tehran, (in Farsi).
[13] Mahmodi, M., 2006, Knowing the impact of physical factors of wind catchers on their thermal behavior, PhD thesis, architecture faculty, Azad University.
[14] Mahmodi, M., and Mofidi S.M., 2008, Analysis of architectural typology of Yazd wind catchers and seeking for practically performed type, Honarhay Ziba, 36(103-105), (in Farsi).
[15] Bahadori, M. 1978. Passive cooling system in Iranian architecture, Scientific American, vol. 238, No 2,pp 144-154.
[16] Karakatsanis, C., Bahadori, M.N., Vickery, B.J. 1986. Evaluation of pressure coefficients and estimation of airflow rates in buildings employing wind towers. Solar Energy; 37(5):363e74.
[17] Cunningham WA, Thompson TL. 1986. Passive cooling with natural draft cooling towers in combination with solar chimney. In: Proceedings of the passive and low energy architecture. Hungarian Ministry of Industry; p. 23–34.
[18] Givoni, B. 1993. Semi-empirical model of a building with a passive evaporative cool tower, Solar Energy 50 (5), 425–434
[19] Alvarez, S., Rodriguez, E., Molina, J.L., 1991. The Avenue of Europe at Expo '92: application of cool towers. In: Alvarez, S., Lopez de Asiain, J., Yannas, S., De Oliviera Fernandes, E. (Eds.), Architecture and Urban Space: 9th PLEA International Conference, Seville, Spain, 24–27 September 1991.
[20] Bahadori M, Pakzad A. 2002. Performance evaluation of new designs of wind tower, ASME 2002 Fluids Engineering Division Summer Meeting, montreal, Quebec, Canada, July 14-18, PN. FEDSM 2002-31247.
[21] Montazeri H, Azizian R. 2008. Experimental study on natural ventilation performance of one-sided wind catcher, Building and Environment 43 2193–2202
[22] Huges, BR. Calautit, JK. Ghani, SA. 2012. The development of commercial wind towers for natural ventilation: A review. Applied Energy 92 (2012) 606–627.
[23] Abouseba M., 2012, PRACTICAL OPTIMIZING OF TRADITIONAL WIND CATCHERS-MULTI PRESSURE TYPE APPROACH, Master's thesis, Faculty of Engineering, University of Ilam, (in Farsi).

13

Simulating an ocean wave power plant with Homer

Jones S. Silva, Alexandre Beluco, Luiz Emílio B. de Almeida

Inst Pesquisas Hidráulicas, Univ Fed Rio Grande do Sul, Porto Alegre, Brazil.

Abstract

In recent years, renewable resources have become an important alternative to increase the capacity of power generation. Among the resources available, the energy in the oceans have attracted more interest and experiencing a period of strong development. Wave energy is the one that is currently closest to reach a stage of technical and economic maturity. In this process, computational tools to facilitate evaluation of the feasibility of power systems including ocean wave power plants are critical. This paper presents the use of software Homer, version Legacy, for evaluation of technical and economic feasibility of hybrid energy systems including contribution of wave energy. Homer is a software designed for the optimization of micro power systems that has a very broad spectrum of applications. A case study in southern Brazil is presented, studying the inclusion of an ocean wave power plant in a wind biodiesel hybrid system and showing the functionality of Homer for this purpose.

Keywords: Ocean energy; Ocean wave energy; Computational simulation; Feasibility study; Software Homer; Southern Brazil.

1. Introduction

Amid a global economic crisis, it still remains necessary to find alternatives to fossil fuels that have technical and economic capabilities for providing energy supplies that are currently needed. In some places, renewable resources are the only available solution rather than a solution environmentally desirable.

Among the resources available, the energy in the oceans have been experiencing in recent years a strong and fast ripening process. Specifically wave energy must shortly reach its technical and economic maturity. There are currently a wide range of designs that allow the conversion of wave energy into electricity.

There is a lot of designs and projects to convert wave energy into electricity. Some of these projects are closer to viability than others. Each year several new developments arise, just as many others are being abandoned [1, 2]. Environmental issues involving the use of wave energy are permanent [3] and should always be seen as a challenge.

In this process, it is important that computational tools can be applied to expedite the process of evaluating technical and economic feasibility for a given power plant. Tools that also allow the study of its insertion in hybrid systems with other resources in most advanced technological stages and the determination of specific costs that must be achieved.

Among the computational tools available for assessment of the feasibility of energy systems, Homer [4, 5] is designed for optimization of micro power systems that presents a wide and interesting range of applications. Among the energy conversion equipment based on renewables, except those that use fuels,

Homer simulates hydro power plants, wind turbines and PV modules.
Above all, Homer is convenient and fast to be programmed, simple to be learned [6] and, even with some limitations, very friendly with regard to the evaluation of results. The version [7] that is available free of charge is sufficient for most applications.
Homer is a very user friendly software and beyond the feasibility assessment of power systems, also allows to study other aspects such as operational details related to failures in the power supply or details of the rules established for energy dispatches
There are several designs being developed for converting wave energy into electricity and the power plants can be installed close to the beach as far away from the coast [2]. Wave energy has a variability that can be compared with wind energy and its use has impact on the system in which it will be connected [8]. Operating in conjunction with reversible hydro power plants can mitigate these effects [9] increasing the capacity factor of plants based on these renewables.
The variability of the energy available in the ocean waves, as well as the variability of other renewable resources may be alleviated if the resources involved present energetic complementarity in time [10, 11]. When one of the energy resources is in its period of minimum availability, the other has higher availability, ensuring the supply of energy to consumers. Wave energy and wind energy are probably not complementary to each other, but can be exploited in complementarity with other renewables.
This technical note shows how Homer can be used to simulate an ocean wave power plant and describes a case study in southern Brazil in which the inclusion of an ocean wave power plant in a wind biodiesel hybrid system is simulated. As suggested in this work, Homer can be used for any design of equipment for energy conversion, with the focus always on the evaluation of the economic feasibility. The case study is not exhaustive and only shows some results.

2. Energy from ocean waves

The power P [W/m] available per unit of length of wave front is given by equation (1) [12], where ρ [kg/m3] is the density of sea water, g [m/s2] is the acceleration due to gravity, Hs [m] is the significant wave height and Te [s] is the wave period.

$$P = \frac{\rho g^2}{64\pi} H_S^{\,2}\, T_e \tag{1}$$

The available energy in ocean waves is presented a more thoroughly manner by reference [13], which also discusses the estimation of energetic potential. Reference [14] discusses the estimation of potential in conditions of scarce data.
For a real plant with lenght L [m] and an overall performance of the plant equal to ηwav [1], the power Pwav [W] which may be available will be given by equation (2). Most devices currently in development involve performances between 30% and 40% [1, 2].

$$P_{wav} = P\, L\, \eta_{wav} \tag{2}$$

Substituting (1) into (2), the power Pwav can be written as shown in (3).

$$P_{wav} = \frac{\rho g^2}{64\pi} H_S^{\,2}\, T_e\, L\, \eta_{wav} \tag{3}$$

This equation will be used to build the model for the description of a wave power plant.

3. Homer, the micropower optimization model

The software HOMER (2009), The Micropower Optimization Model, was developed by National Renewable Energy Laboratory (NREL) and is available for universal access in its version 2.68 beta. HOMER simulates a system for power generation over the time period considered in the project, which may be several years, at intervals of 60 minutes, presenting the results for a period of one year [4, 5].
HOMER includes models for simulation of micro hydro power plants, wind turbines and PV modules, in addition to fuel based generators. The software must be supplied with technical specifications of the

generator sets, the availability of energy resources and characteristics of consumed fuels. The systems simulated by Homer also include various types of loads and some alternatives for energy storage. Homer also has full capacity to studies including the production and consumption of hydrogen.

Figure 1 shows the alternatives offered by HOMER for the simulation of energy systems. HOMER can simulate systems with multiple generators in parallel, with up to ten different designs of gen sets, up to ten different models of batteries. HOMER can also simulate the connection with interconnected energy systems.

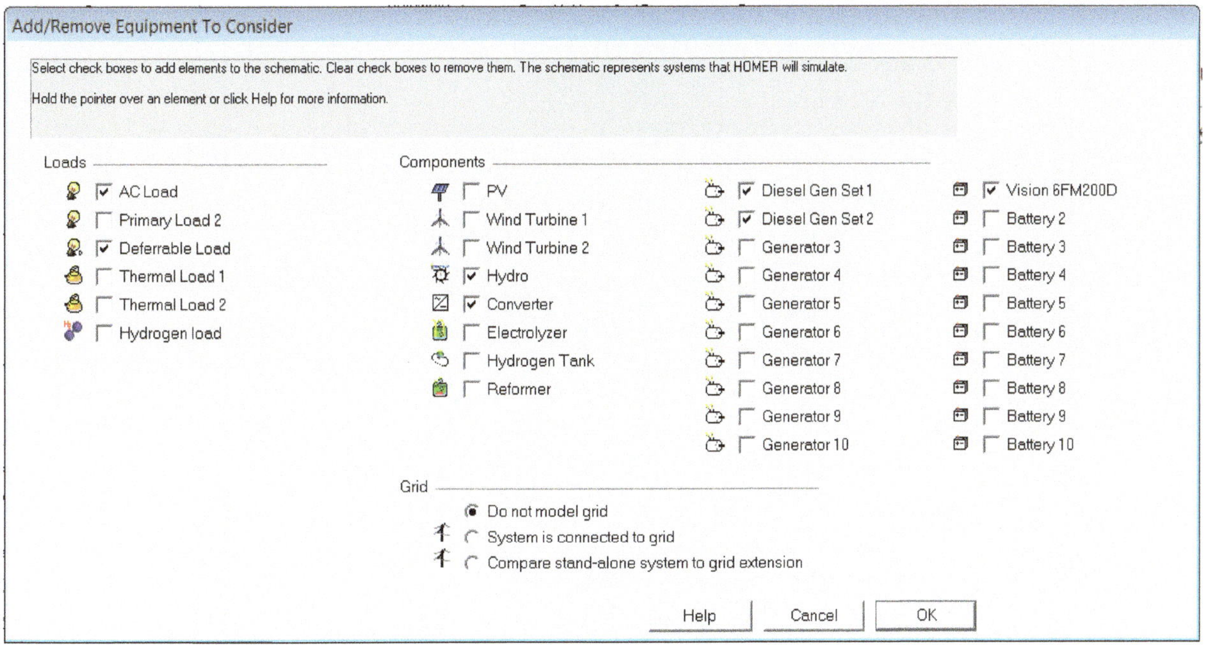

Figure 1. Alternatives offered by HOMER for simulation of hybrid energy systems

The intention of this work is the simulation of an ocean wave power plant, not currently contemplated by HOMER. In the absence of a specific model, then the solution will be obtained improvising with some of the available models.

The most appropriate choice is the model of hydroelectric plants. It should be emphasized however that available height and design flow rate, in the case of simulation of an ocean wave power plant, lose their meaning and become new model parameters.

4. Homer's model for micro hydro power plants

The power in a hydroelectric plant, Phyd [W], is given by equation (4), where γ [N/m^3] is the specific weight of sea water, Q [m^3/s] is the flow rate, H [m] is the available head and η_{hyd} [1] is the total performance of the power plant.

$$P_{hyd} = \gamma \, Q \, H \, \eta_{hyd} \tag{4}$$

The HOMER simulates a "run-of-river" plant, with constant available head [4]. The Figure 2 shows the information that must be supplied to the power plant model used by Homer.

Among the specifications of the turbine, the available head [m], the design flow rate [L/s], minimum and maximum flow ratios, as a percentage of the design flow rate, and efficiency.

For Economic values, capital cost, replacement cost, operation and maintenance cost per year and life time in years.

There is a module for calculating the pressure drop in the pipeline, depending on the pipe length, pipe diameter and pipe roughness.

At the end, there is still the option to include a hydro power plant in all simulated systems or to include only those systems for which prove feasible.

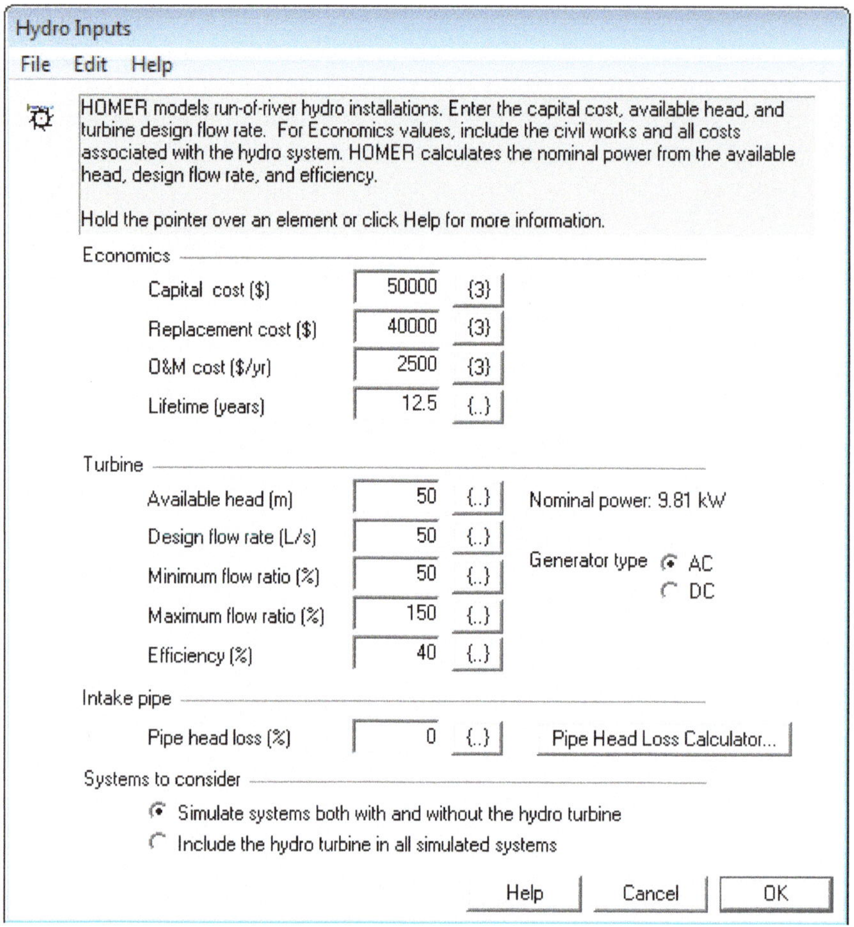

Figure 2. Parameters for the HOMER's model of a micro hydro power plant

5. Adaptation to simulate ocean wave power plants

As Homer does not offer a model that reproduces a wave power plant, it will be necessary to perform this simulation with some of the available models. The most appropriate choice for the characteristics of the adopted model is the model that reproduces hydro electric power plants.

Adjustments will be made simply by determining which values should be entered as height and stream flow in the model used to describe a hydroelectric power plant and will be adapted to describe an ocean wave power plant. Equation (5) shows what will be done.

$$P_{hyd} = P_{wav} \tag{5}$$

Substituting equations (3) and (4) into (5) and defining that ηhyd [1] is equal to ηwav [1] and that H [m] is equal to L [m], it is possible to write (6).

$$Q = \frac{g}{64\pi} H_s^{\,2}\, T_e \tag{6}$$

The flow rate then takes a new interpretation. Since (g/64л) is equal to 0.0488 m/s2, the variable Q [m3/s] can also be expressed by (7).

$$Q = (0.05) H_s^{\,2}\, T_e \tag{7}$$

The flow rate can also be determined by the other side, from the data of available wave energy. Thus, simply divide the available power per meter of wave front by the specific weight of water, as shown in (8). If H is assigned to a constant value, it should appear in the denominator of (8).

$$Q = \frac{1}{\gamma} \frac{P_{wav}}{L \eta_{wav}} \quad (8)$$

6. An application example

The model was developed to study the insertion of a wave power plant in a micro power hybrid system situated not far from the coast. Since there are no commercially available technologies, the study was aimed to determine minimum cost to make it viable. The results can serve as a parameter for building a power plant or to purchase equipment.

The application example is a wind diesel hybrid system located on a small farm in the municipality of Mostardas1, in the state of Rio Grande do Sul. This system has been the subject of a study, conducted by Ponticelli and Beluco [16] for inserting biodiesel and PV modules and will now also be considered for the use of wave energy.

The farm lies between the Atlantic Ocean and the Lagoa dos Patos, in a region of rice cultivation and livestock activities, dedicating themselves to this second activity. The hybrid system is used to supply small mechanical drives and other minor loads. The system also supplies the electrical loads of the farmhouse when the interconnected system failure, but this was not simulated in a specific way. The system is shown in Figure 3(a).

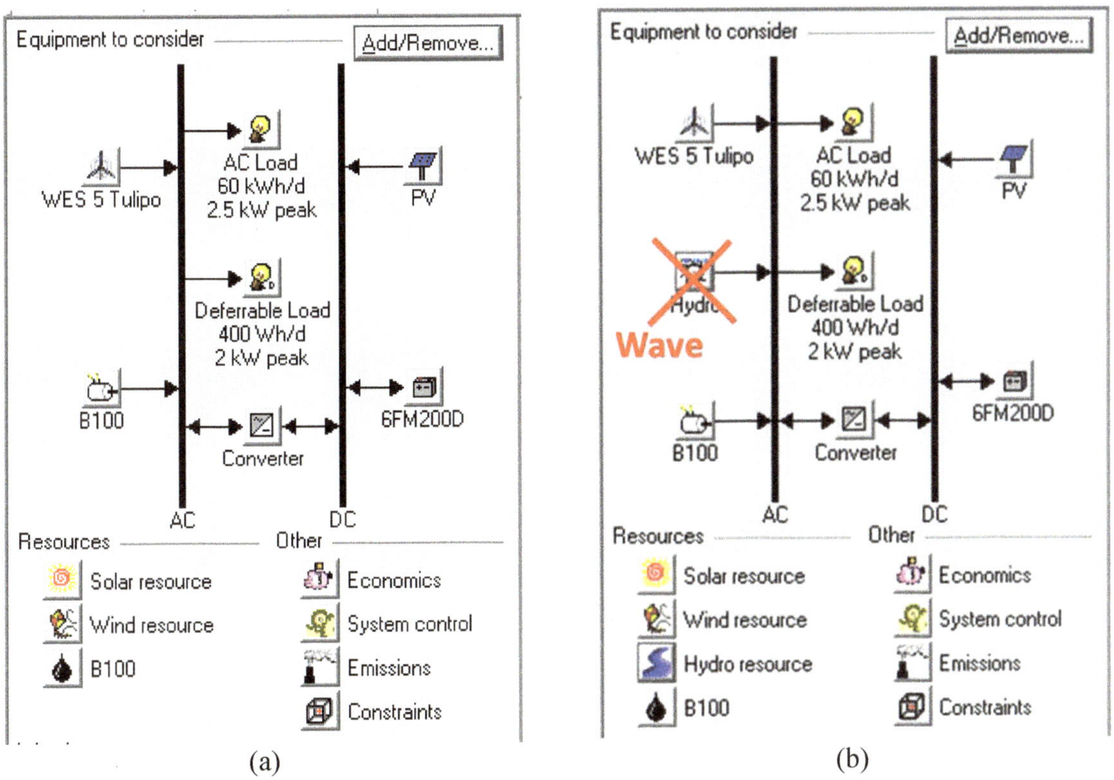

Figure 3. (a) Hybrid system of the Reference [16] and (b) hybrid system including an ocean wave power plant on the right

Loads have consumption value around 105 kWh per day, kept constant by a control system which connects loads for heating water. There is also a water storage system, modeled on Homer as deferrable loads, with consumption of about 400 Wh per day.

The engine generator set is Branco [17], model BD 6500 CF with 10 hp, with 5.5 kVA / 5.0 kW nominal power. The wind turbine WES 5 [18] was employed, with 2.5 kW rated with three rotor blades 5m diameter, with induction generator and IGBT converter. An anemometer connected to a control system allows controlling the speed of the turbine.

1 The city of Mostardas can be located on Google Maps (Google Inc., [15]) at goo.gl/maps/qlC8S.

Figure 4 shows the wind speeds recorded in the first year of operation of the wind turbine, between August 2009 and July 2010. Data were obtained at 25 meters from the ground at approximately the same height as the wind turbine. Figure 5 shows the availability of solar power for the hybrid system, obtained through software Homer with NASA.

Figure 6 shows the instantaneous power in the form of waves available per meter of wave front over a year. These data were obtained with Reference [19] and correspond to the availability of energy throughout the year 1998.

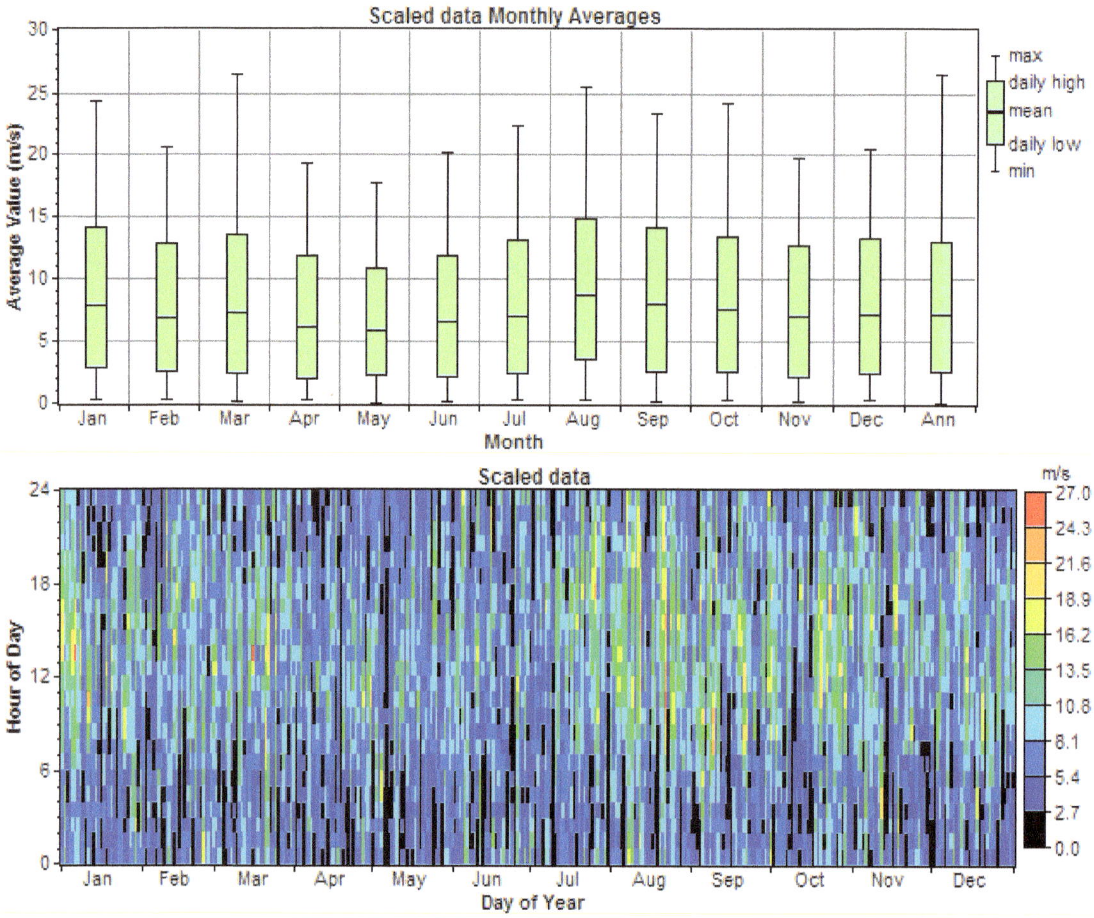

Figure 4. Wind speed data obtained in the first year of operation of the equipment

Figure 7 shows these data as used by software Homer, obtained using equation (17). These data correspond to a wave power plant with a length of 1 meter, simulated with the model of a plant with 25 meters high. The performance of the power plant is considered equal to 100% and hydraulic losses are equated to zero.

Figure 3(b) shows the hybrid system already equipped with the wave power plant. The system includes an AC bus and the other in DC, with the first receiving supplies of wind turbines, diesel generator and wave power plant and the second photovoltaic contribution. The electrical load is connected to the AC bus and battery bank [20] to the DC bus, with the energy flow in both directions between the two buses made possible by the converter device.

Simulations were performed for the following values of the optimization variables: 0, 1, 2, 4 or 8 wind turbines; 0 kW, 5 kW or 10 kW for diesel generator set; 0.00 kW, 0.20 kW, 0.40 kW, 0.80 kW, 1.20 kW, 1.60 kW and 2.00 kW for PV array; 0.00 kW, 0.25 kW, 0.50 kW, 1.00 kW and 2.00 kW for converter capacity and 0, 1, 2, 4 and 8 batteries. Simulations were performed for the following values of the sensitivity inputs: 60 kWh/d for AC Load; 4 m/s, 6 m/s 8 m/s or 10 m/s for wind speed; and 0.40 $/L, 0.60 $/L, 0.80 $/L, 1.00 $/L, 1.20 $/L or 1.40 $/L for pure diesel price. These simulations were repeated to 0 kW, 5 kW and 10 kW for biodiesel generator set, with 0.40 $/L, 0.60 $/L, 0.80 $/L, 1.00 $/L, 1.20 $/L or 1.40 $/L for biodiesel price.

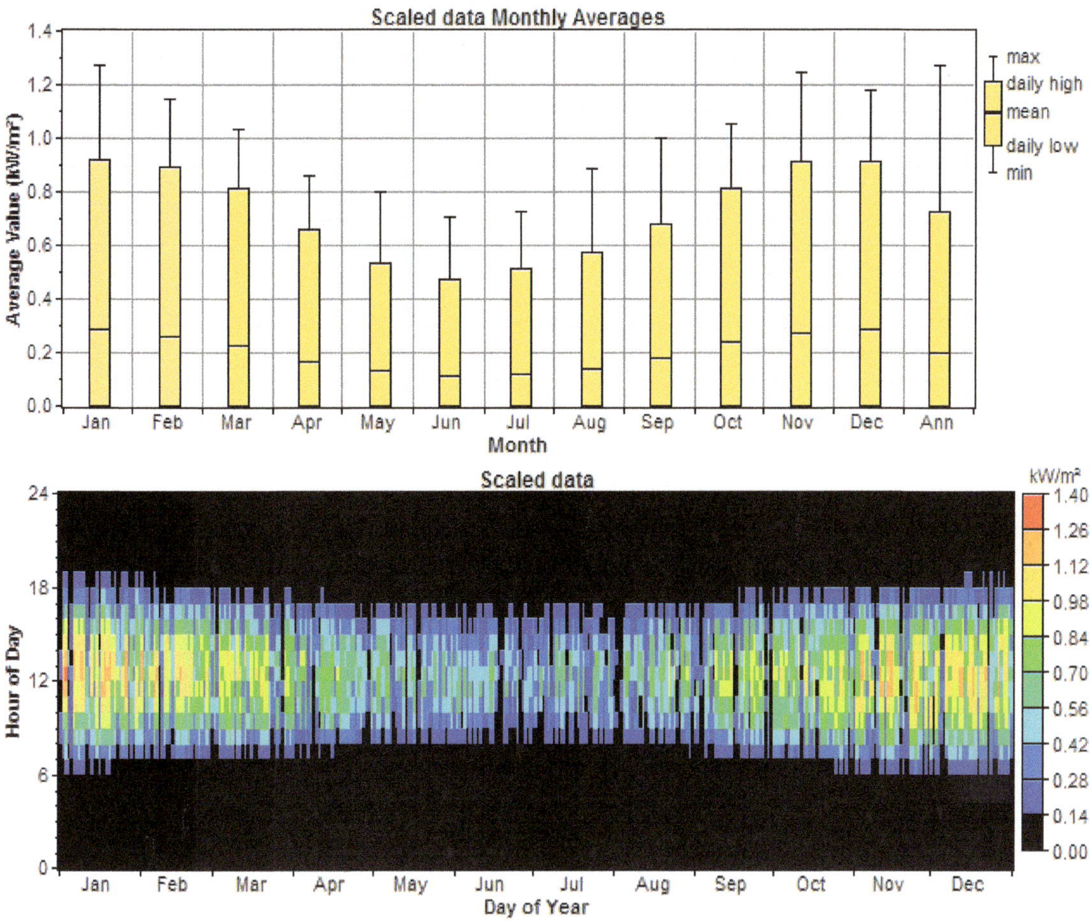

Figure 5. Availability of solar energy obtained by Homer with the databases of NASA

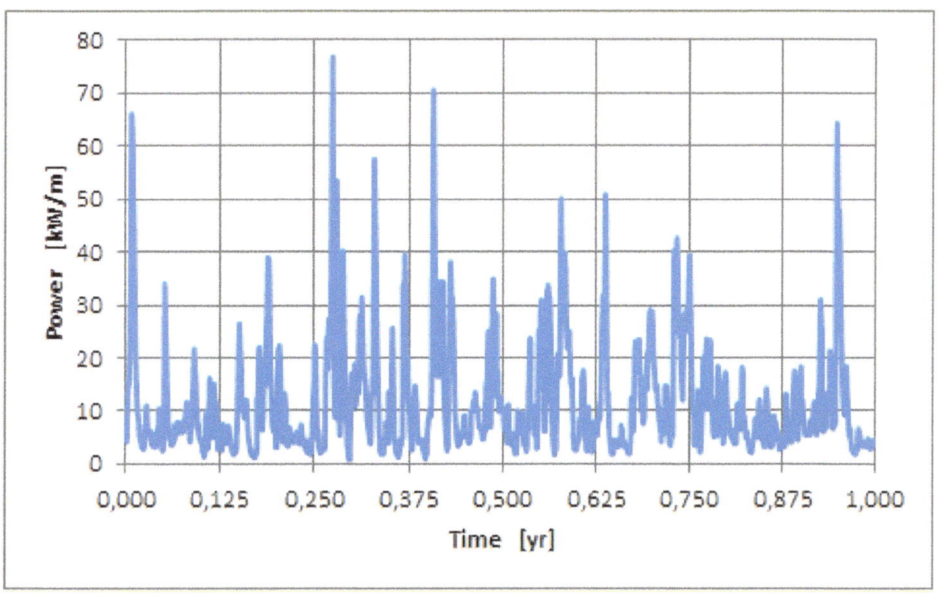

Figure 6. Data series for the wave energetic potential available in the location of the case study

Additionally, these simulations were also repeated to 7,500 $/kW installed, 5,000 $/kW installed, 3,000 $/kW installed, 2,000 $/kW installed and 1,000 $/kW installed for capital cost of wave power plant, always with 80% of this for replacement costs and 5% of this for annual operation and maintenance costs. Finally, the plant was considered with two meters wavefront harnessed and efficiency 40%. All results are presented in Reference [21].

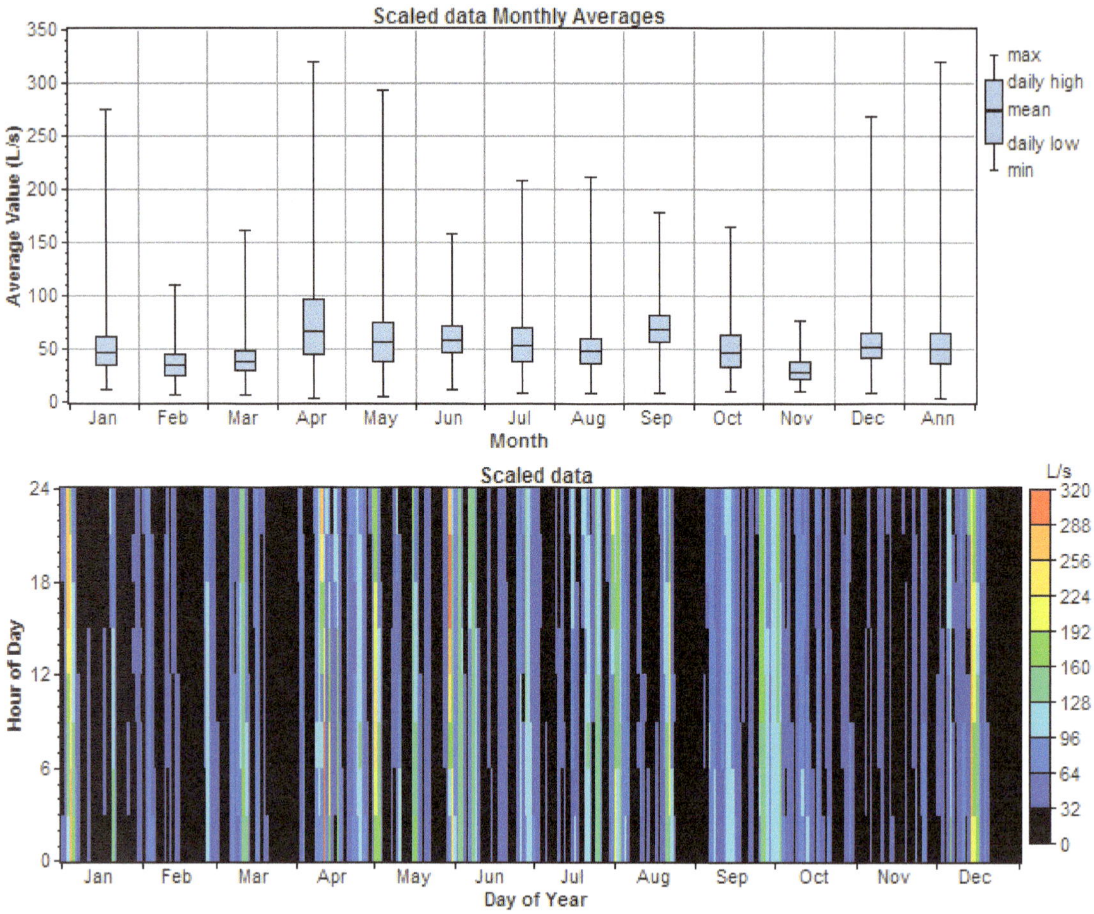

Figure 7. Data for wave energetic potential on-site of the case study, manipulated for Homer

Figure 8 reproduces the optimizations space for the system of Figure 3(a) even without the addition of ocean wave power plant. This result is also valid for the capital cost of the plant over U.S. $ 7,000 per kW installed, without solutions in optimization space that include wave power plant.

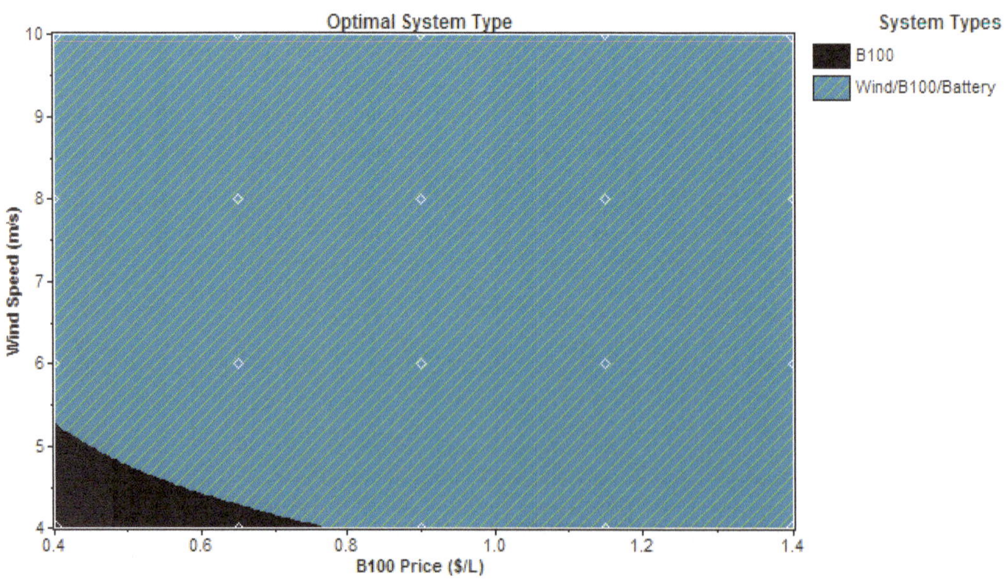

Figure 8. Simulation results with the system of the Figure 3 (a)

Figures 9, 10, 11 and 12 show the optimization spaces for the system of Figure 3(b) respectively with US$ 5,000/kW installed, US$ 3,000/kW installed, US$ 2,000/kW installed and US$ 1,000/kW installed.

In these figures, where it is written "Hydro", should read "Wave". The area corresponding to solutions that include wave power plants appear in the lower right corner of the optimization space, where wind speeds are lower and biodiesel costs higher.

In Figure 9, the brown area corresponds to systems containing wave power plant and biodiesel generator set and the blue area with green stripes corresponds to systems composed of wave power plant, biodiesel gen set, wind turbines and batteries. These two areas will grow in the following figures, as the cost of installation is decreasing. The Figure 12 is dominated by these two areas, meaning that all solutions include wave power plants.

The results indicate that for a capital cost of US$ 2,000/kW installed for converting wave energy, only the systems subject to higher wind speeds do not include wave power plants. For cost of US$ 1,000/kW installed for converting wave energy, all solutions include wave power plants.

Obviously, more studies are needed to evaluate the relationship between dimensions of the wave power plants, their efficiency and capital costs. But these results already show the potential use of Homer as a tool for evaluation of feasibility of systems including wave energy converters.

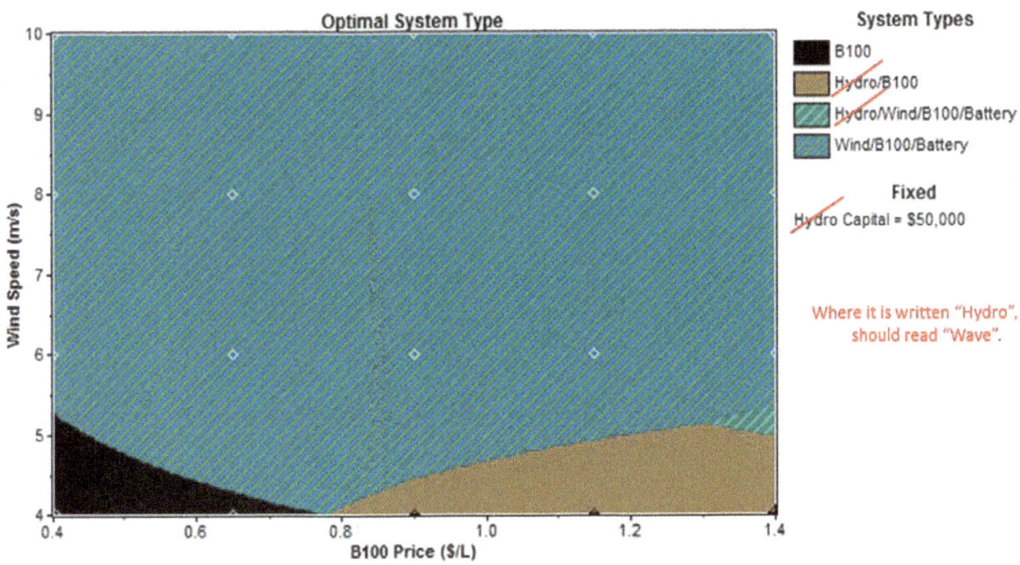

Figure 9. Simulation results with the system of the Figure 3 (b) with capital costs for wave power plant equal to US$ 5,000/kW installed

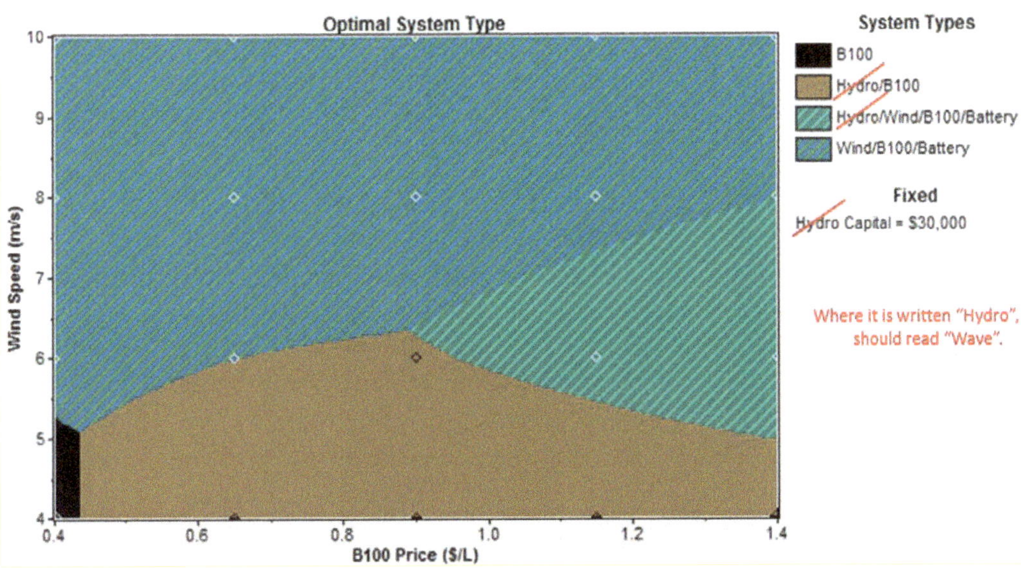

Figure 10. Simulation results with the system of the Figure 3 (b) with capital costs for wave power plant equal to US$ 3,000/kW installed

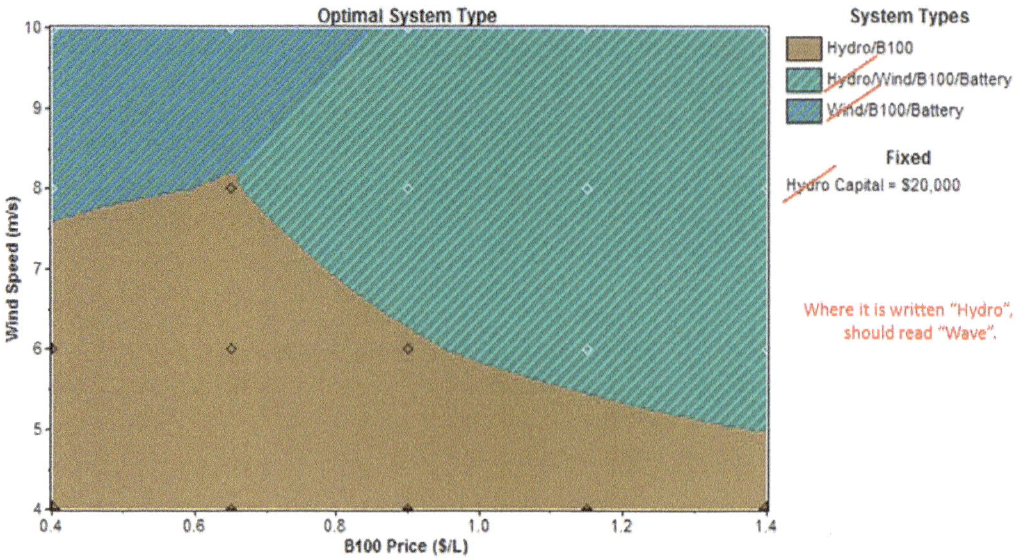

Figure 11. Simulation results with the system of the Figure 3 (b) with capital costs for wave power plant equal to US$ 2,000/kW installed

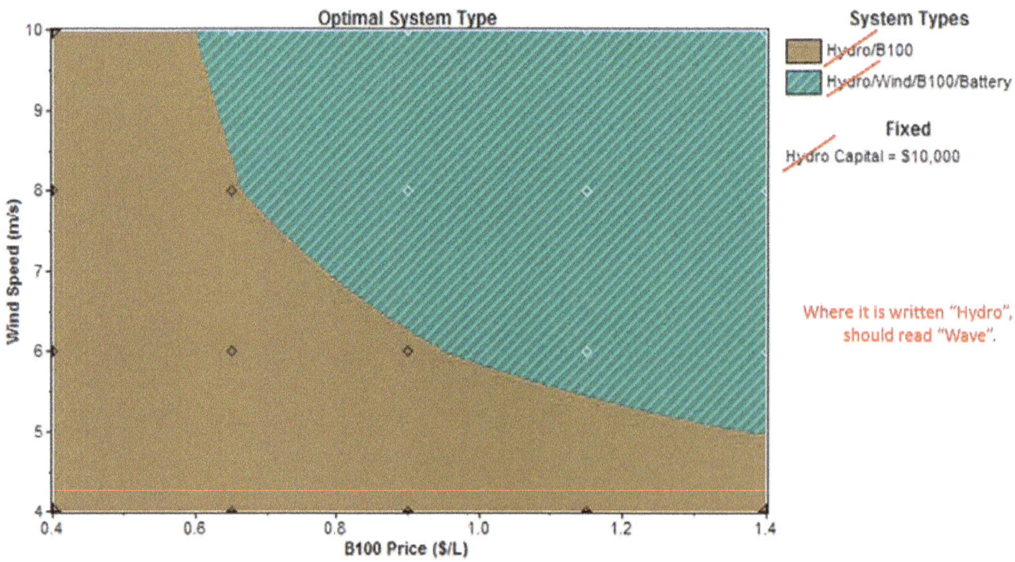

Figure 12. Simulation results with the system of the Figure 3 (b) with capital costs for wave power plant equal to US$ 1,000/kW installed

7. Final remarks

This paper has presented and discussed the use of software Homer for simulating ocean wave power plants. As there is no model for this type of power plant, was proposed to adapt the model of hydroelectric power plants. A case study allowed the discussion of some initial results considering the inclusion of an ocean wave power plant in a wind biodiesel hybrid system.

Homer can be applied in the simulation of any device configuration for converting wave energy. The case study considered an ocean wave power plant exploring two meters of wave front, with 40% efficiency. The results indicated that all the solutions of the optimization space will include ocean wave power plants, when the installation cost to be US$ 1,000 per kW installed.

Acknowledgements

This work was developed as a part of research activities on renewable energy developed at the Instituto de Pesquisas Hidráulicas, Universidade Federal do Rio Grande do Sul. The authors acknowledge the support received by the institution and the receptivity found among the owners of the small farm considered in this study, which do not want her identity disclosed.

References

[1] Bahaj, A.S. (2011) Generating electricity from the oceans. Renewable and Sustainable Energy Reviews, v.15, p.3399-3416.

[2] Falcão, A.F.O. (2010) Wave energy utilization: a review of the technologies. Renewable and Sustainable Energy Reviews, v.14, p.899-918.

[3] Falnes, J. Lovseth, J. (1991) Ocean wave energy. Energy Policy, v.19, n.8, p.768-775.

[4] Lambert, T.W. Gilman, P. Lilienthal, P.D. (2005) Micropower system modeling with Homer. In: Farret, F.A. Simões, M.G. Integration of Alternative Sources of Energy, John Wiley & Sons, p.379-418. ISBN 0471712329.

[5] Lilienthal, P.D. Lambert, T.W. Gilman, P. (2004) Computer modeling of renewable power systems. In: Cleveland, C.J. (ed.) Encyclopedia of Energy, Elsevier, v.1, p.633-647. NREL Report CH-710-36771.

[6] Lilienthal, P.D. Lambert, T.W. Gilman, P. (2011) Getting started guide for Homer Legacy (version 2.68). Homer Energy, Boulder, CO; National Renewable Energy Laboratory (NREL) of the USDOE, Golden, CO; 28p.

[7] Homer Energy (2009) Software HOMER, version 2.68 beta, The Micropower Opyimization Model. Available at www.homerenergy.com.

[8] Polinder, H. Scuotto, M. (2005) Wave energy converters and their impact on power systems. International Conference on Future Power Systems, 9p.

[9] Anagnostopoulos, J.S. Papantonis, D.E. (2008) Simulation and size optimization of a pumped storage power plant for the recovery of wind farms rejected energy. Renewable Energy, v.33, p.1685-1694.

[10] Beluco, A. Souza, P.K. Krenzinger, A. (2008) A dimensionless index evaluating the time complementarity between solar and hydraulic energies, Renewable Energy, v.33, p.2157-2165.

[11] Beluco, A. Souza, P.K. Krenzinger, A. (2012) A method to evaluate the effect of complementarity in time between hydro and solar energy on the performance of hybrid hydro PV generating plants. Renewable Energy, v.45, p.24-30.

[12] Twidell, J. Weir, T. (2006) Renewable Energy Resources. Taylor and Francis, Great Britain, 2nd edition.

[13] Izadparast, A.H. Niedzwecki, J.M. (2011) Estimating the potential of ocean wave power resources. Ocean Engineering, v.38, p.177-185.

[14] Ortega, S. Osorio, A.F. Agudelo, P. (2013) Estimation of the wave power resource in the Caribbean Sea in areas with scarce instrumentation. Case study: Isla Fuerte, Colombia. Renewable Energy, v.57, p.240-248.

[15] Google Inc. (2013) Google Maps. Available at maps.google.com.

[16] Ponticelli, F.A. Beluco, A. (2014) Inclusion of biodiesel and PV modules in a wind diesel hybrid system supplying electrical loads on a small farm. International Journal of Renewable Energy Technology, v.5, n.3, p.229-250.

[17] Caetano Branco Company (2012) Branco Generators Set, www.branco.com.br/us. Accessed on December 28, 2012.

[18] Cell Energy International (2013). WES 5 Tulipo Wind Turbine. Available at cellenergyinternational.co.uk/pdf/WES5_Tulipo_brochure_ds1.pdf. Accessed on July 31, 2013.

[19] Assis, L.E. Beluco, A. Almeida, L.E.S.B. (2014) On the wave energy potential along the southern coast of Brazil. International Journal of Energy and Environment, v.5, n.1, p.59-66.

[20] Vision Battery (2013) Vision 6FM200D Model. Available at www.vision-battery.com. Accessed on Abril 02, 2013.

[21] Silva, J.S. Beluco, A. Almeida, L.E.B. (2013) Simulation results with Homer on the inclusion of an ocean wave power plant in a wind biodiesel hybrid system. Internal report. UFRGS, IPH. Available at www.beluco.net/reports/homer-wave-01.pdf.

14

CFD study of a twisted blade H-Darrieus wind turbine

Rajat Gupta, Rituraj Gautam, Siddhartha Sankar Deka

Department of Mechanical Engineering, National Institute of Technology, Silchar, Assam, PIN-788010, India.

Abstract

In this paper, a two-dimensional Computational Fluid Dynamics (CFD) study of the performance of a H-Darrieus turbine with three twisted blade had been carried out. The chord length of each blade is 5cm and the blade height is considered to be same for all the rotors. A two dimensional (2D) model of the turbine was designed in CATIA V5R19 software and a k-epsilon turbulence closure was adopted with the unstructured mesh generated around the rotor modeled in GAMBIT 2.3.16. The inlet velocities and the rotational speeds are taken from the experimental results and the CFD analysis was carried out in CFD Code-FLUENT 6.3.26. From the CFD analysis, power coefficient (C_p) and torque coefficient (C_t) at three different H/D ratios of 1.13, 1.31 and 1.55 respectively were calculated and compared with available experimental results. The computational analysis showed that the highest values of C_p (0.525) and C_t (0.95) were obtained at H/D ratios of 1.31 and 1.13 respectively. The deviation of computational C_p from experimental C_p was within ±3.08 % and that of computational C_t from experimental C_t was within ±1.106 %. A study of the flow behaviour around the rotor was also carried out using the pressure contours and velocity vectors plots. A maximum pressure drop is obtained for H/D ratio of 1.31 and a vortex reattachment near rear blade of rotor with H/D ratio of 1.31 was observed from the pressure contours and velocity vectors plots. The vortex attachment to the blade of the rotor enhances the lift coefficient of the rotor which helps in improving the power coefficient of the rotor. The comparison between the computational results and previous experimental work is pretty encouraging.

Keywords: CFD; H-Darrieus; TSR; Power coefficient; Velocity contour.

1. Introduction

With the exponential rise in use of fossil fuel there has been a rapid depletion of these non-renewable sources of energy. Besides, increasing pollution & global warming have forced environmentalist and energy researchers to vote against these non-renewable sources as the energy source of future. So there began a widespread research in the field of renewable sources of energy which can help in realizing a secure energy future. It is clean in nature and its abundant availability along with the ease with which it can be converted to other forms of energy.

Wind turbines are of two types, Horizontal axis Wind Turbine (HAWT) and Vertical Axis Wind Turbine (VAWT). The original H-Darrieus wind turbine was originally invented and patented by G.J.M Darrieus, a French aeronautical engineer in the year of 1931 [1]. Two types of Darrieus rotors are mainly available, namely troposkien (Egg beater) Darrieus rotor and H-Darrieus rotor. H-Darrieus is a lift type vertical axis wind turbine consisting mainly of two to three airfoil shaped blades which are attached vertically to the central shaft through support arms. It is self-starting and Omni directional.

Due to the progress in computer technology, Computational Fluid Dynamics (CFD) is now able to simulate the fluid flow at moderate costs and time-to solution. The prospect and success of CFD will therefore depend on the accuracy of the approach, in particular the predictive realms of the employed physical models. The interaction of the wind and the blades of the turbine results in the formation of vortices generated both at the upwind and downwind passage near the wind turbine; which is now possible to study very easily due to CFD else complex & time consuming experiments need to be carried out. Though a considerable amount of research had been carried out in the field of design and analysis of VAWT [2-5] but very few literatures can be found in Wind Turbine Community regarding the CFD approaches in H-Darrieus wind turbines. Considerable improvement in the understanding of VAWT can be achieved through the use of Computational Fluid Dynamics and experimental measurements [6].

This paper had attempted to study the numerical aspects of a modified H-Darrieus wind turbine. The present study is confined to incompressible and steady flow around the wind turbine. Pre-processing was carried out in GAMBIT 2.3.16 software for generating the two dimensional mesh. The governing N-S (Navier-Stokes) equations (the continuity equation and the momentum equation) were solved in FLUENT 6.3.26 and for post processing the TECPLOT software was utilized. The results obtained with unstructured mesh were fairly in agreement with the experimental results with minimum difference between the experimental and CFD results. The use of unstructured grids has two major advantages over structured grids in case of modeling the mesh around a VAWT. First, the unstructured mesh allows for fast and efficient grid generation around highly complex geometries like the rotor of the turbine. Second, appropriate unstructured-grid data structures facilitate the insertion and deletion of points and enable the computational mesh to locally adapt to the flow field solution. This means that a very fine grid can be obtained in the immediate vicinity of the rotor which can correctly model the vortices and pressure variation; hence the numerical dissipation can be avoided.

2. Description of the rotor model

The H-Darrieus rotor was 25cm in height and 5cm in chord length. It was twisted with an angle of 30° at the trailing end to make it self-starting from no load condition. Blade thickness was 5 mm. The blades of the rotor were straight length wise and are 120^0 apart. Three different rotors with different H/D ratios namely 1.13, 1.31 and 1.55 with constant height but with different diameter of the rotor were modeled in CATIA V5R19. The isometric view and top view of a particular rotor model is shown by Figures 1(a) and 1(b) respectively whereas the blade profile of the H-Darrieus rotor is shown by the Figure 1(c). The experimental results were obtained from the experiments conducted by Kakati and Kakati [7] to study the performance of H-Darrieus Wind Turbine rotor in both normal and skewed flow conditions in the wind tunnel present in Mechanical Engineering Department of NIT Silchar. A brief description of the subsonic wind tunnel is provided in the paper [8].

3. Computational methodology

3.1 Computational domain

The computational Fluid Dynamics code FLUENT was utilized for solving the N-S equations and GAMBIT was used for generating the two dimensional unstructured mesh. Figure 2 shows the computational domain, which has the three-bladed rotor along with surrounding four edges resembling the test section of the wind tunnel. The unstructured triangular meshing is used for discretizing the computational domain shown in Figure 3. Velocity inlet and outflow conditions were taken on the left and right boundaries respectively. The inlet and outlet BCs were defined considering the wind speed in inlet at on-design value for each rotor to be fixed [9, 10] so that the calculated torque is function of rotational speed only and thereby simplifying the Cp versus λ comparison. The top and bottom boundaries, which signify the sidewalls of the wind tunnel, had symmetry conditions on them. The Symmetry BC is useful because it allows the solver to consider the wall as part of a larger domain, like a 'free shear slip wall', avoiding in this way the wall effects [11]. The blades, shaft and the support arms were set to moving wall conditions. The H-Darrieus wind turbine blades rotate in the same plane as the approaching wind. Moreover, the for an H-Darrieus wind turbine the geometric properties of the blade cross section are usually constant with varying span unlike that of a Darrieus rotor for which the geometric properties vary with the local radius.

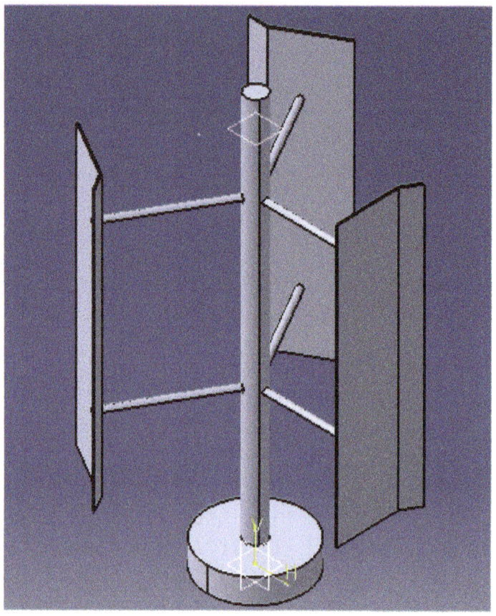

(a) An isometric view of the rotor model

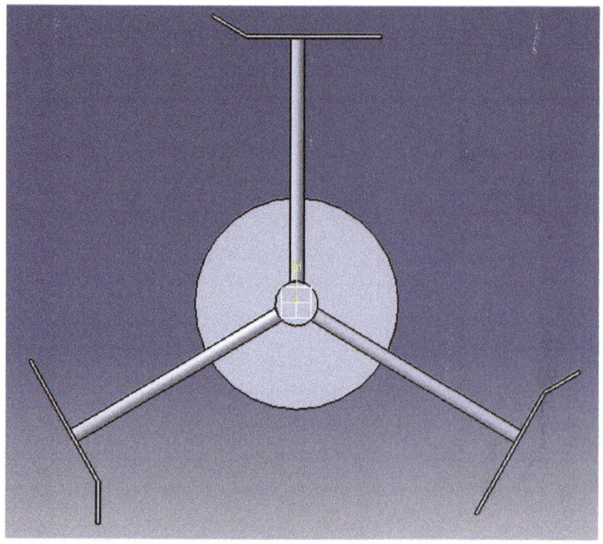

(b) A top view of the rotor model

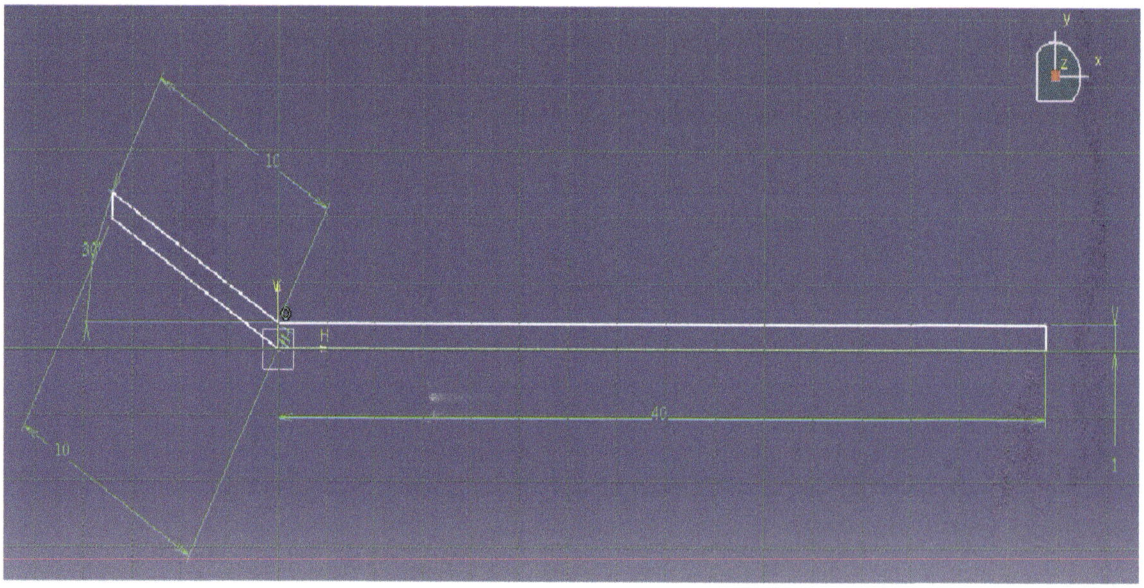

(c) A sketch of the blade profile (all dimensions in mm)

Figure 1. The CAD model of the rotor and the blade profile conceived in CATIA V5R19

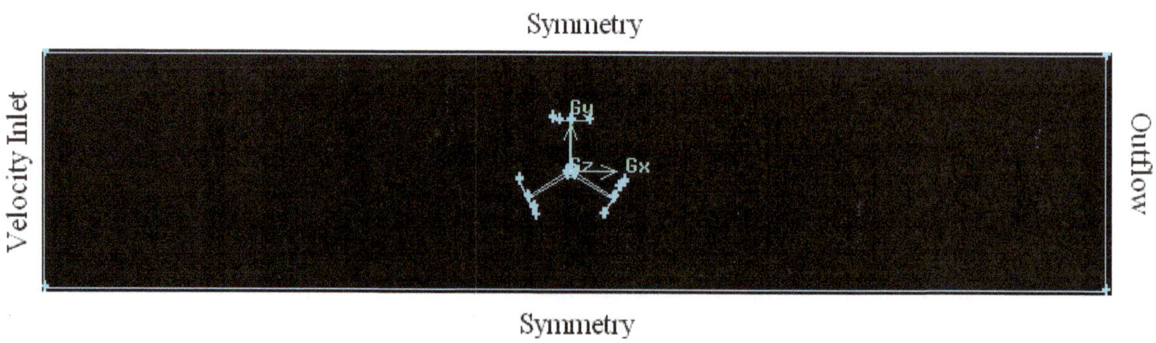

Figure 2. Physical model, boundary conditions and computational domain of the 3-bladed H-Darrieus Rotor for H/D = 1.13

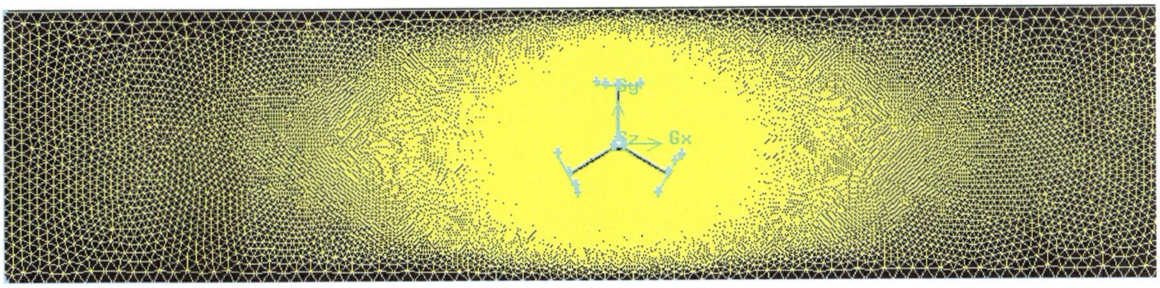

Figure 3. Computational domain after discretization for H/D = 1.13

3.2 CFD solver description

The commercial code FLUENT 6.3.26 was used for performing the CFD analysis for Reynolds number of the order 3×10^5. A steady state segregated solver with absolute velocity and implicit formulation and Green Gauss cell-based gradient option was considered. The FLUENT default segregated solver helps in solving the governing integral equations for continuity and momentum sequentially. For the 2D, steady and incompressible flow the continuity equation is:

$$\frac{\partial U}{\partial X}+\frac{\partial V}{\partial Y}=0 \quad (1)$$

Momentum equation for viscous flow in X direction:

$$U\frac{\partial U}{\partial X}+V\frac{\partial U}{\partial Y}=-\frac{1}{\rho}\frac{\partial P}{\partial X}+\frac{\partial}{\partial Y}\left[\left(\upsilon+\upsilon_t\right)\frac{\partial U}{\partial Y}\right] \quad (2)$$

In the above equation $\frac{\partial U}{\partial X}$, $\frac{\partial V}{\partial X}$ indicates velocity gradient along x and y direction respectively whereas $\frac{\partial P}{\partial X}$ indicates rate of pressure variation along x direction and υ, υ_t represents the dynamic viscosity. As the governing equations are non-linear, several iterations must be performed before a converging solution is obtained. A flow chart (Figure 4) is illustrated how the iterations are carried out. The fluid properties are updated initially based on initialized solution.

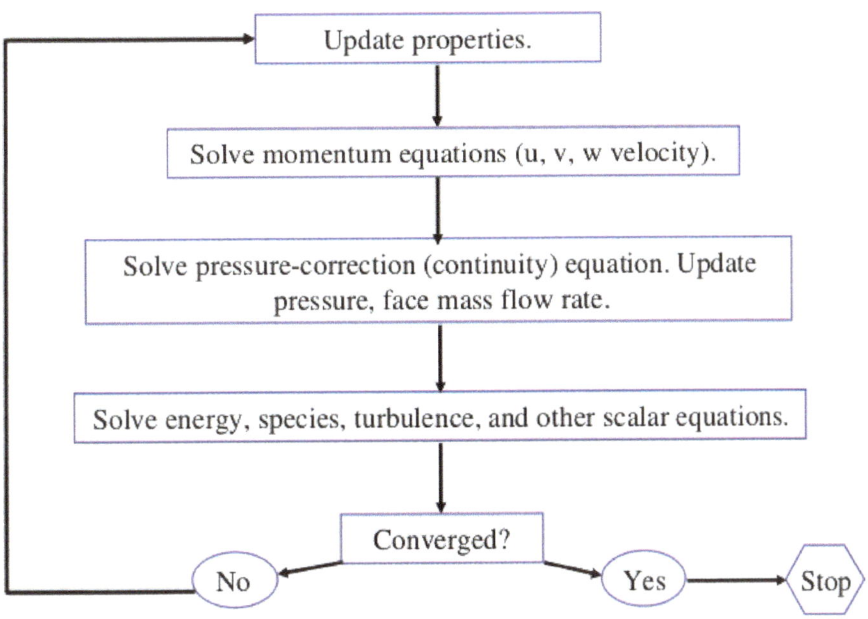

Figure 4. Overview of the segregated solution method

A second order upwind discretization scheme was adopted for momentum discretization and SIMPLE (Semi-Implicit Method for Pressure Linked Equations) algorithm was used for pressure velocity coupling of the flow. When the second order upwind discretization is used the quantities at the cell faces are computed using a multidimensional reconstruction approach as proposed by Barth and Jespersen [12]. The SIMPLE algorithm [13] converts the continuity equation into a discrete Poisson equation for pressure. The differential equations are linearized and solved implicitly in sequence: starting with pressure equation (predictor stage), followed by momentum equations and the pressure correction equation (corrector stage). The equation of the scalars (turbulent quantities) is solved after updating of both pressure and velocity components. The iterations were done to achieve a converged solution. Implicit formulation is used for linearizing the non-linear governing equations to produce a system of equations for the dependent variables for every computational cell. In the segregated method each governing equation is linearized implicitly with respect to that equation's dependent variable. This will result in a system of linear equations with one equation for each cell in the domain. A Gauss-Siedel linear equation solver is used in colligation with an Algebraic Multigrid method to solve the resultant systems of equations for the dependent variable in each cell. Gradients are needed for computing secondary diffusive terms, velocity derivatives as well as for constructing values of all scalar variables at the cell faces. Green- Gauss Cell based had been considered for computing the gradients. The turbulence in the flow was modeled using a standard k-ε viscous model with standard wall functions was chosen. The standard wall functions give reasonably accurate predictions for the majority of high-Reynolds-number, wall-bounded flow & is also accurate for far field flows. The standard wall functions are based on the proposal by Launder and Spalding [14]. The two equation k-ε turbulence model is the simplest one and is widely used to solve the two separate transport equations to allow the turbulent kinetic energy and its dissipation rate to be independently determined. The model is reasonably accurate and computationally less expensive. The governing equations were integrated over each control volume to construct discretized algebraic equations for dependent variables. The transport equations for k and ε in the standard k- ε model [11] are as given below:

Turbulent kinetic energy (k) equation:

$$\frac{\partial}{\partial t}(\rho k)+\frac{\partial}{\partial x_i}(\rho k u_i)=\frac{\partial}{\partial x_j}\left[\left(\mu+\frac{\mu_t}{\sigma_k}\right)\frac{\partial k}{\partial x_j}\right]+P_k-\rho\varepsilon-Y_M \tag{3}$$

For dissipation (ε):

$$\frac{\partial}{\partial t}(\rho\varepsilon)+\frac{\partial}{\partial x_i}(\rho\varepsilon u_i)=\frac{\partial}{\partial x_j}\left[\left(\mu+\frac{\mu_t}{\sigma_\varepsilon}\right)\frac{\partial\varepsilon}{\partial x_j}\right]+C_{1\varepsilon}\frac{\varepsilon}{k}(P_k)-C_{2\varepsilon}\frac{\rho\varepsilon^2}{k} \tag{4}$$

Model constants:

$C_{1\varepsilon}=1.44$, $C_{2\varepsilon}=1.92$, $C_\mu=0.09$, $\sigma_k=1.0$, $\sigma_\varepsilon=1.3$

Modeling turbulent viscosity:

$$\mu_t=\rho C_\mu\frac{k^2}{\varepsilon} \tag{5}$$

Production of k due to mean velocity gradients:

$$P_k=-\rho\overline{u_i'u'}\frac{\partial u_j}{\partial x_i} \tag{6}$$

In these equations, P_k represents generation of turbulent kinetic energy due to mean velocity gradients, the turbulent Prandtl Numbers σ_k and σ_ε is for k and ε respectively. All the variables including turbulent kinetic energy k, its dissipation rate ε were shared by each fraction of fluid volume.
The philosophy behind adopting RANS (k- ε model) is that in it the Navier-Stokes equations are decomposed into time-averaged and fluctuating components. This yields a turbulence simplification that reduces computational demands from the LES/DES, and still achieves acceptable accuracy for complex flow. The details of the solution strategy adopted is provided in Appendix.

3.3 Grid independence test

The correctness of the computational results depends upon the resolution of the grid. However very fine resolution of the mesh results in increased computational cost leading to higher time consumption and money. It can be seen that with gradually refining of the mesh (or improved grid density) the variation in the results obtained computationally is very less after a while it becomes nil. At this point a limit of the refinement of the mesh is reached and further refinement would not have any effect in the results. In the analysis C_d had been chosen as the parameter to obtain the Grid Independence Limit. Grid refinement is carried out until a steady value of C_d is obtained. From the grid independence test it can be clearly seen that the value of C_d is constant beyond 68369 nodes (Figure 5). So the mesh with 121657 nodes & skewness less than 0.5 is being considered for the final simulation.

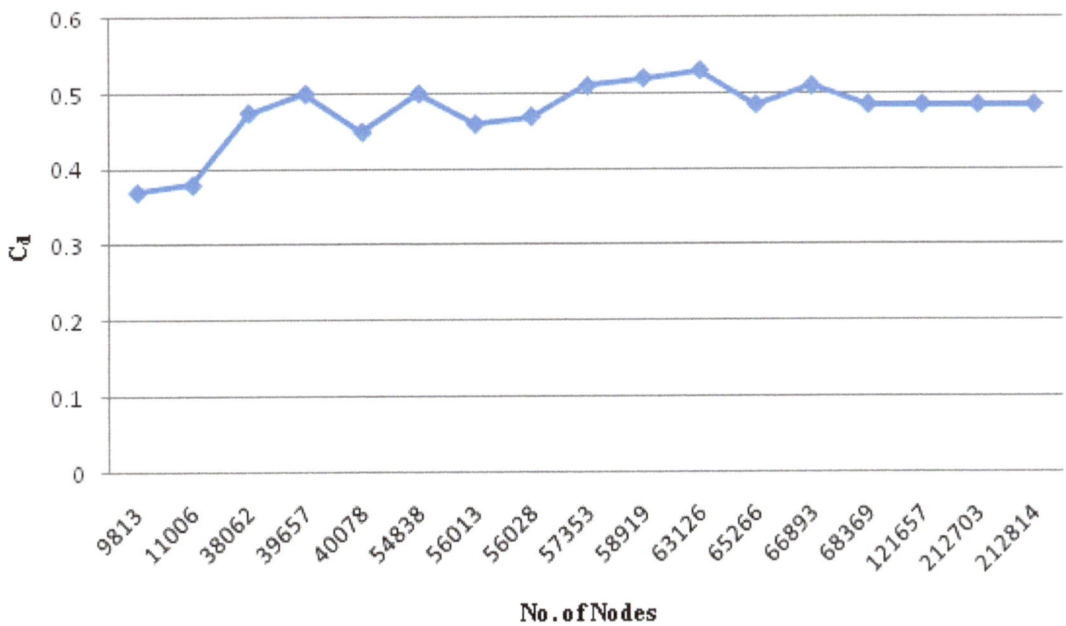

Figure 5. Grid independence test carried out for the 3-bladed H-Darrieus rotor

4. Analysis of results

The experimental and computational C_p and C_t with respect to tip speed ratio and the percentage deviations of the computational results from the experimental results for various H/D ratios are calculated as below [15].

$$C_t = \frac{T}{\frac{1}{4}\rho A V_{free}^{2} D} \tag{7}$$

$$C_p = \frac{T\omega}{\frac{1}{2}\rho A V_{free}^{3}} \tag{8}$$

Now putting the value of torque from equation (7) in (8)

$$C_p = \frac{1}{2} C_t \frac{D\omega}{V_{free}} \tag{9}$$

Tip Speed Ratio (TSR) is given by

$$\lambda = \frac{\text{Tip Speed of Blade}}{\text{Wind Speed}} = \frac{C_p}{C_t} = \frac{\pi DN}{60V_{free}} \tag{10}$$

In the above equations, the values of C_t (torque coefficient) and C_p (power coefficient) are known after FLUENT analysis. The difference of the experimental and computational results had been calculated in the form of standard deviation from the data set for each set of H/D ratios. The standard deviation formula used for power coefficient is

$$\sigma^2 = \frac{1}{n-1}\left[\sum_{i=1}^{n}\left(C_{p_i} - \overline{C_p}\right)\right] \tag{11}$$

where,

$$\overline{C_p} = \frac{1}{n}\sum_{i=1}^{n} C_{p_i} \tag{12}$$

Here σ is the standard deviation and n is the number of data taken.

A comparison of the experimental and computational values of C_p and C_t for different values of H/D ratios is shown by Figures (6, 8, 10, 12, 14 and 16). The standard deviation of the computational C_p and C_t values from the experimental ones is shown by Figures (7, 9, 11, 13, 15 and 17). Finally Figures 18, 19 show the maximum values of C_p and C_t, both computational and experimental for the three different H/D ratios.

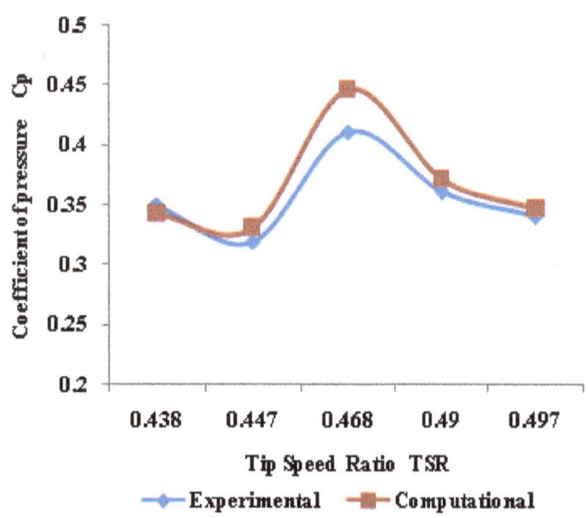

Figure 6. Variation of C_p with TSR for H/D=1.13

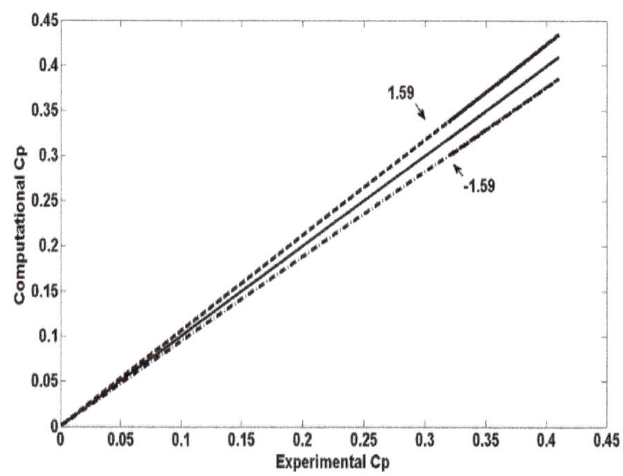

Figure 7. Deviation of computational C_p from experimental C_p for H/D=1.13

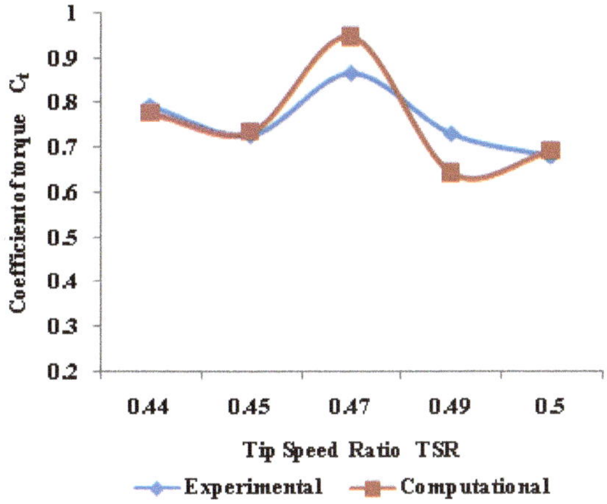

Figure 8. Variation of C_t with TSR for H/D=1.13

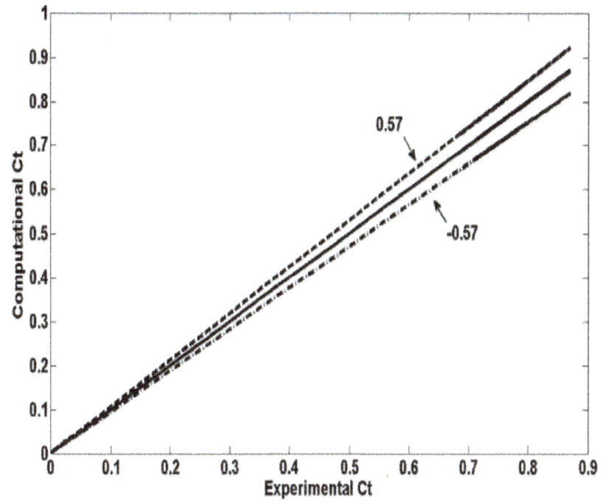

Figure 9. Deviation of computational C_t from experimental C_t for H/D=1.13

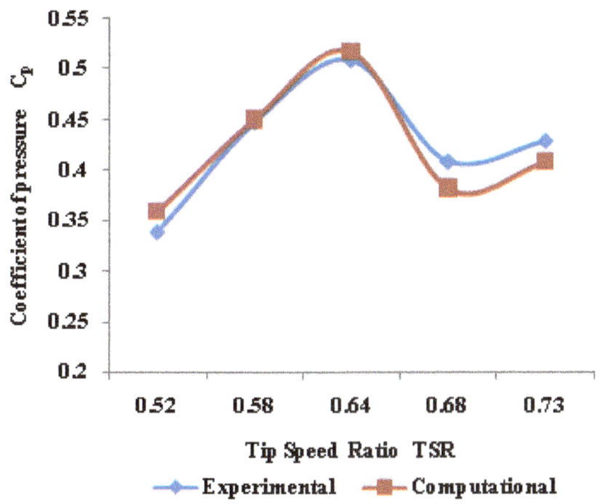

Figure 10. Variation of C_p with TSR for H/D=1.31

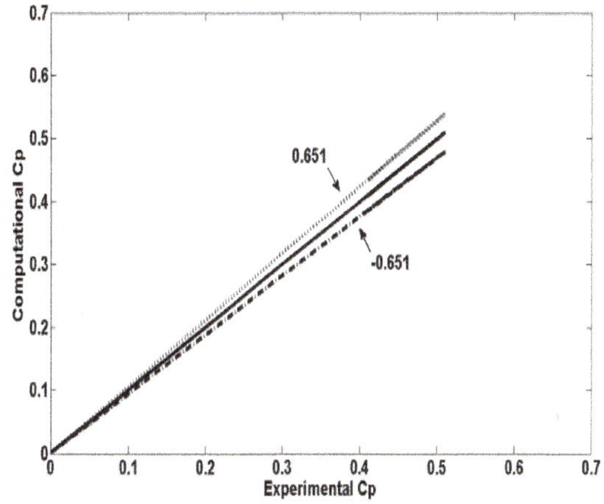

Figure 11. Deviation of computational C_p from experimental C_p for H/D=1.31

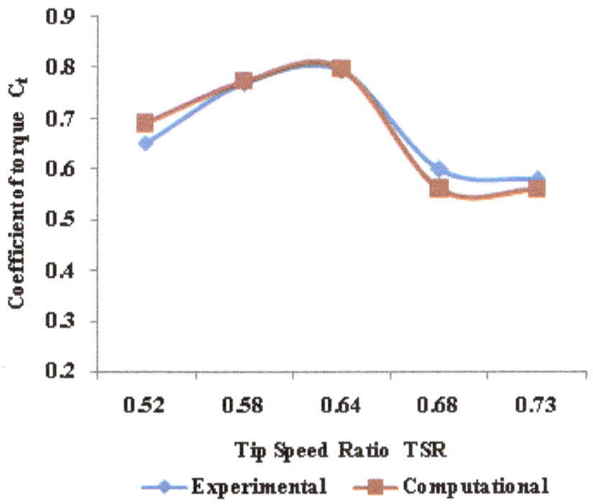

Figure 12. Variation of C_t with TSR for H/D=1.31

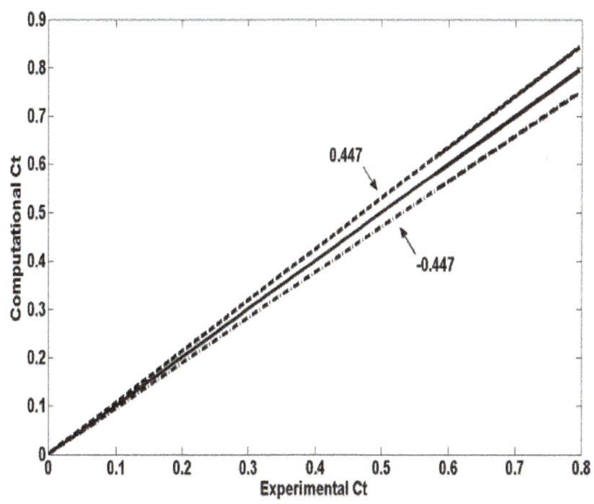

Figure 13. Deviation of computational C_t from experimental C_t for H/D=1.31

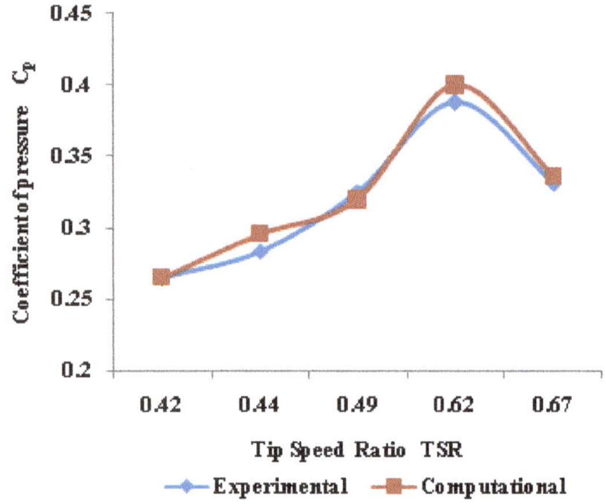

Figure 14. Variation of C_p with TSR for H/D=1.55

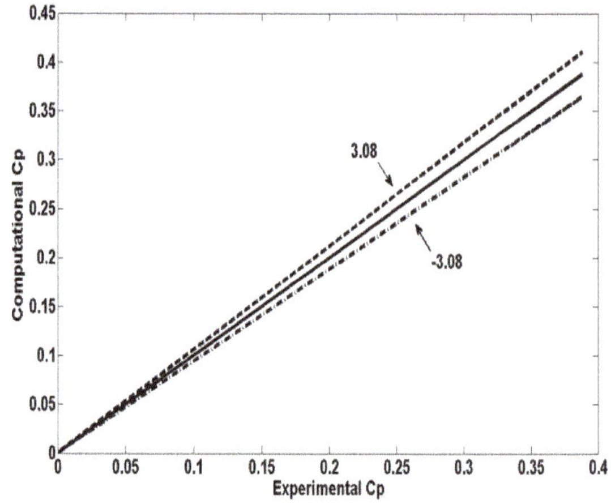

Figure 15. Deviation of computational C_p and experimental C_p for H/D=1.55

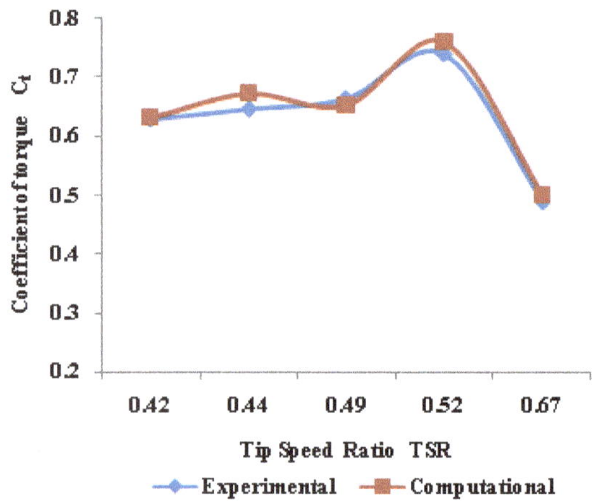

Figure 16. Variation of C_t with TSR for H/D=1.55

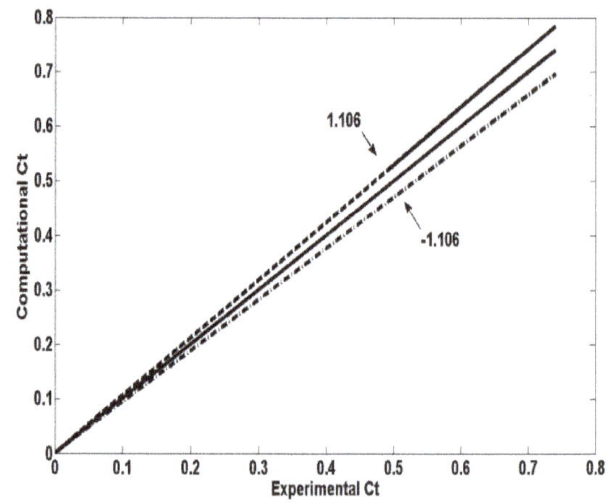

Figure 17. Deviation of computational C_t and experimental C_t for H/D=1.55

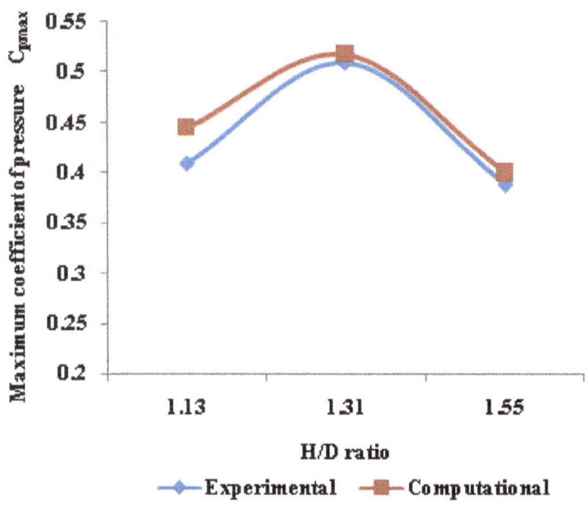

Figure 18. Comparison between experimental and computational C_p for different H/D ratios

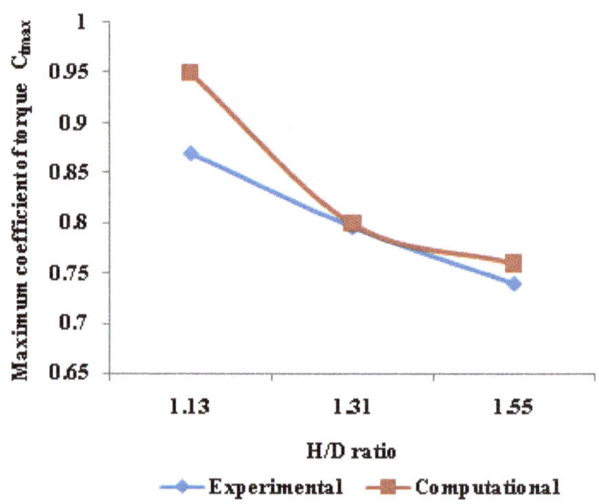

Figure 19. Comparison between experimental and computational C_t for different H/D ratios

4.1 Comparison of experimental and computational results

From Figures 6 and 8, it is found that, for H/D ratio 1.13, the maximum C_p obtained is 0.445 at a TSR of 0.477, and the maximum C_t obtained is 0.95 at a TSR of 0.471. And for this H/D ratio, the standard deviation (Figures 7 and 9) of computational C_p from experimental C_p is 1.59% and that of computational C_t from experimental C_t is 0.57%. From Figures 10 and 12, it is found that, for H/D ratio 1.31, the maximum C_p obtained is 0.518 at a TSR of 0.643, and the maximum C_t obtained is 0.80 at a TSR of 0.652. And the standard deviation (Figures 11 and 13) of computational C_p from experimental C_p is 0.651% and that of computational C_t from experimental C_t is 0.447%. From Figures 14 and 16, it is found that, for H/D ratio 1.55, the maximum C_p obtained is 0.40 at a TSR of 0.526, and maximum C_t obtained is 0.76 at a TSR of 0.533; the standard deviations (Figures 15 and 17) of C_p and C_t are 3.08% and 1.106% respectively. From Figure 18, it can be found that the maximum computational C_p of 0.518 and maximum experimental C_p of 0.51 occur for H/D ratio of 1.31. Also from Figure 19, the maximum computational and experimental C_t comes out to be 0.95 and 0.87 respectively for H/D ratio corresponding to 1.13. From Figures 6-19 it can be clearly be interpreted that there is a close concurrence of the computational and experimental results.

4.2 Pressure contours and velocity vectors analysis for different H/D ratios

Post processing of the results after solving the governing equations is obtained in Tecplot 360. From the Figures 21, 23 and 25 it can be observed the velocity magnitude vectors and Figures 20, 22 and 24 represents the pressure contours of the three bladed H-Darrieus rotor for three different H/D ratios, namely 1.13, 1.31 and 1.55. The flow is mainly affected downstream of the rotor, while the flow upstream of the rotor almost remains unaffected except for the regions very near to the rotor. The velocity vectors show that the flow is accelerated while passing over the rotor blades. The velocity at the blade tips is almost 1.8 times higher compared to the input velocity on the extreme left of the computational domain. This velocity gradient generates the power stroke of the blades during its clockwise rotation. In case of wind turbines the highest aerodynamic torque at is developed at the blade tips and the high velocity magnitude at the blade tips ensures augmentation in aerodynamic torque production. Further, the velocity vectors show that, with the increase of H/D ratio, the velocity magnitude difference from inlet up to the rotor increases up to a certain H/D ratio and then decreases meaning loss of performance for the turbine with increase of H/D ratio. As is evident from Figure 18 it can be seen that the C_p value increases initially with increase of H/D ratio from 1.13 to 1.31 and then decreases with further increase of H/D ratio from 1.31 to 1.55. For example at H/D ratio of 1.13, there is a velocity drop of 16.71m/s which increases to 25.67 m/s for H/D ratio of 1.31 but when the H/D ratio of 1.55 is considered the velocity drop decreases to 11.9 m/s. This fact is further supplemented with the observation that maximum static pressure drop of 1990 Pa is observed for H/D ratio of 1.31 as shown in Figure 22 and minimum static pressure drop for H/D ratio of 1.55 around 1395Pa as shown in Figure 24. This occurrence enhances the aerodynamic torque of the rotor and improves the power extraction of the rotor of H/D ratio 1.31 to maximum. This is in agreement with the work done on two bladed H-Darrieus rotor by Gupta et al. [16] where it was shown that the power extraction of the two bladed H-Darrieus rotor increases with increasing H/D ratio to 1.0 and beyond that there is a fall in power coefficient. From the velocity vectors plot it is pretty evident that vortices are formed close to the rotor. These clockwise rotating vortices formed on the rear side of the rotor gradually grows in size for increasing H/D ratio up to 1.31 but it becomes smaller on approaching H/D ratio of 1.55. Another interesting observation is that for the blade downstream of the flow the vortices are separated for H/D ratios of 1.13 and 1.55 but are attached to the blade in case H/D ratio of 1.31. Also not much variation in velocity can be found in the vortex region between the approaching blades in case of H/D ratio of 1.31. However a variation in velocity is observed for H/D ratio of 1.13 and 1.55.

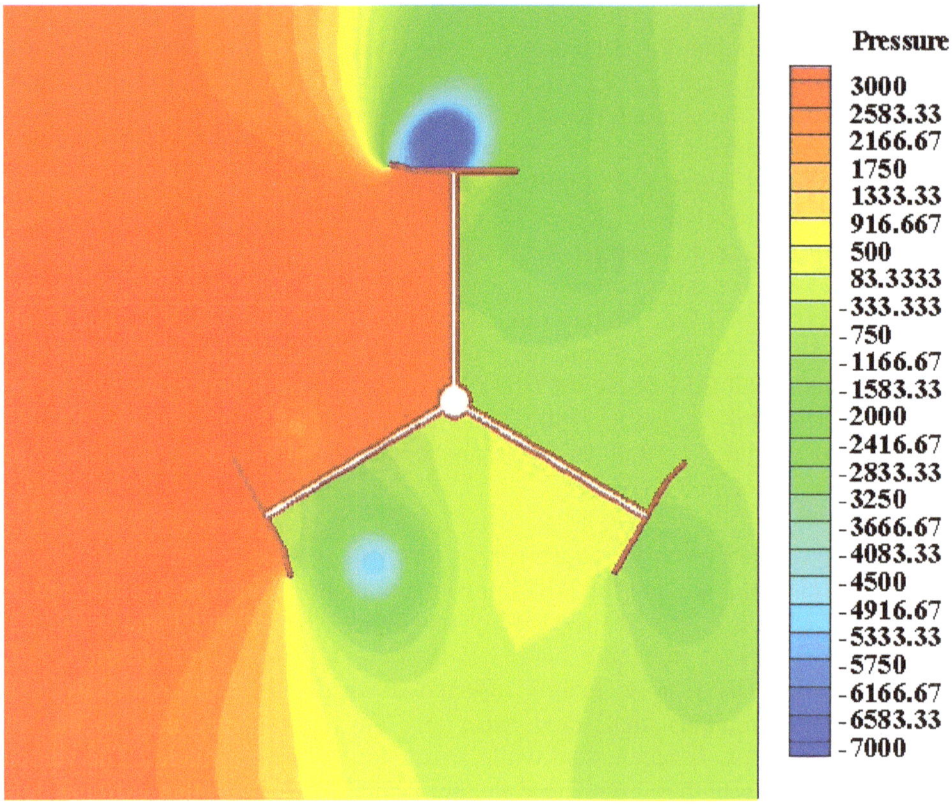

Figure 20. Pressure (Pa) contours for H/D=1.13

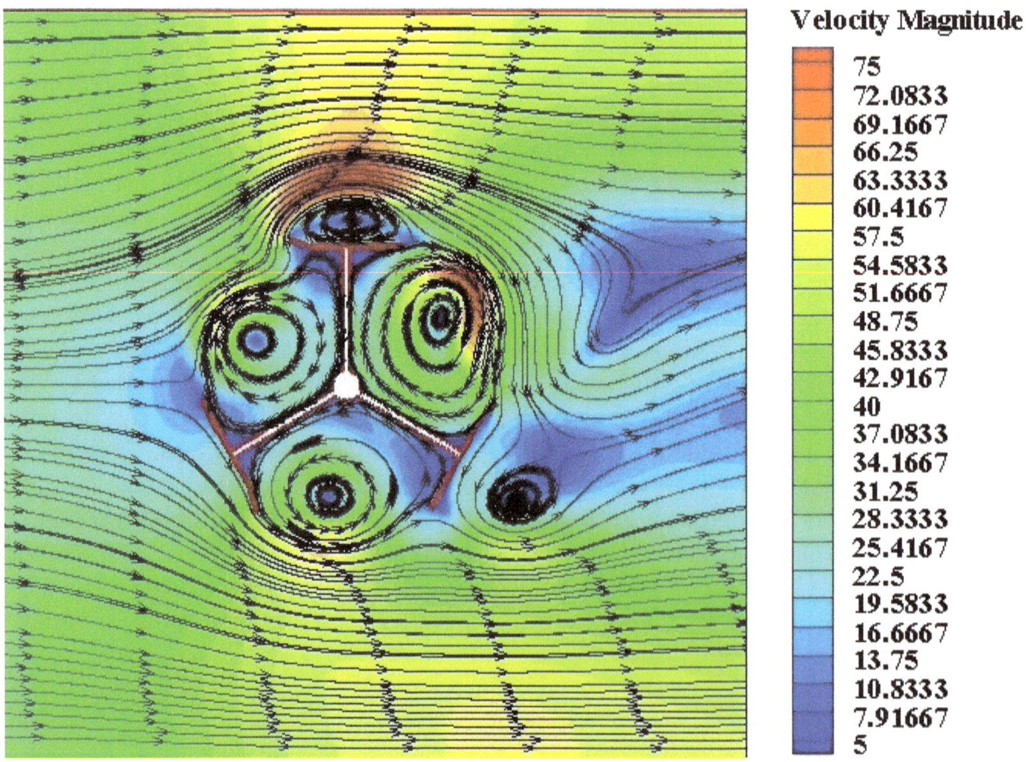

Figure 21. Contours of velocity magnitude (m/s) along with streamlines for H/D=1.13

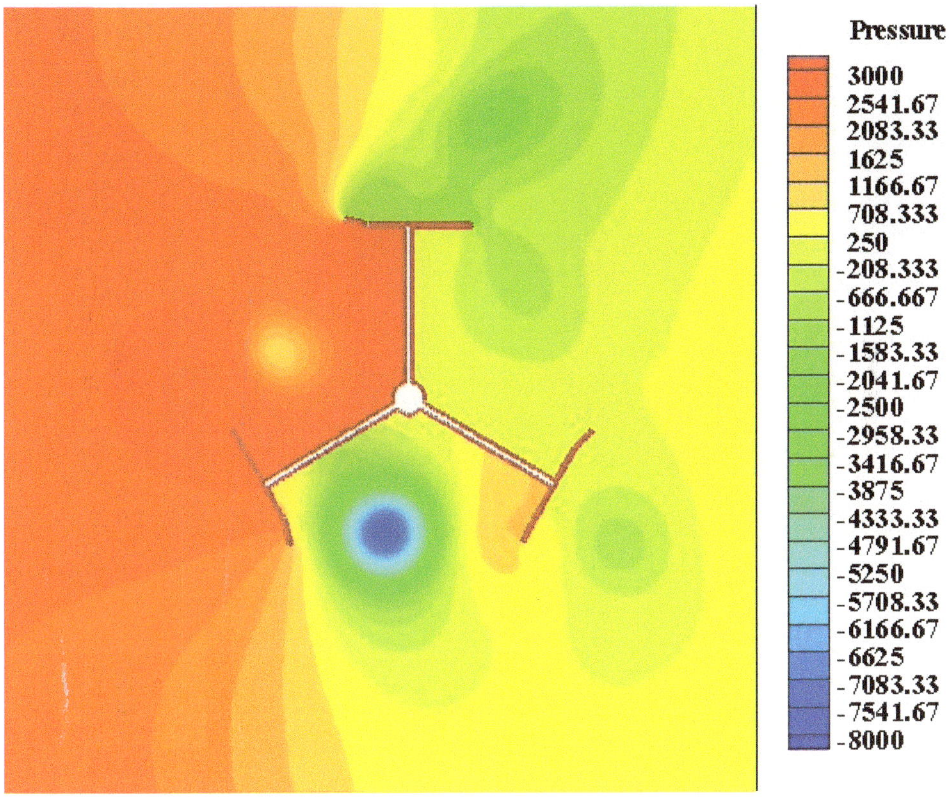

Figure 22. Pressure (Pa) contours for H/D=1.31

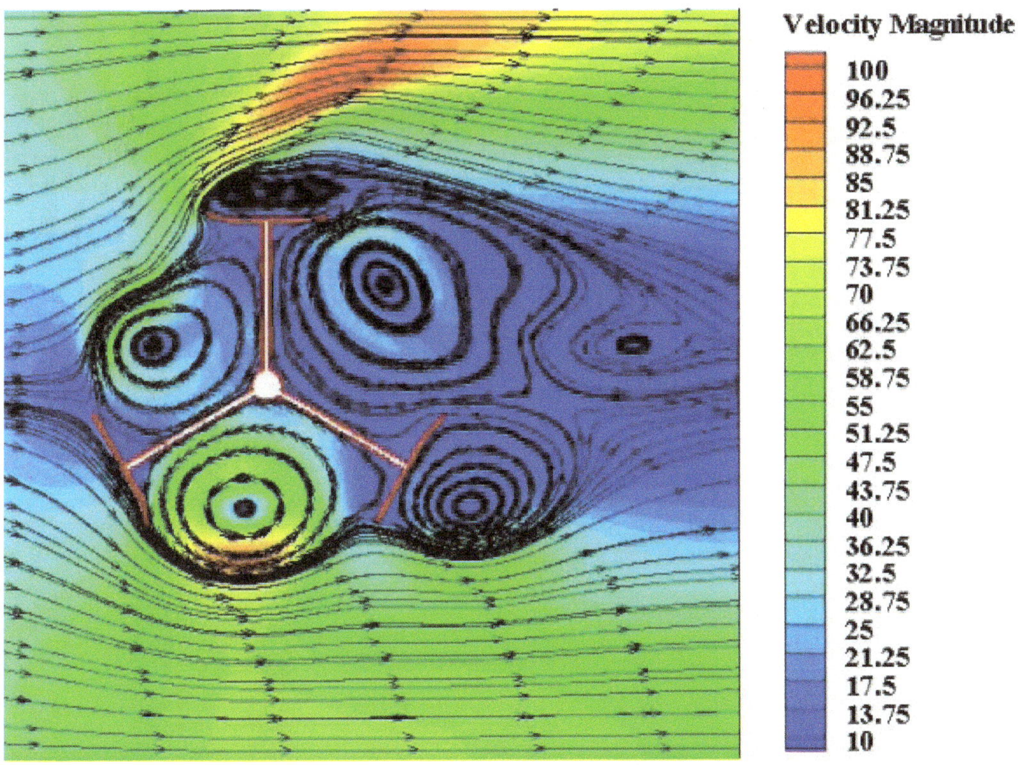

Figure 23. Contours of velocity magnitude (m/s) along with streamlines for H/D=1.31

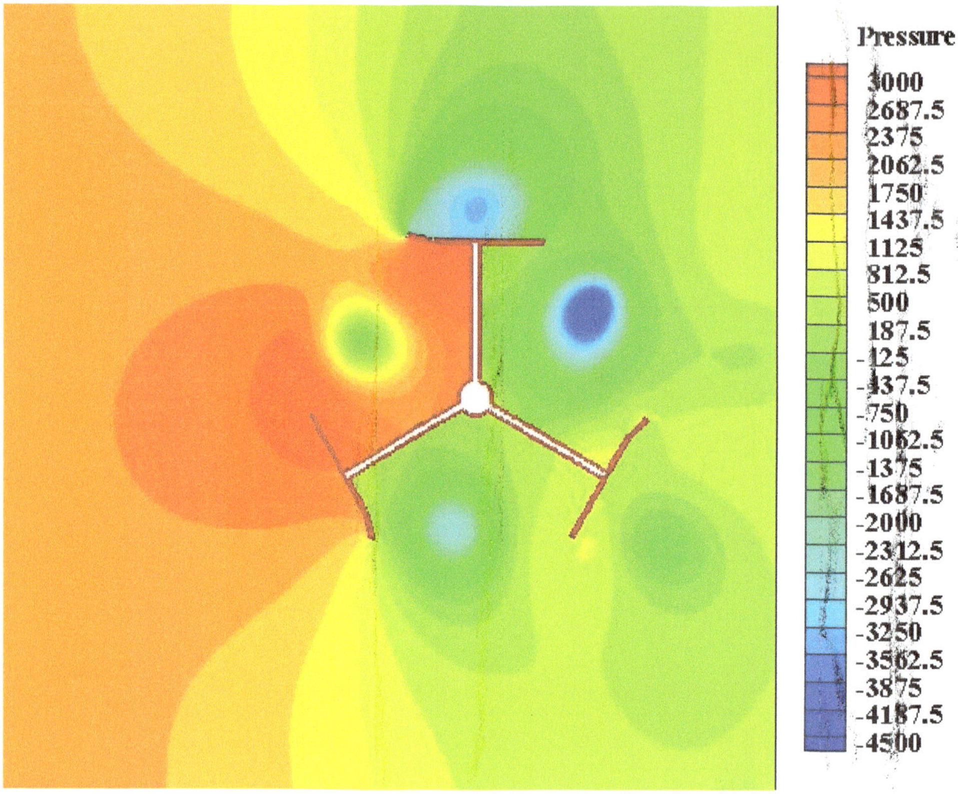

Figure 24. Pressure (Pa) contours for H/D=1.55

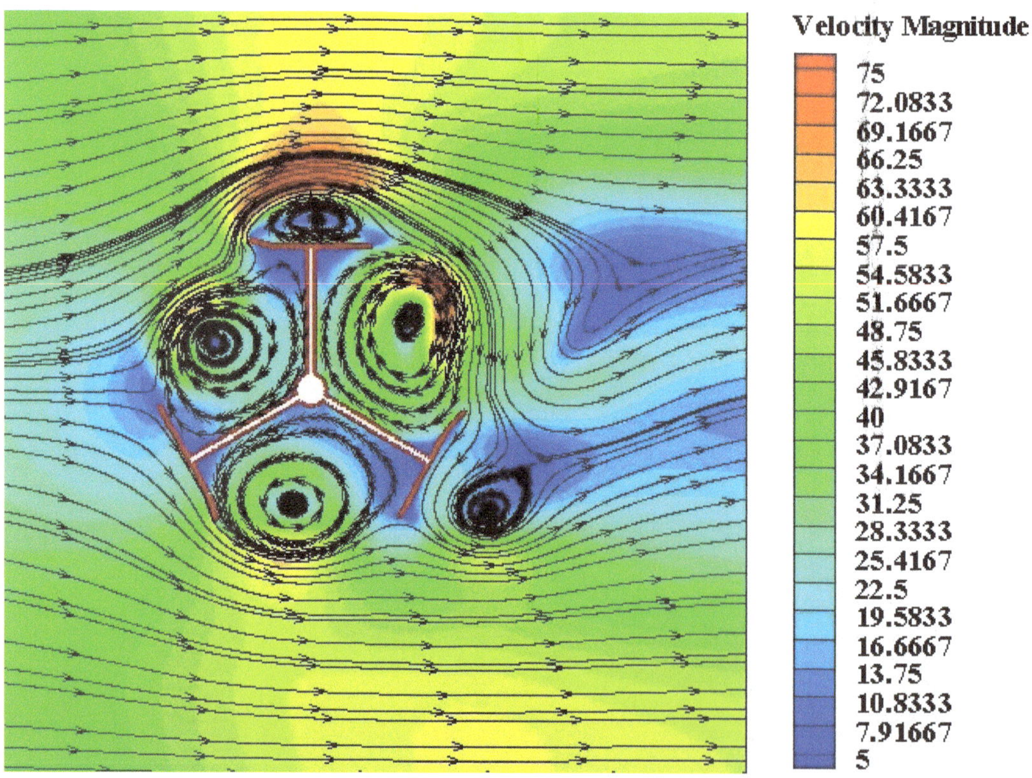

Figure 25. Contours of velocity magnitude (m/s) along with streamlines for H/D=1.55

5. Conclusion

The power coefficients and torque coefficients of the H-Darrieus rotor were evaluated numerically using CFD for three different H/D ratios. A comparative study was followed between the experimental and computational results. It was found that there was a close conformity between the computational and experimental results. The deviation of computational C_p from experimental C_p was within ±3.08%, whereas the deviation of computational C_t from experimental C_t was within ±1.106%. These findings highlight the potential of CFD as a tool for the performance measurement of vertical axis wind rotor and go a step forward in establishing the fact that the standard k- ε turbulence model can predict the flow behaviour in steady state analysis of VAWTs with a 2D rotor model (for a Re 3×10^5 of the flow) within ±3.08%. From the contour plot analysis, it is observed that velocity at the blade tips is quite high compared to the input velocity. With increase in H/D ratio from 1.13 to 1.31 the velocity drop across the rotor increases and becomes the maximum at H/D of 1.31 thereby suggesting maximum performance at this H/D ratio. But beyond H/D ratio of 1.31 the velocity drop and hence the power extraction decreases as being clearly evident from the rotor with H/D ratio of 1.55. The k epsilon model had also been able to closely predict the formation of the vortices near the blade. In case of the rotor with H/D ratio of 1.31, an attachment of the vortices is seen near the rear blade on the downstream side of the rotor. These are very strong vortices and an extremely low pressure region is formed in the centre of the vortices. The generation of the vortices improves in the lift coefficient of the rotor. Due to the increased lift coefficient an increment in the power coefficient of the rotor with ratio of H/D ratio is seen.

Appendix:

Solution specifications

Solver:

- Solver: Pressure Based
- Time: Steady
- Space: 2D
- Formulation: Implicit

Viscous Model:

- Model: Standard k-epsilon (k-ε)
- Near-Wall Treatment: Standard Wall Functions

Material: Air ($\mu = 1.7894\times10^{-5}$ kg/m-s, $\rho = 1.225$ kg/m^3)

Operating conditions: Atmospheric Pressure (1.0132 bar)

Boundary Conditions:

- Inlet: Velocity Inlet
- Sides: Symmetry
- Rotor Blades: Moving Wall
- Outlet: Outflow

Solution controls:

- Pressure Velocity Coupling: SIMPLE
- Under Relaxation Factor: 0.7 (Momentum)
- Discretization: Momentum (Second Order Upwind) for pressure velocity coupling, Pressure (Standard)

Solution Initialization: Inlet condition

Residual Monitors: 0.001 (continuity, x-velocity, y-velocity, k, ε)

Force Monitors: Drag Coefficient

Inlet Velocity: 31.3 m/s

References

[1] G.J.M Darrieus, Turbine having its rotating shaft transverse to the flow of the current, US Patent no. 1.835.018, 1931.
[2] S. Mertens and E. de Vries, "Small Wind Turbines: Driving Performance", Renewable Energy World, 11: Issue 3, 2008.
[3] A. Iida, K. Kato, A. Mizuno, in Numerical Simulation of Unsteady Flow and Aerodynamic Performance of Vertical Axis Wind Turbines with LES: Proceedings of the 16th Australian Fluid Mechanics Conference, Crown Plaza, Gold Coast, Australia, 2-7 December 2007.
[4] D.P. Coiro and C. de Nicola, in Prediction of Aerodynamic Performance of Airfoils in Low Reynolds Number Flows: Proceedings of the Low Reynolds Number Aerodynamics Conference, Notre Dame, Indiana, USA, 5-7 June,1989.
[5] L. P. Fernando, M. J. Pablo, A vortex model for Darrieus turbine using finite element techniques, Journal of Renewable Energy, pp 1–18, vol.24, 2001.
[6] R. Howell, N. Qin , J. Edwards, N. Durrani, Wind tunnel and numerical study of a small vertical axis wind turbine, Renewable Energy Journal, 35, 412-422 (2010).
[7] J.Kakati and H.Kakati, Study of the performance of H-Darrieus Wind turbine in both normal aned skewed flow conditions, B.Tech Thesis, NIT Silchar, India, 2007.
[8] R. Gupta, R. Das, K. K. Sharma, Experimental study of a Savonius-Darrieus wind machine: Proceedings, International Conference on Renewable Energy for Developing Countries, Washington DC, USA, 2006.
[9] Bravo R., Tullis S., Ziada S. Performance Testing of a Small Vertical-Axis Wind Turbine. Mechanical Engineering Department, McMaster University
[10] Takao M., Kuma H., Maeda T., Kamada Y., Oki M., Minoda A. A Straight-bladed Vertical Axis Wind Turbine with a Directed Guide Vane Row - Effect of Guide Vane Geometry on the Performance - Journal of Thermal Science Vol.18, No.1 (2009) 54–57
[11] Ferziger J.H., Peric M. Computational methods for fluid dynamics. Springer.
[12] T. J. Barth and D. Jespersen, The design and application of upwind schemes on unstructured meshes, Technical Report AIAA-89-0366, AIAA 27th Aerospace Sciences Meeting, Reno, Nevada, 1989.
[13] S.V. Patankar, Numerical Heat Transfer and Fluid Flow. Hemisphere, Washington, D.C, 1980. ISBN 0-89116-522-3.
[14] B. E. Launder and D. B. Spalding, The Numerical Computation of Turbulent Flows, Computer Methods in Applied Mechanics and Engineering, 3pp. 269-289, 1974.
[15] R.Gupta, A.Biswas, K.K.Sharma, Comparative study of a three-bucket Savonius rotor with a combined three-bucket Savonius–three-bladed Darrieus rotor, Journal of Renewable Energy, (2008).
[16] R. Gupta, S. Roy, A. Biswas, Computational fluid dynamics analysis of a twisted airfoil shaped two-bladed H-Darrieus rotor made from fibreglass reinforced plastic (FRP), International Journal of Energy and Environment, ISSN 2076-2909 (Online) Volume 1, pp.953-968 (Issue 6, 2010).

15

Fabrication and characteristics of 8YSZ/Ni functionally graded materials by applying spark plasma sintering procedure

M. S. EL-Wazery

Department of Production Engineering and Mechanical Design, Faculty of Engineering, Menoufiya University, Shebin El-Kom, Egypt.

Abstract
Functionally graded materials (FGM) in the form of layered structure consisting of yttria stabilized zirconia (YSZ) and nickel were fabricated by spark plasma sintering procedure. The relative density, linear shrinkage and Vickers hardness of each layer of graded materials were measured. The microstructure and the composition of these components were studied. The results obtained show that functionally graded materials produced by spark plasma sintering exhibited a low porosity level and consequently fully dense specimens. Also, the results demonstrate that the composition and microstructure of YSZ/Ni FGM have the expected gradient distribution. There are no macroscopic distinct interfaces in YSZ/Ni FGM due to the gradient change in components. This good continuity of microstructure can eliminate the disadvantage of traditional macroscopic interface in YSZ/Ni joint, and reflect the design idea of FGM. Vicker's hardness of YSZ/Ni is lower than that of pure zirconia (YSZ) and increases by increasing the relative density of the layer of YSZ/Ni.

Keywords: Functionally graded materials (FGM); Yttria stabilized zirconia (YSZ); Spark plasma sintering (SPS); Relative density and Microstructure.

1. Introduction

The concept of functionally graded materials (FGM), first proposed in the mid-1980s in Japan [1], is based on the development of compositional and/or microstructural gradients within the material leading towards graded properties in one, two or three dimensions [2, 3]. Compared to monolithic and composite materials, these structures offer the possibility of improving performance of components under demanding technological applications, e.g. in fuse-large surfaces of spacecraft, thermal and environmental barrier coatings for turbines and engines, super-hard cutting tools or artificial bones [4]. The zirconia/nickel FGMs can be designed to reduce thermal stresses and take advantage of the heat and corrosion resistances of ceramic and the mechanical strength, high toughness, good machinability and bonding capability of metals without severe internal thermal stresses. YSZ/Ni FGM are an ideal solution to the problem of metal–ceramic bonding [5-7].

Spark plasma sintering (SPS) is a newly developed process which makes possible sintering high quality materials in short periods by charging the intervals between powder particles with electrical energy. 8-mol% yttria-stabilized zirconia (8YSZ) is widely chosen as the electrolyte in solid oxide fuel cells (SOFCs) because of its excellent high temperature ionic conductivity and excellent mechanical properties

[8-10]. The influence of YSZ content and sintering temperature on microstructures and mechanical properties of Al_2O_3/YSZ composites was investigated by Menga et al. [11]. All samples could be fully densified at a temperature lower than 1400°C. Vickers hardness and fracture toughness of composites increased with increasing YSZ content, and the samples containing 10 wt.% of YSZ had the highest Vickers hardness of 18GPa. The mechanical properties of Si_3N_4–Al_2O_3 FGM joints with 15 layers for high-temperature applications were investigated [12]. Scanning electron microscopy (SEM) observations of fracture surfaces indicated the absence of any glassy phase at the triple points. This result was quite contrary to the previously reported 20-layer Al_2O_3/Si_3N_4 FGM samples where three-point bend testing revealed severe strength degradation at high temperatures. Shahrjerdi et al. [13] have demonstrated the functionally graded metal-ceramic composite fabricated via pressureless sintering. The pure metallic and ceramic components were Titanium (Ti) and Hydroxyapatite (HA), which were located at the ends of a cylindrical specimen. The properties of all FGM products were characterized by shrinkage, optical microscope (OM), SEM, energy dispersive spectrometry (EDX) and hardness test. The grade of the FGM material was proven by results from all recorded measurements, as well as linearity of shrinkage. The microstructure and mechanical properties of PSZ/NiCr FGMs fabricated by powder metallurgy were investigated experimentally. It was found that hardness increases and ductility decreases with the increase of partially stabilized zirconia (PSZ), which is attributed to the variation of the matrix phase from the metal to the ceramic [14]. Zhou and Li [15] have introduced the inverse homogenization for FGM microstructure design, where layered periodic base cells (PBC) were topologically optimized individually for specific graded properties. To ensure connectivity between different PBCs, heat sinks were prescribed on the PBC boundaries for maximizing their conduction. Lia et al. [16] have fabricated YSZ–FGM–NiCr joints by hot pressing process and studied their thermal properties by thermal cycling test and shear testing. The joint showed good thermal stability and good oxidation resistance up to 1000°C. Also, Mishina et al. [17] studied the fabrication method of another constituent, PSZ/AISI316L FGMs by applying SPS procedure for use in joint prostheses and their mechanical and biotribological properties were evaluated through fracture toughness, bending strength, and Vicker hardness studies.

In this article, FGMs consisting of yttria stabilized zirconia (YSZ) and Ni were fabricated by spark plasma sintering procedure. The relative density, linear shrinkage and Vickers hardness of the FGM were studied. The compositional and microstructure analysis of all samples sets was examined after the sintering process through X-ray diffraction (XRD) and scanning electron microscopic (SEM) with energy dispersive X-ray analysis (EDAX). Vickers hardness was measured by performing indentations at a load of 5 kg within a dwell time of 15 s on the synthesized samples.

2. Expermental procedure

2.1 Materials

Yttria stabilized zirconia (8 mol. % Y_2O_3) powder (TOSOH, Japan) and Ni powder (MERK, USA) were used as the raw powders to fabricate the FGMs by utilizing the powder metallurgy technique. The mean particle sizes of powders were 0.5 μm for the YSZ and 15 μm for Ni, respectively. The YSZ and Ni powders were mixed in weigh ratios of 10:0, 9:1, 8:2, 7:3, 6:4 and 5:5, respectively. The composition of each layer in graded material is shown in Figure 1, and each mixture was suspended in alcohol, milled for five hrs by horizontal ball miller. The ratio of powder to ball was 1:2 and the milling speed was 100 r.p.m. 30wt. % ethyl alcohol was added to the mixtures powders to cover the media (balls). After milling, the mixed powders were sieved using a 125 μm steel mesh to remove the remaining agglomerates. 10wt. % ethyl alcohol, as a lubricant, was added to the powders to form pure ceramic and intermediated layers. Then, the mixture powders were dried in a furnace at 70 °C for 24 hrs. The powder blends were put to form a non-graded composition or were layered so as to form graded composition in a rectangular hot work tool steel die with 35 mm length and 10 mm width. The powder compacts were pressed up to 0.5 MPa at room temperature for one minute. Hydraulic press was used to compact the powder layers once at the same pressure. The die and punch set-up used to produce the non-graded composites (NGC) and FGM are reviewed in our paper [18, 19]. Using the same pressing conditions, a larger thickness (6.5 mm) FGM green compact was also formed but under lower pressure (0.1MPa) during stacking the layers, which would be used for mechanical property testing.

In the SPS process a green compact of the FGM was placed inside the plasma electrode, Figure 2. It applies large DC pulsed current (3 kA) directly to the green compact of FGM layers. Large DC pulsed current is considered to generate sparks between the powder particles and eventual repeat of momentary ultra-high temperature (1400 °C) under vacuum for one hour. The high frequency transfers and disperses the heat phenomena throughout the FGM specimen, resulting in a rapid and thorough heat distribution, high homogeneity and especially strong bonds between particles.

100 wt. % YSZ	Layer 1
90 wt. % YSZ + 10 wt. % Ni	Layer 2
80 wt. % YSZ + 20 wt. % Ni	Layer 3
70 wt. % YSZ + 30 wt. % Ni	Layer 4
60 wt. % YSZ + 40 wt. % Ni	Layer 5
50 wt. % YSZ + 50 wt. % Ni	Layer 6

Figure 1. Composition distribution model of YSZ/Ni FGM

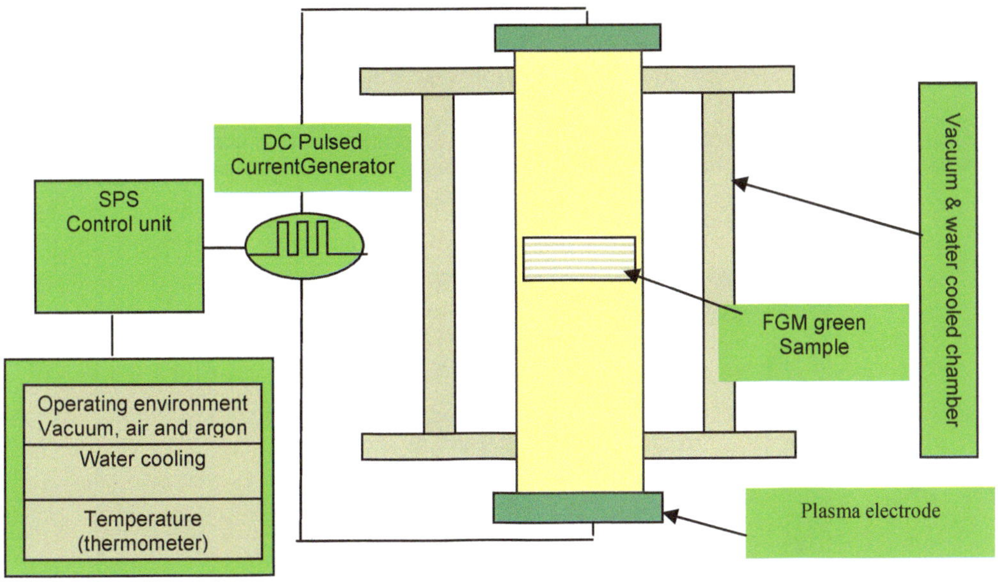

Figure 2. SPS system configuration

2.2 Materials characterization

The actual density and the theoretical density of the sintered specimens were measured by using the water immersion method (Archimedes method) and the rule of mixture, respectively. The radial shrinkage of the NGC was measured by the thermo-mechanical analyzer (TMA-50 Shimadzu-Japan). The composition and microstructures of each layer of the graded materials specimens were examined by X-ray diffraction (XRD) and the scanning electron microscope (SEM-JSM 5400) with energy dispersive X-ray analysis (EDAX). A hardness tester (HSV-40) was used to measure the Vickers hardness of each layer in the YSZ/Ni FGM.

3. Results and discussion

3.1 Microstructure and XRD

Figure 3a shows the microstructure of the fabricated YSZ/Ni FGM. It has six layers: a pure YSZ layer and five YSZ/Ni composite layers containing 10, 20, 30, 40 and 50 % of Ni, Figure 3b. The white and gray regions exhibit the Ni and YSZ phases, respectively. The white area in each composite shows the nickel phase in a grayish YSZ matrix. The white area increased with increasing Ni contents from 0 to 50 wt.% and no distinct boundaries are exhibited. It can be seen that metal particles are homogeneously distributed into the ceramic matrix and there is no evidence of agglomerates. As the particle size of the zirconia powder was considerably smaller than that of the Ni powders, zirconia particles would have entered gaps between Ni particles on the fabrication. As a result, the Ni particles are dispersed in the zirconia matrix in the material with high content of Ni (40%Ni and 50%Ni).

(a)

(b)

Figure 3. (a) The micrographs of the FGM layers (gray part is nickel in YSZ); (b) SEM images of the layers of the YSZ/Ni FGM

The X-ray diffraction patterns (XRD) from the surface of the cross-section for the NGCs (YSZ/40% Ni) after the spark plasma sintering under vacuum at 1400 °C for one hour are shown in Figure 4. No new phase was formed between zirconia and Ni during sintering process. No other impurities above the detection limit have been detected.

3.2 Densification behavior

In the powder metallurgy process, the pores can be treated as the initiate defects. The relative densities of the layers in homogenous specimens and the YSZ/Ni FGM compacts are shown in Figure 5. Spark plasma sintering is an effective method to produce full dense specimens. It is noted that relatively high density (in the range of 0.90–0.99) for each layer in the FGM was obtained after the spark plasma sintering stage. The relative density has a maximum value (99 %) for the pure YSZ and a minimum value (89.87%) for the homogeneous specimen with 50 % nickel. Among the layers, layer 1 has the highest relative density, which is mainly attributed to a small initial particle size (0.5μm) of zirconia selected for making the ceramic layer (pure ceramic layer) of the model structures. The higher relative density of zirconia is mainly due to its higher melting point, higher sinterability and hence higher sintering temperature for attaining fast densification than nickel. Among the five composite layers, the relative density decreases with increasing the content of nickel and the level of porosity is increased. Such effect can be attributed to the decrease in sintrability of Ni as well as its large particle size (15 μm). The FGMs exhibit a low porosity level and a relatively high relative density 99% by SPS procedure. The high value of the relative density of the YSZ/Ni FGM is related to the high densification, high homogeneity, rapid temperature increase and good bonding between the layers of FGM.

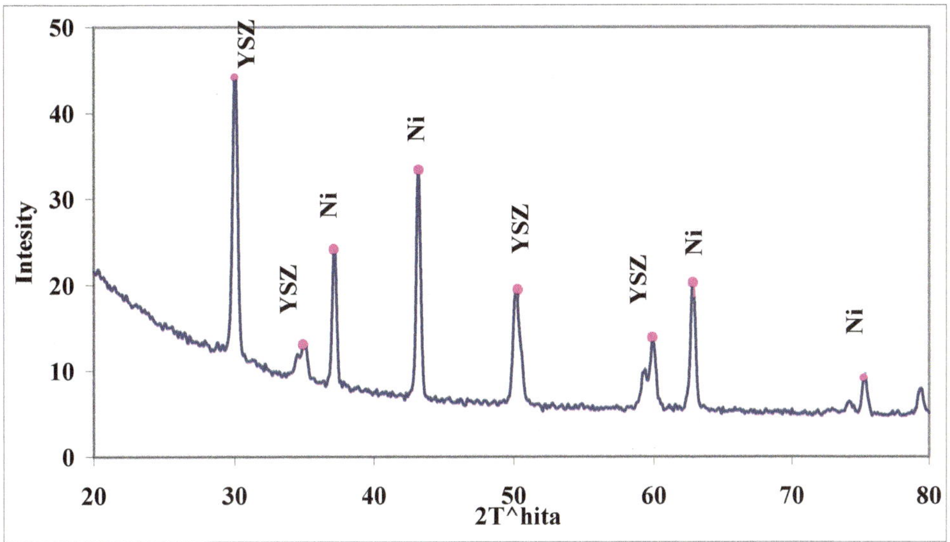

Figure 4. XRD patterns from the surface of the cross-section of the NGCs (YSZ/40%Ni)

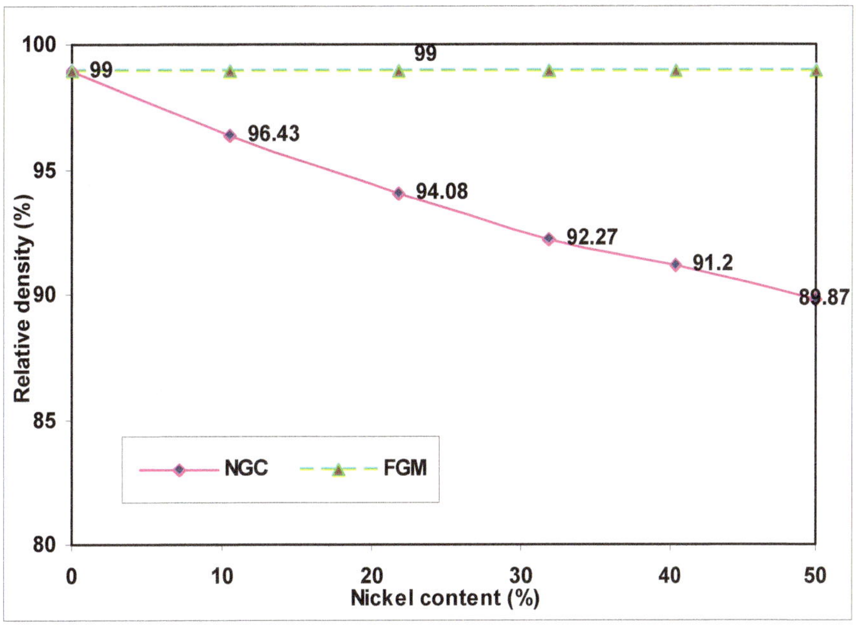

Figure 5. The relative densities of the non-graded composite and FGM specimens

An important factor, which should be considered when analyzing the sinterability of YSZ/Ni FGM is the accurate prediction of the change in thermo-mechanical properties as shrinkage at each temperature. The thermal expansions coefficient of the Ni and the YSZ were (α_{Ni} =15.4 x10^{-6} °C^{-1} and α_{YSZ} = 7 x10^{-6} °C^{-1}), respectively [20]. The TMA was carried out on the layers of the FGM in air with a heating rate of 5°C/min up to 1400 °C. To indirectly estimate the linear shrinkage of the layers in the FGMs fabricated by a spark plasma sintering procedure, the shrinkages of the NGC corresponding to each layer were measured by the thermo mechanical analyzer (TMA-50, Shimadzu-Japan) and the results were shown in Figure 6. It could be seen that the maximal value of linear shrinkage among the layers specimens is 19%

for 10%Ni/YSZ. Among the other NGC layers, the linear shrinkage decreases with increasing the content of nickel and reached the minimum linear shrinkage (7%) at 50% Ni. This is because the lower sinterability of nickel compared to YSZ.

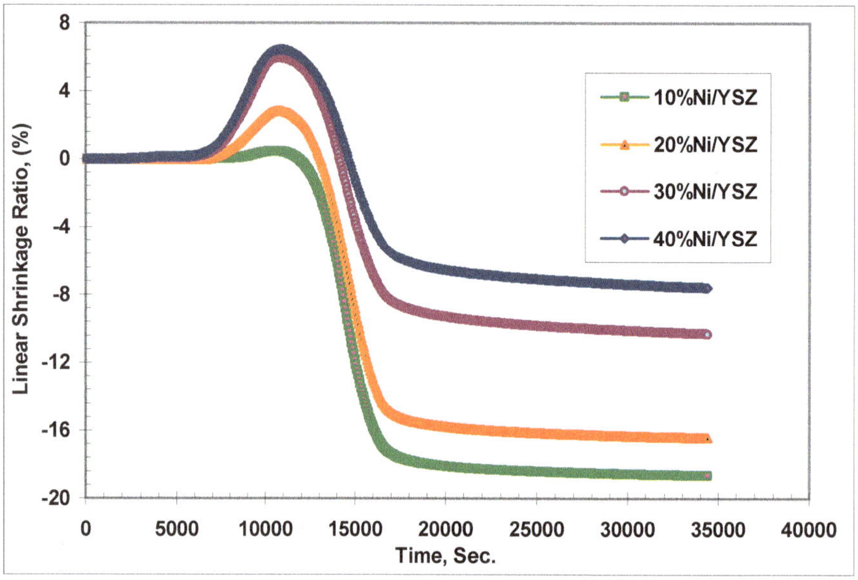

Figure 6. Relation between linear shrinkage and time of the FGM layers

3.3. Mechanical properties

The Vickers hardness of the YSZ/Ni FGM was measured using a hardness tester (HSV-40) on a polished surface under a load of 5 kg at 30 sec. The average of five indentations in each layer at the center and was used to calculate the hardness values of the layers. The Vickers hardness has a maximum value (15GPa) for the pure YSZ and a minimum value (5GPa) for the homogeneous specimen with 50 % nickel (Figure 7).

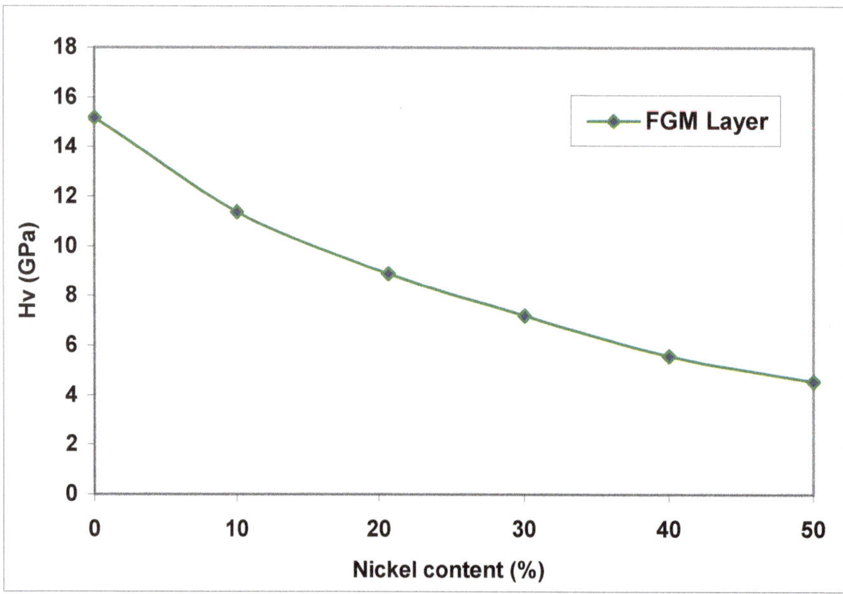

Figure 7. Vickers hardness of the layers in YSZ/Ni FGM

Also, it should be noted that the vickers hardness decreased with increasing nickel content. This is mainly attributed to the pores in intermediate layers in the FGM after sintering stage, where the pores were increased in the FGM layers with increasing nickel content from 0% to 50% Ni and the relative density was decreased also. Furthermore, hardness was reduced as the transformation from pure ceramic layer (YSZ) to intermediate layers (mixtures of zirconia and nickel powders) took place.

4. Conclusions

The spark plasma sintering (SPS) seems to be an effective method to obtain the full dense specimens. The FGMs exhibit a low porosity level and high relative density (99%). There are no macroscopic distinct interfaces in YSZ/Ni FGM due to the gradient change in components. This good continuity of microstructure can eliminate the disadvantage of traditional macroscopic interface in YSZ/Ni joint, and reflect the design idea of FGM. The linear shrinkage of the NGC was reduced with an increase in nickel content. There are more dimples in the microstructure resulted from agglomeration of nickel (metal) leaved from place in 50% NGC. Vicker's hardness of YSZ/Ni is lower than that of pure zirconia (YSZ) and increases by increasing the relative density of the layer of YSZ/Ni.

References

[1] Yamanouchi, M., Hirai, T., and Shiota, I., Proceedings of the First International Symposium on "Functionally Gradient Materials", FGM Forum, Tokyo, 1990, Japan.

[2] Neubrand, A., and Rodel, J., "Gradient materials: an overview of a novel concept", Z Metallkd 88 (1997), 358–371.

[3] Kieback, B., Neubrand, A., Riedel, H., "Processing technique for functionally graded material" Material Science Engineering and Structure 2003; 362:81.

[4] Miyamoto, Y., Kaysser, A., Rabin, B., Kawasaki, A., "Functionally graded materials: design processing and applications", Kluwer, Dordrecht, the Netherlands, Ford R G, 1999.

[5] Wosko, M., Prazmowska, J., "AIII-BV(N) Photodetectors with functionally graded active area " Proceedings of the Symposium on Photonics Technologies for 7th Framework Program Wroclaw, 12-14 October 2006.

[6] Kawasaki, A., Watanabe, R., "Thermal fracture behavior of metal/ceramic functionally graded materials ", Engineering Fracture Mechanics 69 (2002), 1713–1728.

[7] Ruys, J., Popov, B., Sun, D., Russell, J., Murray, J., " Functionally graded electrical/thermal ceramic systems" Journal of the European Ceramic Society 21 (2001), 2025–2029.

[8] Matula, G., Jardiel, T., Jimene,z R., Levenfeld, B., Várez, A., "Microstructure, mechanical and electrical properties of Ni-YSZ anode supported solid oxide fuel cells" International Journal of Materials Science and Engineering, 32 (2008), 21-25.

[9] Rajeswari, K., Buchi, M., Hareesh, U., Srinivasa, Y., Das, R., Johnson, R., "Studies on ionic conductivity of stabilized zirconia ceramics (8YSZ) densified through conventional and non-conventional sintering methodologies"' Ceramics International 37 (2011), 3557–3564.

[10] Yu, J., Park, W., Lee, S., Woo, K., "Microstructural effects on the electrical and mechanical properties of Ni-YSZ cermets for SOFC anode. Power Sources 163 (2008), 921- 926.

[11] Menga, F., Liua, C., Zhangb, F., Tiana, Z., Huanga, W., "Densification and mechanical properties of fine-grained Al2O3–ZrO2 composites consolidated by spark plasma sintering" Journal of Alloys and Compounds 512 (2012), 63–67.

[12] Lee, S., Lemberg, A., Cho, G., Roh, Y., Ritchie, O., " Mechanical properties of Si3N4–Al2O3 FGM joints with 15 layers for high-temperature applications" Journal of the European Ceramic Society 30 (2010), 1743–1749.

[13] Shahrjerdi, A., Mustapha, F., Bayat, M., Sapuan, S., Majid, A., "Fabrication of functionally graded hydroxyapatite-titanium by applying optimal sintering procedure and powder metallurgy" International Journal of the Physical Sciences Vol. 6(2011), 2258-2267.

[14] Wu, Li., Jin, X., Sun, Y., Guo, L., "Microstructure and mechanical properties of ZrO2/NiCr functionally graded materials" , Materials Science and Engineering A 509 (2009) 63–68.

[15] Zhou, S. and Li, Q., "Microstructural design of connective base cells for functionally graded materials", Materials Letters 62 (2008) 4022–4024

[16] Lia, J., Zenga, X., Tanga, N., Xiaob, P., " Fabrication and thermal properties of a YSZ–NiCr joint with an interlayer of YSZ–NiCr functionally graded material" Journal of European Ceramic Society 23 (2003), 1847–1853.

[17] Mishina, H., Inumaru, Y., Kaitoku, K., "Fabrication of ZrO2/AISI316L functionally graded materials for joint prostheses" , Materials Science and Engineering A 475 (2008) 141–147.

[18] EL-Wazery, S. M., EL-Desouky, R. A., Hamed A. O., Mansour A. N. and A. A., Hassan, " Fabrication and mechanical properties of zirconia/nickel functionally graded materials" International Journal of Advanced Materials Research, Vols. 463-464(2012), 463-47.

[19] A. R. EL-Desouky, M. S. EL-Wazery,, "Mixed Mode Crack Propagation of Zirconia/ Nickel Functionally Graded Materials " International Journal of Engineering (IJE), Volume 26, No. 8, (August) 2013, PP: 885-894, Citation Indexed: SCOPUS, Available Online , www. ije.ir.

[20] Jingchuan, Z., Zhongda, Y., Zhonghong, L., Jian, L., " Microstructure and thermal relaxation of ZrO2/Ni functionally graded materials" Trans. Nonferrous Met. Soc. China, 6 (1996), 94-99.

16

Wind energy potential of Marmara region in Turkey

Oktay Arikan[1], Evren Isen[2], Cagri Kocaman[3], Bedri Kekezoglu[1]

[1] Department of Electrical Engineering, Yildiz Technical University, Esenler, Turkey.
[2] Department of Electrical & Electronics Engineering, Kirklareli University, Kirklareli, Turkey.
[3] School of Civil Aviation, Ondokuz Mayis University, Samsun, Turkey.

Abstract
Turkey is one of the developing countries in the world. According to economical, industrial and population growth, electrical energy demand has valuably increased in the last decade. The continually increase of electrical energy demand and global sensitivity to environmental pollution raise the importance of renewable energy sources. Wind energy has become more popular than other renewable energy sources owing to its advantages such as supplying great power, being sustainable and clean energy resource. In this study, wind energy potential analysis of Marmara region which is one of the Turkey's geographical regions is presented. Due to its high energy demand, crowded population, large industrial areas and attractive wind energy potential, this paper focused on the Marmara region. Installed wind energy conversion system (WECS) power is 923.65 MW and available WECS power is 46,996.28 MW in Marmara region. It is obviously seen that only 1.96 % of the existent capacity is utilized. A valuable increase in installed capacity could provide an attractive opportunity to decrease the import energy resources, transmission losses and greenhouse gas (GHG) emission of Turkey.

Keywords: Wind energy; Marmara region; Renewable energy sources; Turkey.

1. Introduction

Limited reserves of fossil fuels and their negative effects on the environment have led institutions, organizations and governments to look for new technologies. Rise in the prices of fossil fuels, environmental pollution and undesired climate changes began to affect the world [1, 2]. Besides, it is known that renewable electrical energy, which is obtained from natural resources, has negligible contribution to environmental pollution [3, 4]. Therefore, many countries trend to renewable energy sources such as solar, wind, biogas and geothermal. Also, especially at the rural areas, renewable energy has become an economical electrical power resource for most of the countries. When compared with the cost of providing electrical energy from national network by using transmission lines, establishment of renewable energy systems are more economical for these regions. Also, the numbers of renewable energy suppliers are increasing consistently with the developments in this technology.

In the last decade, limited reserves, harmful effects on environment and high costs of fossil fuels have accelerated usage of renewable energy systems [5-8]. Wind energy has become more popular than other renewable energy sources owing to its advantages such as supplying great power, being sustainable and clean energy resource. Developments on wind energy systems bring requirements for solving the fast and accurate project producing problems. Accurate and reliable wind data is very important during the planning and designing of a wind energy system. Development of new wind projects in many countries are adversely affected from lack of accurate wind data and reliability studies. This data is obtained by giving the authority to governments or private institutions [7].

Installed wind turbine capacity in the world is 296.255 GW in the middle of 2013 and installed capacity is aimed to be 318 GW until the end of the year. While China takes the first place in the use of wind energy with 80.824 GW installed wind power, United States of America and Germany are in the second and third, respectively [9].

The installed wind power capacity of European Union (EU) is 106.040 GW by the end of 2012 and 6.3% of the total electrical energy requirement was provided from wind energy in 2011. It is targeted to generate 20% of electricity demand in the Europe from WECSs at 2020. When installed wind power within the borders of the European Union is examined, Germany, Spain and United Kingdom are at the first three places [10].

Social, economic and industrial developments of Turkey increase the electrical energy demand. The primary energy sources for electricity generation in Turkey are coal (28.4%), natural gas (43.6%), hydropower (24.2%), geothermal + wind (2.8%), liquid oil (0.7%) and others (0.3%) in 2012. Total installed power capacity is 64,044.10 MW in Turkey according to 2014 data [11]. Installed power capacity of Turkey according to production type at the end of 2013 is given in Figure 1 [12].

As large oil and natural gas reserves do not exist in the country, most of these requirements are largely met by imports [3, 13]. There are efficient energy resources such as coal, hydraulic, solar, thermal, wind, biomass, etc. in Turkey. Although, national hydro and lignite resources are mainly used for electrical energy production, this production is insufficient to supply electrical energy requirements of the country. Dependence on foreign resources (natural gas and liquid oil) increases due to the growing electrical energy demands [1, 13].

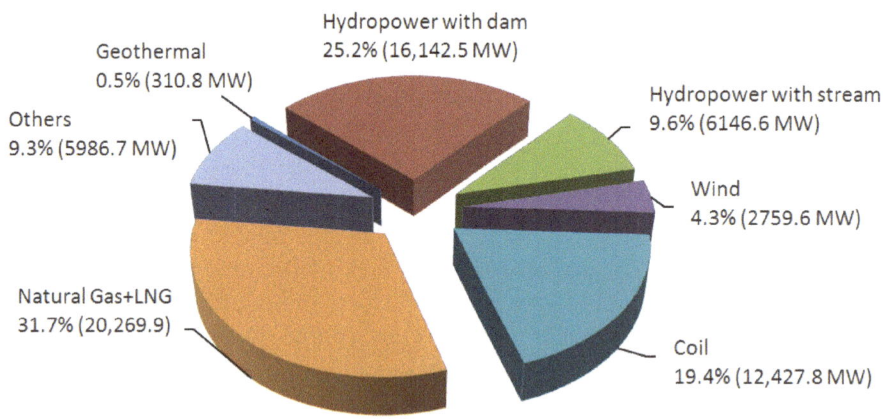

Figure 1. Installed power capacity according to production types

Nowadays, energy status of Turkey has some fundamental problems such as dependence on foreign sources, environmental pollution, inefficient energy usage and low electrical energy production from renewable energy sources [3, 14]. Installation of renewable energy systems is required for obviating dependence on foreign resources and reducing the percentage of fossil fuels at the primary energy consumption of Turkey [3].

Usage of renewable energy sources and technological developments are important for sustainable and environmental economic development of Turkey. Therefore, low utilization percentage of renewable energy sources, especially wind and solar energy is promoted by new regulations.

Besides electrical energy demands, one of the most important problems that affects the humankind is carbon emission. Many cities try to reduce carbon emission, and determine future targets for carbon emission levels as given in Table 1 [15].

Although wind energy conversion systems produce low carbon emission during production and installation, this value can be neglected compared with the fossil fuels [16]. The effects of carbon emission to atmosphere and climate change make governments realize the importance of requirement low carbon emission, thus they started to give economic incentives for this propose [17].

The most crowded and industrialized region in Turkey is Marmara region. Therefore, carbon emission increases in the region, continuously. High wind energy potential is an important advantage to reduce the CO_2 emission for this region.

Table 1. Long-term carbon reduction targets for cities of the world

City	Target year (base year)	Reduction target (%)
Copenhagen	2025 (2005)	Carbon neutral
London	2025 (1990)	-60%
Boston	2050 (1990)	-80%
Melbourne	2020 (1996)	Carbon neutral
Sydney	2050 (1990)	-70%
Toronto	2050 (1990)	-80%
Stockholm	2050	Carbon neutral

Geographical location of Turkey has advantages for the usage of renewable energy sources. The cumulative installed wind power is over 2 GW in 2012 and approximately 3.5% of energy production is obtained from wind energy in Turkey [18]. When the geographical regions of Turkey are taken into consideration, it can be recognized that Marmara region has the highest average wind speed and the largest wind energy potential. Also, this region takes the first place with 923.65 MW in terms of installed wind power capacity. The annual average wind data shows that Marmara, Aegean and Southeast Anatolia Region are suitable for wind energy utilizations [1].

In this study, wind energy potential of Marmara region in Turkey is investigated. The reasons for selection of this region are its high energy demand, crowded population, large industrial areas and attractive wind energy potential.

Additionally, large hydropower plants of Turkey placed at the eastern part of the country. Transferring electrical energy from the hydropower plants to the Marmara Region, which is located at western part and have highest population density and industrialization, brings transmission losses.

Also, remarkable amount of primary electrical energy sources of Turkey are depends on fossil fuels. According to 2011 data, percentages of gas, coal and liquid oil are 43.6%, 28.4% and 0.7%, respectively [11]. Most of these resources are imported and have negative effect on Turkish economical development. Taking into account all of these factors, it is quite obvious that utilization of wind energy potential of Marmara Region would have favorable effect on decreasing transmission losses and imported primary energy resources. Moreover, generated electricity from an environmental source will have a major contribution on reducing the greenhouse gas (GHG) emission.

2. Wind energy in Turkey

Depending on industrialization and population growth, electrical energy demand is continually increasing in Turkey. As conventional energy resources used for electrical energy generation (hydro, coal, etc.) are insufficient, energy dependence of Turkey to foreign countries is increasing.

Rise of energy demand, dependence to foreign countries for energy resources and environmental pollution problems are cause of rapidly increasing interest in renewable energy at Turkey. The investigations show that Turkey has large renewable energy sources. Therefore, researches and investments at this field are increasing in recent years.

The investments in Turkey show that wind energy is the distinguished renewable energy source. Installation of wind energy conversion systems (WECS) started in 1990s and continues rapidly in Turkey. Having high power potential and fast installation are the main reasons for the popularity of WECSs. The development of WECSs in Turkey is shown in Table 2 [19].

The data given in Table 2 show that power production from wind energy started in the late 1990s and continues without interruption in recent years. The investments are raised particularly after 2006 and the largest investment was made in 2007.

The wind map of Turkey for 10 m height is given in Figure 2. Turkey is divided into seven regions geographically and as seen from the figure, the average wind speed above 3 m/s occurs in four regions. Marmara region is particularly suitable for installation of WECS.

Table 3 shows the annual average wind speed and wind power values for geographical regions of Turkey [6, 18, 19]. It reveals that Marmara region has the highest annual average wind speed, annual average wind density. Marmara, South-eastern Anatolia and Aegean regions are at the first three ranks when compared with other regions in terms of the wind intensity.

Table 2. Development of WECSs in Turkey depending on years

Year	Cumulative installed power (MW)	Rate of change (%)	Cumulative rate of change (%)
1998	8.7	0.0	0.0
1999	8.7	0.0	0.0
2000	18.9	117.2	117.2
2001	18.9	0.0	117.2
2002	18.9	0.0	117.2
2003	20.1	6.3	131.0
2004	20.1	0.0	131.0
2005	20.1	0.0	131.0
2006	51	153.7	486.2
2007	146.3	186.9	1581.6
2008	363.7	148.6	4080.5
2009	791.6	117.7	8998.9
2010	1329.15	67.9	15,177.6
2011	1805.85	35.9	20,656.9
2012	2312.15	13	26,476.43

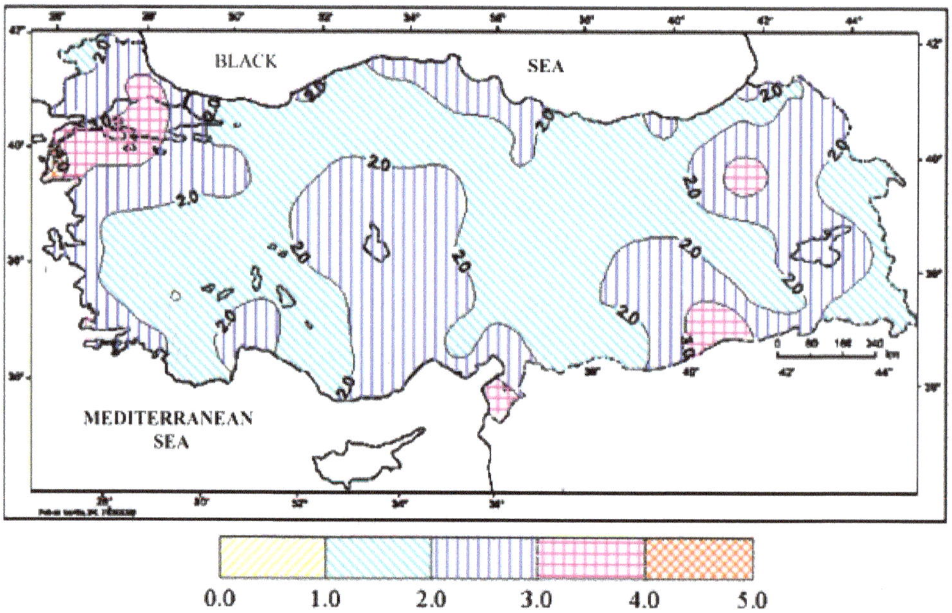

Figure 2. Wind map of Turkey for height of 10m [20]

Table 3. Wind energy potential in Turkey

Region	Annual average wind speed (m/s)	Annual average wind density (W/m^2)	Installed wind power (MW)	Available wind power (MW)
Marmara	3.29	51.91	923,65	46,996.28
SouthEastern(SE) Anatolia	2.69	29.33	-	2630.88
Aegean	2.65	23.47	852	29,228.96
Middle (M) Anatolia	2.46	20.14	72	7824.72
Mediterranean	2.45	21.36	384.50	13,286.08
Black sea	2.38	21.31	80	14,302.64
Eastern (E) Anatolia	2.12	13.19	-	2983.92

Analysis of the annual energy demand in Turkey shows that approximately 40% of the energy is consumed in Marmara region [21]. Maximum usage of wind energy potential in this region will provide beneficial effects, such as reducing transmission losses, decreasing fossil fuel utilization and importing primary energy sources, increasing power capacity of the region...etc.

Status of installed, under constructed and licensed WECSs according to the regions of Turkey are shown in Figure 3 [19]. It is clearly seen that Marmara region stands out in terms of under constructed and licensed WECSs. Due to high wind power potential and energy demand, WECS investments increase in this region. It is the reason why Marmara region is investigated in this paper.

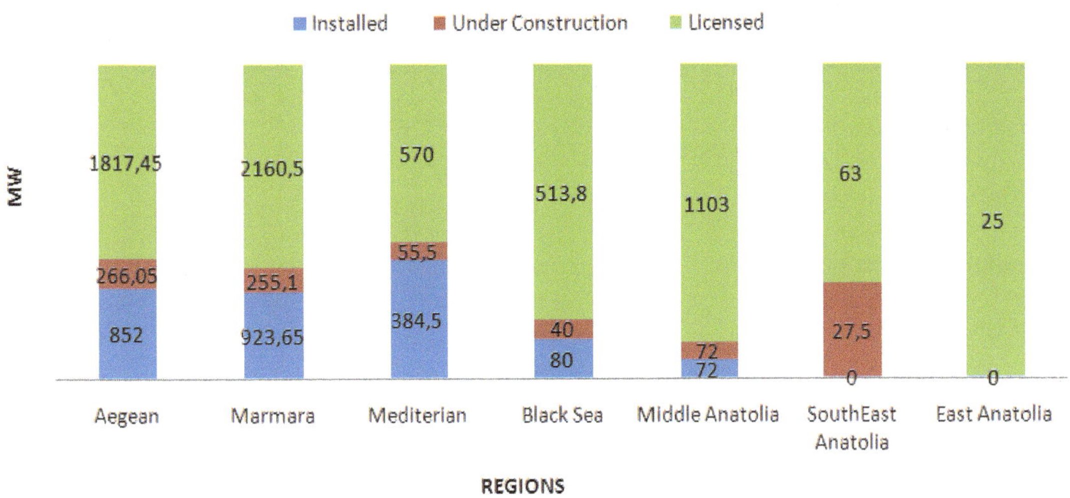

Figure 3. Status of WECSs in seven regions of Turkey

3. Wind Energy Potential in Marmara Region

Marmara region has the highest population density and developed industrialization in Turkey. Therefore, Marmara takes the first place in terms of energy needs. When Turkey's electricity generation is taken into consideration, it is obviously seen that considerable part of production is carried out at the large hydropower and fossil fuel power plants. This fact shows that the electrical energy demand of Marmara region have to be supplied from the hydropower plants or fossil fuel power plants. Both of these options have objections for the economical, sustainable and environmental growing of Turkey.

Large hydropower plants takes place at the eastern part of the country and electrical energy transmission is provided with long transmission lines to the Marmara region which takes place at the western location. Therefore, transmission losses are unavoidable. Also, hydropower potential is inadequate to provide the energy demand in Turkey.

Fossil fuel power plants consume natural gas, liquid oil and coal for electricity generation. Due to the lack of oil, gas and efficient coil reserves, Turkey is importing fossil fuels from other countries. Additionally, generating electrical energy brings economical and environmental problem together. Therefore, renewable energy sources have a great importance for Turkey. Nowadays, wind energy is one of the most popular alternative energy sources with the supports of the government.

Marmara region have the highest available wind power potential and this region is candidate for being the country's most important wind energy production area. As Marmara region has the largest electrical energy demand, significant wind energy potential is important for meeting the electricity requirements. In addition, providing the electrical energy from wind potential will bring an important decrease for import primary energy sources and transmission losses.

There are eleven cities in the border of Marmara region. Wind data for each city are given in Table 4 [18]. It is known that required average wind speed is assumed 7 m/s for economical wind energy conversion system. Table 4 is presenting the appropriate areas and power potentials in accordance with the wind speed data at 50 m height.

It can be clearly seen that all cities of the region are suitable for economical WECS establishment. However, particularly the cities of Balıkesir and Çanakkale have higher wind potential among the other cities. Balıkesir is the only city that has more than 9.5 m/s wind speed with 16.72 km^2 suitableareas. The most proper city for economical WECS installation is Balıkesir with 2,765.5 km^2. The second one

Çanakkale has 2602.52 km^2 sufficient area and 13,012.58 MW power potential. The cities such as Istanbul and Kocaeli that have more industrialization and population density are insufficient to provide energy demand.

Table 4. The wind potential of the cities in Marmara region

Cities	Wind speed (m/s)	6.8-7.5	7.5-8.1	8.1-8.6	8.6-9.5	>9.5
Balıkesir	Area (km^2)	1511.4	850.96	284.51	115.23	3.34
	Power (MW)	7557.1	4254.8	1422.6	576.16	16.72
Bilecik	Area (km^2)	61.73	-	-	-	-
	Power (MW)	308.64	-	-	-	-
Bursa	Area (km^2)	683.6	85.22	5.23	2.29	-
	Power (MW)	3418	426.08	26.16	11.44	-
Çanakkale	Area (km^2)	863,7	802.99	761.09	174.74	-
	Power (MW)	4318.5	4015	3805.4	873.68	-
Edirne	Area (km^2)	578.24	115.78	-	-	-
	Power (MW)	2891.2	578.88	-	-	-
İstanbul	Area (km^2)	832.91	2.48	-	-	-
	Power (MW)	4164.6	12.4	-	-	-
Kırklareli	Area (km^2)	572.66	43.22	-	-	-
	Power (MW)	2863.3	216.08	-	-	-
Kocaeli	Area (km^2)	15.57	-	-	-	-
	Power (MW)	77.84	-	-	-	-

Wind maps of Balıkesir and Çanakkale which cities have the most wind power potential in Marmara region are given for 50m height in Figure 4. As obvious in the figure, particularly shorelines have great energy potential. Although the areas that have high average wind speed has spread wide part of Çanakkale, this case could not be seen in Balıkesir.

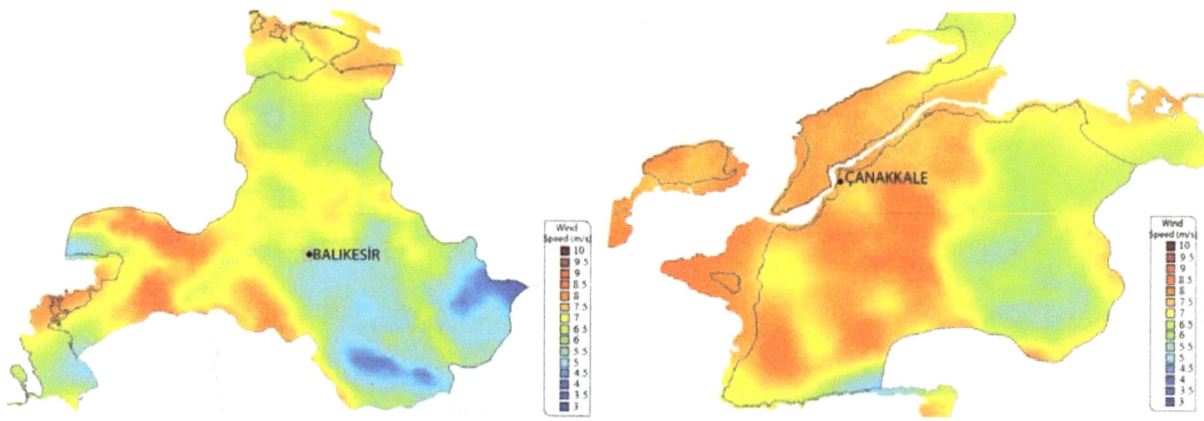

Figure 4. Wind maps of Balıkesir and Çanakkale [18]

The power capacities of installed, licensed and available WECSs are given in Table 5 [18, 19]. Based on available power and installed WECS power, Balıkesir takes the first place and Çanakkale is the second most appropriate city for wind power applications. It is remarkable that in five cities (Bursa, Kırklareli, Kocaeli, Sakarya, Yalova) there are no installed WECS.

When the received WECS licenses are investigated, it is recognized that Marmara region will be more popular for WECS installations in the near future. The locations of installed and licensed WECSs in Marmara region are shown in Figure 5.

Despite the potential to increase the WECS with licensed power plants, it is obvious that rate of usage capacity is still very low. Installed WECS power is 1.96% of the available power. Also, if it is assumed that the licensed plants are constructed, capacity will be only the 6.56% of the regions potential. This reality shows that the Marmara region have an attractive inappropriate wind energy resources.

Table 5. Wind power potential of cities in Marmara region

City	Installed WECS power (MW)	Available WECS power (MW)	Licensed power (MW)	Rate of capacity usage (%)
Balıkesir	616.1	13,827.36	338	4.45
Bilecik	40	308.64	120	12.96
Bursa	-	3881.68	59	-
Çanakkale	133.7	13,012.58	407	1.03
Edirne	15	3470.08	51	0.43
İstanbul	90.05	4177	382.5	2.16
Kırklareli	-	3079.38	247	-
Kocaeli	-	77.84	336	-
Sakarya	-	2	70	-
Tekirdağ	28.8	4626.6	84	0.62
Yalova	-	533.12	66	-
TOTAL	923.65	46,996.28	2160.5	1.96

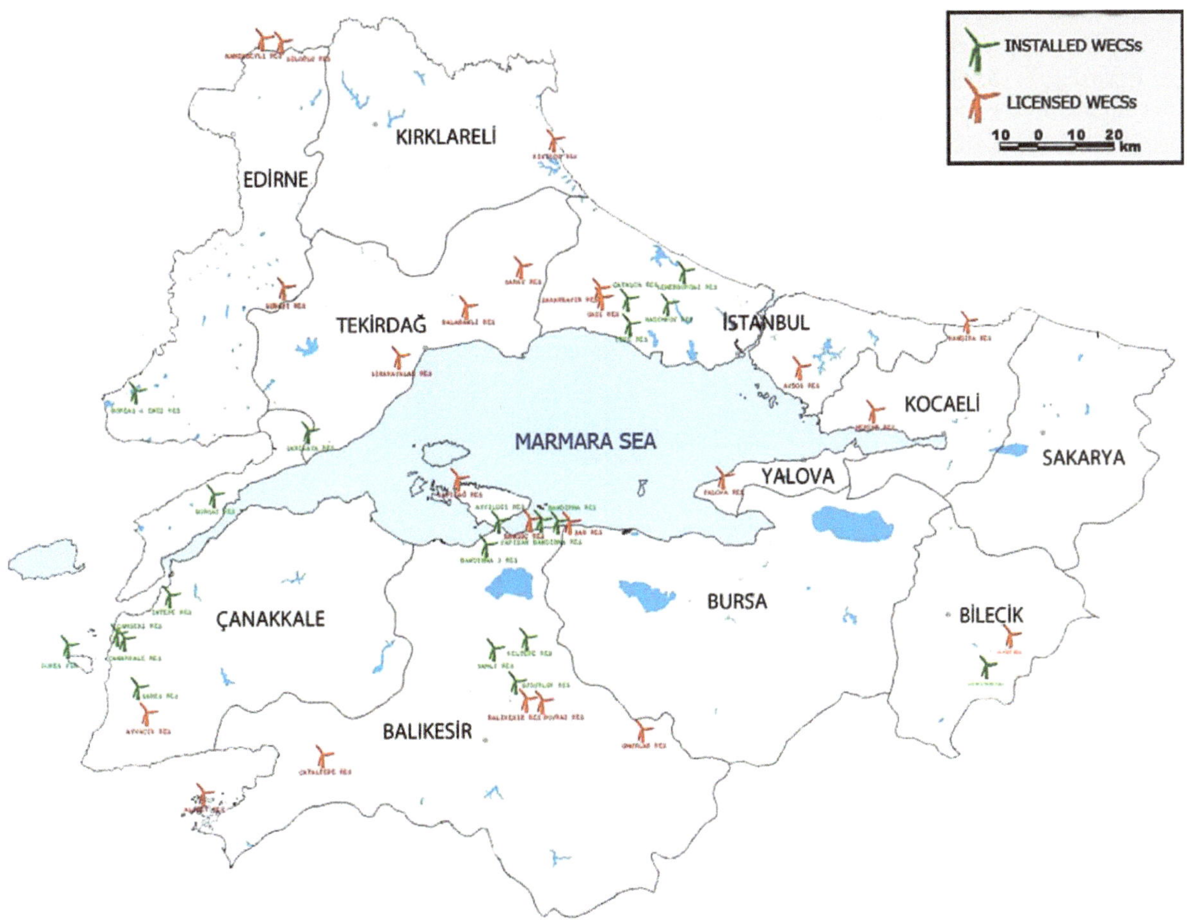

Figure 5. Installed and licensed WECSs in Marmara region [18]

The details of installed WECSs are given in Table 6 [19]. The first installed WECS project BORES is realized in Çanakkale-Bozcaada. This system was installed in 2000 and includes 17 wind turbines each has 0.6 MW power. The last installation was carried out in Balıkesir in 2012/2013 with 50 MW power of WECSs. Table 6 shows that installed power capacities of WECs and power capacity of turbines are increasing in recent years. That is because of the technological developments at wind turbine instruction and growing popularity of wind energy at this region.

Table 6. The installed power values of WECSs in Marmara region

Location	Project name	Installed Power (MW)	Turbine Brand	Turbine Power (MW)	Year of Installation
Çanakkale	Bozcaada RES	10.2	Enercon	0.6	2000
İstanbul	Tepe RES	0.85	Vestas	0.85	2006
Balıkesir	Bandırma-2 RES	35	Ge+Nordex	1.5+2.5	2006/2012
İstanbul	Sunjüt RES	1.2	Enercon	0.6	2008
İstanbul	Çatalca RES	60	Vestas	3	2008
Balıkesir	Bandırma-3 RES	25	Nordex	2.5	2008
Edirne	Boreas 1 Enez RES	15	Nordex	2.5	2008
Balıkesir	Şamlı RES	113.4	Vestas	3	2008/2010
Çanakkale	İntepe RES	30.4	Enercon	0.8	2009
Çanakkale	Bozcaada RES	10.2	Enercon	0.6	2009
Balıkesir	Keltepe RES	20.7	Enercon	0.9	2009
Balıkesir	Ayyıldız RES	15	Vestas	3	2009
İstanbul	Saray RES	4	Enercon	2	2009
Balıkesir	Bandırma RES	60	Vestas	3	2009/2010
Balıkesir	Çataltepe RES	16	Enercon	2	2010
Tekirdağ	Sarıkaya RES	28.8	Enercon	2+2+0.8	2010
Çanakkale	Burgaz RES	14.9	Enercon	0.8+0.9	2010
Çanakkale	SaRES	22.5	Ge	2.5	2010/2011
Çanakkale	Çamseki RES	20.8	Enercon	2+0.8	2011
Çanakkale	Ay RES	5	Vestas	1.8	2011
Balıkesir	Şah RES	93	Vestas	3	2011
Bilecik	Metristepe RES	40	Nordex	2.5	2011
Balıkesir	Susurluk RES	45	Nordex	2.5	2012
Çanakkale	Çanakkale RES	29.9	Siemens	2.3	2012
Balıkesir	Balıkesir RES	143	Ge	2.75	2012
Balıkesir	Poyraz RES	50	Enercon	2	2012/2013

4. Conclusion

In recent years, according to the increasing demand for electricity, decreasing fossil fuel reserves and environmental pollution, great importance is given to the investigation and investments for renewable energy sources in the world. The investments on construction of WECSs are rapidly increasing in Turkey where produces large amount of electrical energy from fossil fuels (oil, natural gas,..) and imports most of this fuels from foreign countries.

Marmara region investigated in this study is one of the seven regions in Turkey and it has share of 40% by 46,996.28 MW available wind power capacity of total capacity in Turkey. Due to the amount of population and developed industry, Marmara region has the highest requirements in terms of electric energy usage (62,531.068 MW). Also, Marmara has the highest average wind speed by 3.29 m/s in seven regions. This study reveals that Marmara region where has high average wind speed and areas with glamorous wind power, is the most suitable region for efficient wind energy production in Turkey. Especially Balıkesir and Çanakkale cities outshine with high wind speed potential.

Despite the vast electrical energy need, only 1.96 % of the available wind power capacity of this region is utilized. Large increases in capacity utilization are expected to occur in the near feature. This is a great opportunity for investors.

In this paper, it is clearly presented that related to the government supports and private enterprise investments; higher wind power energy can be achieved in this region. Installing new plants would significantly decrease energy dependence of Turkey on foreign resources. Furthermore, the installation of WECS would minimize the installation cost of transmission lines and reduce the transmission losses in Marmara region where large energy consumers exist.

Acknowledgment

Authors would like thank to the Yildiz Technical University for full financial support of project namely "Reliability Analysis of Small Scale Hybrid Wind and Solar Energy Systems: A Case Study of Davutpasa Campus", Project No: 2012-04-02-KAP04.

References

[1] Kose, R. An evaluation of wind energy potential as a power generation source in Kütahya, Turkey Energy Conversion and Management 2004, 45,1631-41.

[2] Erturk, E. The evaluation of feed-in tariff regulation of Turkey for onshore wind energy based on the economic analysis Energy Policy 2012, 45, 359-67.

[3] Ilkilic, C. Wind energy and assessment of wind energy potential in Turkey Renewable and Sustainable Energy Reviews 2012, 16 (2), 1165-73.

[4] Ucar, A., Balo, F. Evaluation of wind energy potential and electricity generation at six locations in Turkey Applied Energy 2009, 86 (10), 1864-72.

[5] Duic, N., Alves, L.M. Chen F., da Graca Carvalho, M. Potential of Kyoto protocol clean development mechanism in transfer of clean energy technologies to small island developing states; case study of Cape Verde Renewable and Sustainable Energy Reviews 2003, 7 (1), 83-98.

[6] Ilkilic, C., Turkbay I. Determination and utilization of wind energy potential for Turkey Renewable and Sustainable Energy Reviews 2010, 14 (8), 2202-2207.

[7] Radics, K., Bartholy, J. Estimating and modelling the wind resource of Hungary Renewable and Sustainable Energy Reviews 2008, 12 (3); 874-82.

[8] Kose, F., Aksoy, M.H., Ozgoren, M. An Assessment of Wind Energy Potential to Meet Electricity Demand and Economic Feasibility in Konya, Turkey International Journal of Green Energy 2014, 11 (6), 559-576.

[9] World Wind Energy Association 2013 Half-year report. http://www.wwindea.org/webimages/Half-year_report_2013.pdf (accessed on March 15, 2014).

[10] The European Wind Energy Association Annual Report 2012, United in tough times. http://www.ewea.org/fileadmin/files/library/publications/reports/EWEA_Annual_Report_2012.pdf (accessed on March 15, 2014).

[11] Turkish Electricity Transmission Corporation, TEIAS Annual Report 2012. http://www.teias.gov.tr/FaaliyetRaporlari/Faaliyet2012/TEIASfaayilet2012INGLIZCE.pdf (accessed on March 15, 2014).

[12] Turkish Electricity Transmission Corporation. http://www.teias.gov.tr/yukdagitim/ kuruluguc.xls (accessed on March 15, 2014).

[13] Ucar, A., Balo, F. Investigation of wind characteristics and assessment of wind generation potentiality in Uludag-Bursa, Turkey Applied Energy 2009, 86 (3), 333-39.

[14] Yumak, H., Ucar T., Yayla, S. Wind Energy Potential on the Coast of Lake Van International Journal of Green Energy 2012, 9 (1), 1-12.

[15] Zhang, L., Feng Y., Chen, B. Alternative Scenarios for the Development of a Low-Carbon City; A Case Study of Beijing, China Energies 2011, 4 (12), 2295-2310.

[16] Saidur, R., Rahim, N.A., Islam, M.R., Solangi, K.H. Enviromental impact of wind energy Renewable and Sustainable Energy Reviews 2011, 15 (5), 2423-2430.

[17] Hardisty, P.E., Clark, T.S., Hynes, G.R. Life cycle Greenhouse Gas Emissions from Electricity Generation; A Comparative Analysis of Australian Energy Sources Energies 2012, 5(4), 872-897.

[18] Republic of Turkey Ministry of Energy and Natural Resources- General Directorate of Electrical Power Resources. http://www.eie.gov.tr/yenilenebilir/document/Turkiye_Isletmedeki_RES_Temmuz%282012%29.pdf (accessed on March 15, 2014).

[19] Turkish Wind Energy Association. Turkish Wind Energy Statistics Report January 2013 http://www.tureb.com.tr/dosyalar/Turkiye_Ruzgar_Enerjisi_istatistik_Raporu_2013_Ocak.pdf (accessed on March 15, 2014).

[20] Kaygusuz, K. Wind energy status in renewable electrical energy production in Turkey Renewable and Sustainable Energy Reviews 2010, 14(7), 2104-2112.

[21] Republic of Turkey Energy Market Regulatory Authority, Elektrik ve Doğalgaz Piyasa Yatırım Verileri http;//www2.epdk.org.tr/data/index.htm (accessed on March 15, 2014).

17

Development of a thermal resistance model to evaluate wellbore heat exchange efficiency

Albert A. Koenig[1], Martin F. Helmke[2]

[1] ARB Geowell, 100 Four Falls Corporate Center Suite 215, West Conshohocken, PA 19428, USA.
[2] West Chester University, Department of Geology and Astronomy, 207 Merion Science Center, West Chester, PA, 19383, USA.

Abstract
A new model is proposed to simulate conduction of heat between a pipe loop in a geoexchange system and the ground. The approach employs the thermal resistor technique coupled with a conduction shape factor modified by an occultation factor. The model is compared to available data and demonstrates suitable agreement with previous studies. The model facilitates a parametric study of borehole resistance as a function of geometry and thermal conductivity of the components. By spacing the legs of the loop against the borehole and increasing the pipe size, the study shows that one can maximize the wellbore heat transfer using a moderate (1.2 W/mK) thermal conductivity grout. This study further demonstrates that improved well construction techniques could increase the efficiency of most closed-loop geothermal systems by 10 percent.

Keywords: Ground-source heat pump; Borehole thermal resistivity; Heat transfer model.

1. Introduction

Contemporary ground-source heat pump (geothermal) systems rely on efficient transfer of heat between the borehole and surrounding geologic medium. Typical closed-loop systems in the United States employ a circuit of high-density polyethylene (HDPE) pipe embedded within a borehole filled with an aqueous grout mixture. Previous studies [1] have documented that grout thermal conductivity and the proximity of the pipe to the borehole wall strongly influence the rate of heat transfer. Increased heat transfer efficiency improves system performance and decreases installation costs due to a reduction in borehole length.

This study proposes a thermal resistor network model to evaluate the thermal resistance of geoexchange wells. The method is straight-forward, computationally efficient, and flexible. Previous studies have employed numerical techniques [2-4] that are rigorous but are time-consulting to construct. Other investigators have proposed analytical solutions [5-7] that are readily applied, but are restricted to a limited number of pipe geometries. The method proposed by this study uses a straight-forward analytical solution with the advantage of a lessrestrictive borehole geometry.

2. Model development

The transfer of heat between the geoexchange pipe and surrounding geologic medium is governed by the thermal properties of carrier fluid, pipe, grout, and geologic medium; pipe location, and the dimensions of pipe and the borehole (Figure 1). A model must consider the resistance contributions from the fluid (typically water) flowing inside the two pipes (supply and return joined at the base of the wellbore by a U-bend), through the pipe walls, and through the medium (typically grout) that fills the space between the pipes and surrounding bore wall.

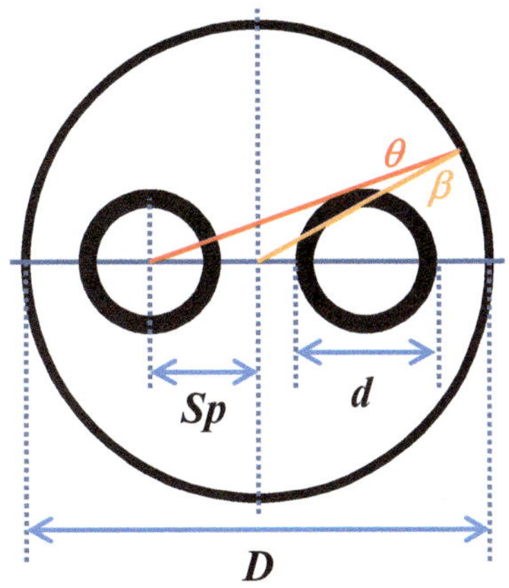

Figure 1. Cross-section geometry of a typical closed-loop heat exchange well. Dimensional parameters D, d, and Sp refer to the borehole diameter, pipe exterior diameter, and pipe spacing, respectively. The occultation angles θ and β are also shown

Thermal transport within the system may be represented by an equivalent network of resistors (Figure 2), where the temperatures are designated T_1, T_2, and T_b for the supply and return legs and borehole wall, respectively. The thermal resistance of each pipe, R_{p1} and R_{p2}, are composed of convection at the inside wall and conduction through the pipe wall. The shunt resistance, R_s, addresses heat transfer between the pipes through the medium, while R_b defines the thermal resistance between pipe surfaces to the surrounding wall.

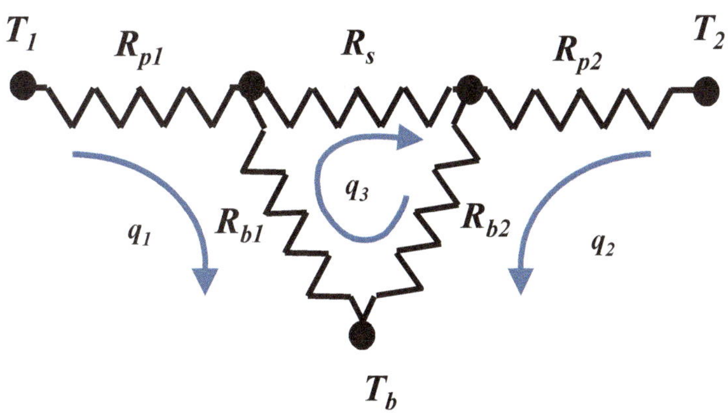

Figure 2. Resistance network model representing wellbore heat transfer

The temperature of the elements within the network model are considered unique, such that $T_1 > T_2 > T_b$ for the cooling season and $T_b > T_2 > T_1$ during heating. The three governing equations that describe the resistive network are

$$(T_1 - T_b) - (q_1 - q_3)R_{b1} - q_1 R_{p1} = 0 \quad (1)$$

$$(T_2 - T_b) - (q_2 + q_3)R_{b2} - q_2 R_{p2} = 0 \quad (2)$$

$$q_3 R_s + (q_3 + q_2)R_{b2} + (q_3 - q_1)R_{b1} = 0 \quad (3)$$

where q is heat flux. Assuming $R_{p1} = R_{p2}$ and $R_{b1} = R_{b2}$ (symmetric case in which pipe is equidistant from the borehole center), Eq.(1) through (3) may be solved simultaneously assuming steady-state conditions to reveal

$$q_1 = \frac{\Delta T_{avg}}{R_p + R_b} + \frac{\Delta T}{2R_3} \tag{4}$$

$$q_2 = \frac{\Delta T_{avg}}{R_p + R_b} - \frac{\Delta T}{2R_3} \tag{5}$$

$$q_3 = \frac{\Delta T}{R_{shunt}} \tag{6}$$

where $\Delta T = T_1 - T_2$, $T_{avg} = (T_1 + T_2)/2$, $\Delta T_{avg} = (T_{avg} - T_b)$, $R_3 = (R_p + R_b) - 2R_b/(R_s/R_b + 2)$, and $R_{shunt} = (R_p + R_b)(R_s/R_b + 2) - 2R_b$.

The pipe thermal resistance includes convective heat transfer from the carrier fluid and heat conduction through the pipe wall, defined as

$$R_p = \frac{1}{\pi d_i h_{cv}} + \frac{\ln(d_o / d_i)}{2\pi k_p} \tag{7}$$

where subscripts o and i denote outside and inside pipe diameters, h_{cv} is the convection coefficient for water flowing inside the pipe, and k_p is the thermal conductivity of the pipe wall.

Thermal transport between the pipes and bore wall must consider the 2-dimensional geometry of the system and the thermal interference, or obscuration, created by the companion pipe. Applying the shape-factor method for 2-dimensional, steady-state heat conduction [8, 9], the thermal resistance between single pipe and borehole wall may be calculated as:

$$R_1 = \frac{\cosh^{-1}\left(\dfrac{D^2 + d^2 - 4x_1^2}{2Dd}\right)}{2\pi k_g f_{occ}} \tag{8a}$$

$$R_2 = \frac{\cosh^{-1}\left(\dfrac{D^2 + d^2 - 4x_2^2}{2Dd}\right)}{2\pi k_g f_{occ}} \tag{8b}$$

where D is borehole diameter, d is pipe diameter, and k_g is the grout thermal conductivity. The variables x_1 and x_2 are the distances between the borehole center and center of pipes 1 and 2, respectively. In the special case where $x_1 = x_2$, these parameters are equivalent to pipe spacing, Sp, and the distance between the pipes is $2Sp$.

Equation (8) may be used to simulate a single off-centered pipe conducting to an unobscured surrounding wall. In a two-pipe system, conductive interference increases the net thermal resistance. To address this obscuration effect, an occultation factor, f_{occ}, is applied to the equation that considers the spacing and dimensions of pipes within a cylindrical borehole:

$$f_{occ} = (1 - \beta / \pi) \tag{9}$$

$$\theta = \sin^{-1}\left(\frac{d}{2(x_1 + x_2)}\right) \tag{10}$$

$$\beta = \theta + \sin^{-1}\left(\frac{d}{2D}\right) \tag{11}$$

The final thermal resistance, R_b, may be calculated by

$$R_b = \frac{R_1 R_2}{R_1 + R_2} + \frac{R_p}{2} \tag{12}$$

The borehole thermal resistance may be used to calculate the heat transfer efficiency of a wellbore and predict seasonal ground response to an operation geoexchange system.

3. Results and discussion

3.1 Typical wellbore

The proposed model was employed to simulate a geoexchange borehole similar to those typically installed in the United States. Common drilling techniques in the US produce a borehole with a diameter of 152 mm (6 inches) filled with a neat bentonite grout with a k of 0.78 W/m K, and round, high density polyethylene (HDPE) tubing with dimensions established in Table 1. In this simulation, the supply and return pipe legs are assumed to be equally-spaced within the borehole, resulting in anSp of 32.3 mm. Applying equations (7) through (12) results in a calculated R_b of 0.17 m K/W.

Table 1. Calculation of thermal resistance for a typical closed-loop borehole

Variable	Value
Model input parameters	
D, borehole diameter	152 mm
Sp, pipe spacing	32.3 mm
d, pipe exterior diameter	42 mm
d_i, pipe interior diameter	34 mm
t, pipe wall thickness	4 mm
k_p, pipe thermal conductivity	0.4 W/m K
h_{cv}, thermal exchange coefficient	1100 W/m^2 K
k_g, grout thermal conductivity	0.78 W/m K
Model output, this study	
R_p, pipe thermal resistivity	0.084 m K/W
R_b, borehole thermal conductivity	0.17 m K/W

The model was compared to approaches employed by previous studies to validate results (Figure 3). All simulations included the resistive effect of fluid heat transfer and thermal conduction through the pipe wall according to eq. (7). Simulations demonstrate that the proposed model is in general agreement with previous approaches, showing a reduction in R_b as pipe spacing increases. Predicted R_b from eq. (12) is most similar to (within 5 percent of) the results from Sharqawy et al. [4], who employed a detailed 2-dimensional finite element solution. The Sharqawy et al. study, similar to the current approach, focused on heat transfer through the borehole and assumed a constant temperature boundary condition at the borehole wall. Hellström [6] applied a semi-infinite boundary condition that accounted for a variable thermal field along the borehole perimeter. This results in a predicted R_b that is 14-percent greater than this study's approach. Finally, the semi-empirical findings from Remund [1] are 25-percent higher than this study. However, Remund acknowledged his results were higher than predicted by models due to grout voids, uncertainty of pipe placement, and borehole wall irregularities that are inherent in the field.

3.2 Sensitivity analysis

Equations (7) through (12) were used to evaluate the causal relationship between the dependent variable R_b and independent variables k_g, Sp, t, and d (Figure 4a, 4b, 4c, and 4d, respectively). All other variables were consistent with the base case summarized in Table 1. Increasing the thermal conductivity of grout dramatically decreases R_b. However, increasing k_g above 1.2 W/m K appears to provide little additional benefit. Moreover, increasing k_g has the disadvantage of increasing thermal shunting between legs of the u-loop. Increasing pipe spacing can reduce R_b by half. Most geothermal boreholes installed in the United States do not utilize spacers, allowing the pipes to twist which minimizes pipe spacing. We recommend that pipes be pushed against the borehole wall to increase efficiency. Reducing pipe wall thickness improves thermal transfer across the pipe wall. However, reducing t below 3 mm would likely weaken the pipe and cause it to collapse at depth. Another option would be to increase the diameter of the pipe. An increase in pipe diameter increases the surface area and corresponding heat flux. Increasing d is limited by the physical properties of the pipe and the space available within the borehole. The combined benefit of optimizing these parameters yields a significant reduction in R_b. Simulation of the "best reasonable case" (k_g = 1.2 W/m K, Sp = 43 mm, t = 3 mm, and d = 60 mm) reveals an R_b of 0.05 m K/W.

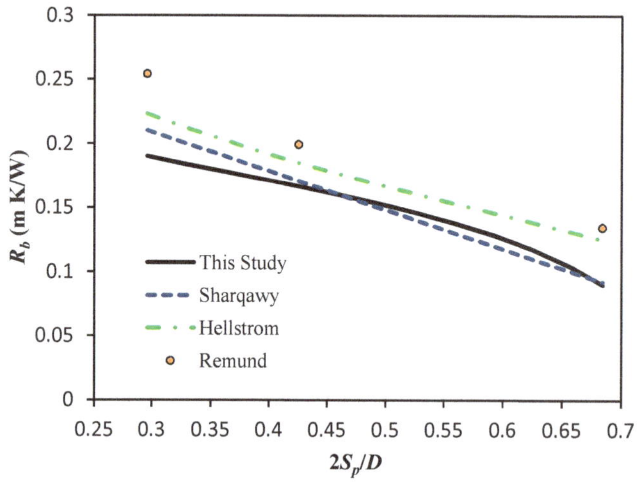

Figure 3. A comparison of effective borehole resistance (R_b) predicted by this study, a finite-element model (Sharqawy [4]), an semi-infinite analytical model (Hellström [6]), and empirical observations (Remund [1])

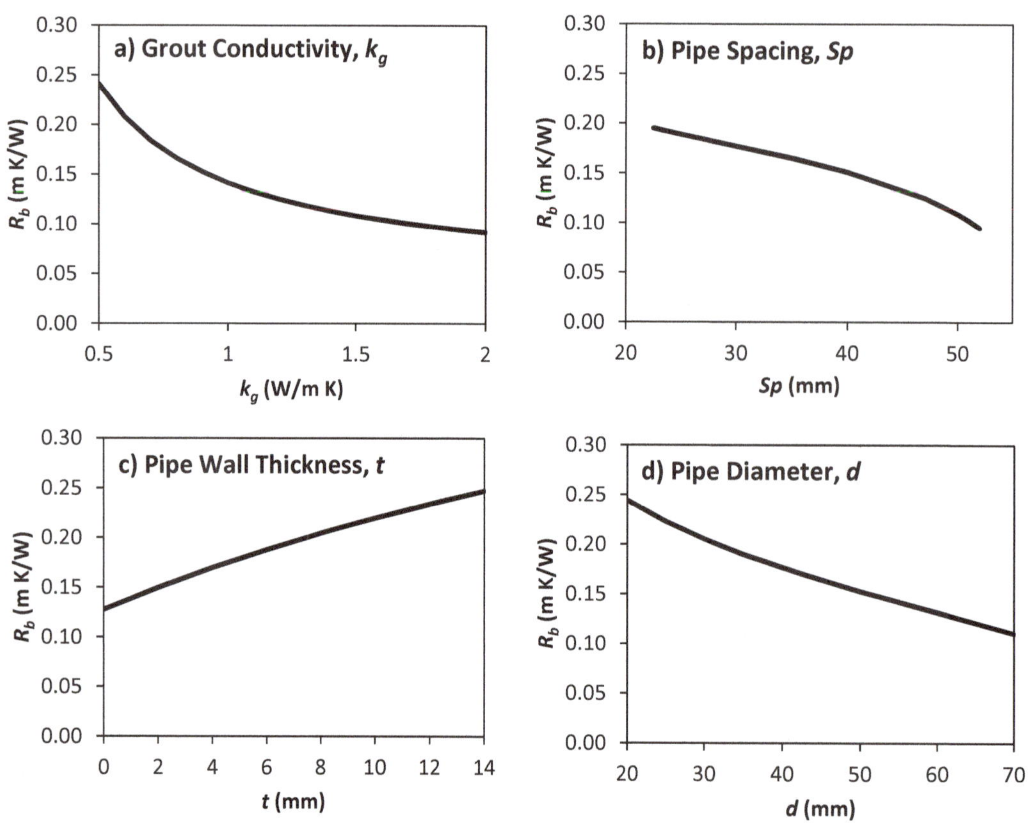

Figure 4. Borehole thermal resistance predicted by the model as a function of grout thermal conductivity, pipe spacing, pipe wall thickness, and pipe diameter

3.3 Importance of R_b

Previous studies have emphasized the importance of minimizing R_b. Comparatively few, however, have quantified the overall benefit of reducing borehole thermal resistance in cases where heat transfer may be limited by the thermal conductivity of the surrounding geologic medium (k_r). To test the relative importance of R_b, we simulated a borehole using the spatial dimensions in Table 1 with a 2-dimensional finite difference model[10]. Well efficiency, E_c, was computed by comparing the response of borehole with resistance R_b to a borehole with zero thermal resistance as a function of k_r. Representative values of R_b ranged from 0 to 0.3 m K/W as suggested by this study. Thermal conductivity was adjusted to represent coal, limestone, saturated sand, granite, and quartzite (Table 2).

Table 2. Thermal conductivity, k_r, and the critical borehole thermal resistance, R_b, for a variety of representative geologic materials

Formation Type	Thermal conductivity, k_r (W/m K) [11]	Critical borehole thermal resistance, R_b (m K/W)
Coal	0.3	0.31
Limestone	1.3	0.07
Sand	2.6	0.04
Granite	4	0.02
Quartzite	8	0.01

Well efficiency diminishes as R_b and k_r increase (Figure 5). This results because a geoexchange well installed within a high-k_r formation is more likely to be limited by thermal resistance within the borehole. A significant (>1 percent) reduction in well efficiency is predicted if R_b increases above a critical value (column 3 in Table 2). This demonstrates that for most geologic media (with the notable exception of coal), thermal exchange is likely to be limited by borehole thermal resistance. This emphasizes the benefit of reducing R_b under most geologic situations. Moreover, the model shows that improved construction of the geoexchange piping could improve the overall efficiency of a system installed in typical geologic materials (limestone, sand, and granite) by 10 percent.

The relationship between well efficiency and the product of R_b and k_r is well-defined (Figure 5b). For the typical well geometry investigated by this study, E_c may be calculated from the relation

$$E_c = e^{(aR_b k_r)} \tag{13}$$

where a = -0.105. This equation fits the simulated data well, with an R^2coefficient of 0.9999.

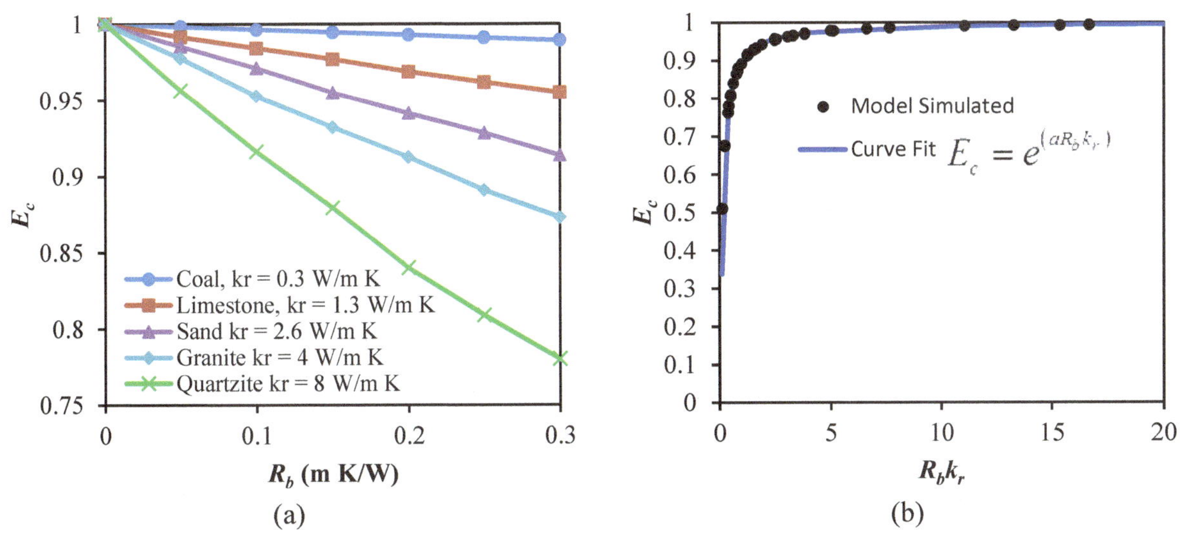

Figure 5.Thermal well efficiency (E_c) as a function of borehole thermal resistance (R_b) and the thermal conductivity of the surrounding geologic formation (k_r)

3.4 Shunting

The thermal shunt between the legs of the pipe loop is calculated from eq. (6), where R_s is developed from the pipe to pipe shape factor analysis in a surrounding cylinder from [8] where:

$$R_s = \cosh^{-1}\left[2(2Sp/d)^2 - 1\right] \tag{14}$$

As shown in Figure 6, the loop's ability to dissipate heat is compromised by the thermal shunt, which exists between supply and return legs of the loop. The vertical axis is a measure of the heat shunted between legs of the loop to the heat applied or extracted at the surface. When the pipes are in close proximity, the shunted heat transfer increases, and the performance of the loop is diminished. Alternatively, when the pipes are spread far apart (contacting the perimeter of the bore wall, for instance), shunting is minimized and the loop performance is maximized. Figure 6 shows that the thermal shunt is exacerbated for close spacing (small *Sp*) and for larger grout conductivity, k_g, while the least shunting is for legs that are separated widely and for low grout conductivity (i.e., neat bentonite).

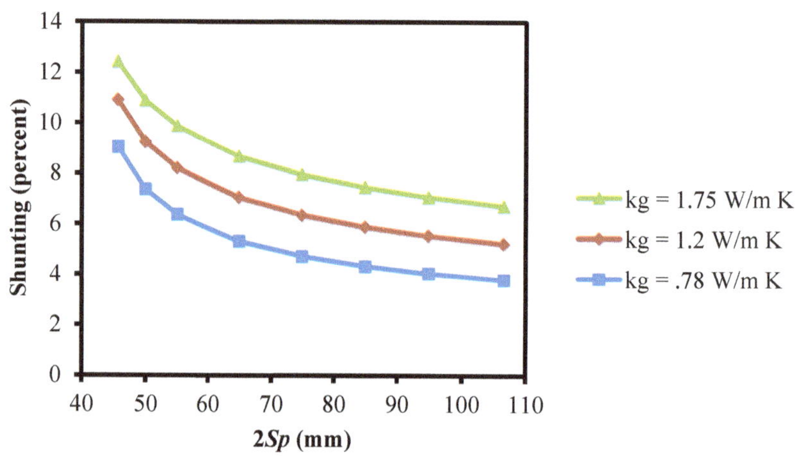

Figure 6. Loop thermal shunt between legs as function of spacing and grout conductivity

In practice, closely spaced U-bends and installation practices that allow the pipes to twist and contact one another over the pitch of the twist will suffer the consequences of greater thermal shunting. This has the effect of reducing the effective length of the loop, since the applied heat never makes it fully down hole, but instead raises the temperature of the upcoming water, so that (T_1-T_2), or the effective loop heat rejection, is smaller than what it would be with no shunting.

4. Conclusions

The analytical solution developed by this study to calculate borehole thermal resistance is computationally efficient, represents a variety of borehole geometries, and is readily accessible to installers because it requires no additional software. The model is particularly useful for predicting the benefits of optimizing borehole geometry and properties of the pipe and grout. The model compares well with previous studies, although actual R_b is likely to be somewhat greater than predicted by any model due to problems with grouting, pipe placement, and drilling techniques encountered in the field. Previous studies have emphasized the importance of increasing k_g as a means of reducing R_b; however, increasing k_g alone may cause unwanted thermal shunting between pipe legs. Numerical simulations demonstrate that reducing R_b could increase well thermal exchange efficiency by over 10 percent for most geologic materials. Future efforts to improve performance of geoexchange wells should consider maximizing pipe spacing, pipe surface area, and minimizing pipe wall thickness in addition to increasing k_g, which could significantly improve efficiency.

References

[1] Remund C.P. Borehole thermal resistance: Laboratory and field studies. ASHRAE Transactions 2009, 105(1), 439-445.

[2] Liao Q., Zhou C., Cui W., and Jen T. Effective borehole thermal resistance of a single u-tube ground heat exchanger. Numerical Heat Transfer 2012, 62, 197-210.

[3] Sagia Z., Stegou A., and Rakopoulos C. Borehole resistance and heat conduction around vertical ground heat exchangers. The Open Chemical Engineering Journal 2012, 6, 32-40.

[4] Sharqawy M.H., Mokheimer E.M., and Hassan M.B. Effective pipe-to-borehole thermal resistance for vertical ground heat exchangers. Geothermics 2009, 38, 271-277.

[5] Gu, Y. and O'Neal, D. Development of an equivalent diameter expression for vertical u-tubes used in ground-coupled heat pumps. ASHRAE Transactions 1998, v. 104, n. 2, 9 p.

[6] Hellström, G. Ground heat storage, thermal analyses of duct storage systems. Doctoral thesis 1991, Department of Mathematical Physics, University of Lund, Lund, Sweden.

[7] Philippe M. Bernier M., Marchio D. Sizing calculation spreadsheet vertical geothermal borefields. ASHRAE Journal 2010, 7, 20-28.

[8] Andrews R.V. Solving conductive heat transfer problems with electrical-analog shape factors, Chemical Engineering Process, 1955, 51(2), 67 p.

[9] Holman J.P. Heat Transfer, Sixth Edition. McGraw-Hill Publishers, New York, NY, 676 p.

[10] Harbaugh, A. Banta, E., Hill, M., and McDonald, M. MODFLOW-2000, the U.S. Geological Survey modular ground-water model. Open-File Report 00-92 2000, U.S. Geological Survey, 121 p.

[11] Banks, D. An Introduction to Thermogeology: Ground Source Heating and Cooling 2008, Blackwell Publishing, Oxford, UK, 339 p.

18

Natural convection mass transfer hydromagnetic flow past an oscillating porous plate with heat source in a porous medium

S. S. Das[1], S. Mishra[2], P. Tripathy[3]

[1] Department of Physics, KBDAV College, Nirakarpur, Khurda-752 019(Odisha), India.
[2] Department of Physics, Christ College, Mission Road, Cuttack-753 001(Odisha), India.
[3] Department of Physics, Centurion University, Paralakhemundi, Gajapati-761 211(Odisha), India.

Abstract
This paper analyzes the effect of mass transfer on natural convection hydromagnetic flow of a viscous incompressible fluid through a porous medium past an oscillating porous plate in a porous medium with heat source. The governing equations of the flow field are solved analytically and the expressions for velocity and temperature of the flow field, skin friction τ and the heat flux in terms of Nusselts number N_u are obtained. The effects of the important flow parameters such as magnetic parameter M, permeability parameter K_p, Grashof number for heat and mass transfer G_r, G_c, Schmidt number S_c, heat source parameter S and the Prandtl number P_r on the velocity and temperature of the flow field are to be discussed with the help of figures. It is observed that a growing magnetic parameter M retards the magnitude of the velocity of the flow field at all points due to the action of the Lorentz force on the flow field. The heat source parameter S has an accelerating effect on the magnitude of the velocity of the flow field at all points. The effect of growing Grashof number for mass transfer G_c and the permeability parameter K_p is to enhance the velocity (absolute value) of the flow field at all points. An increase in Schmidt number S_c is to increase the magnitude of the velocity of the flow field at all points. A growing rarefaction parameter R enhances the magnitude of the velocity of the flow field at all points.

Keywords: Natural convection; Mass transfer; Hydromagnetic flow; Porous medium; Oscillating plate; Heat source.

1. Introduction

Hydromagnetic flow through a porous medium with heat and mass transfer is gathering momentum day by day in view of its possible applications to geophysical sciences, astrophysical sciences and also in industry. The study of fluctuating flow is important in paper industry and many other technological fields. In view of these applications several researchers have given much attention towards fluctuating flows of viscous incompressible fluids past an infinite plate.

The nature of vertical natural convection flow resulting from the combined buoyancy effects of thermal and mass diffusion effects was analyzed by Gebhart and Pera [1]. Georgantopoulos *et al.* [2] estimated the effect of free convection and mass transfer on the hydro-magnetic oscillatory flow past an infinite vertical porous plate. Hossain and Begum [3] discussed the effect of mass transfer and free convection on the flow past a vertical plate. Bejan and Khair [4] studied the heat and mass transfer effects by natural convection in a porous medium. Hossain and Begum [5] discussed the effect of mass transfer on the unsteady flow past an accelerated vertical porous plate with variable suction. Raptis and Perdikis [6] analyzed the oscillatory flow through a porous medium in presence of free convection. Sattar [7] reported the free and forced convection boundary layer flow through a porous medium with large suction, Chamkha [8] studied the hydromagnetic three-dimensional free convection flow on a vertical stretching surface with heat generation/absorption.

The effect of combined heat and mass transfer hydromagnetic flow by natural convection from a permeable surface embedded in a fluid saturated porous medium was analyzed by Chamkha and Khaled [9]. Nagraju *et al.* [10] discussed the simultaneous radiative and convective heat transfer in a variable porosity medium. The problem of heat and mass transfer in MHD flow of a viscous fluid past a vertical plate under oscillatory suction velocity was studied by Singh and his co-workers [11]. Hayat *et al.* [12] discussed the flow of a visco-elastic fluid on an oscillating plate. Jain and Gupta [13] have reported the unsteady hydromagnetic thermal boundary layer flow past an infinite porous surface in the slip flow regime. Singh and Gupta [14] studied the MHD free convective mass transfer flow of a viscous fluid through a porous medium bounded by an oscillating porous plate in slip flow regime.

Sharma and Singh [15] estimated the unsteady MHD free convective flow and heat transfer along a vertical porous plate with variable suction and internal heat generation. Das and his associates [16] studied the mass transfer effects on MHD flow and heat transfer past a vertical porous plate through a porous medium under oscillatory suction and heat source. Das *et al.* [17] reported the hydromagnetic convective flow past a vertical porous plate through a porous medium with suction and heat source. Natural convection unsteady magnetohydrodynamic mass transfer flow past an infinite vertical porous plate in presence of suction and heat sink was studied by Das and his team [18] Recently, Das and his co-workers [19] analyzed the hydromagnetic mixed convective mass diffusion boundary layer flow past an accelerated vertical porous plate through a porous medium with suction by finite difference scheme.

The study reported herein analyzes the effect of mass transfer on natural convection hydromagnetic flow of a viscous incompressible fluid in a porous medium past an oscillating porous plate with heat source. The governing equations of the flow field are solved analytically and the expressions for velocity and temperature of the flow field, skin friction τ and the heat flux in terms of Nusselts number N_u are obtained. The effects of the important flow parameters such as magnetic parameter M, porosity parameter K_p, Grashof number for heat and mass transfer G_r, G_c, Schmidt number S_c, heat source parameter S and the Prandtl number P_r on the flow field are to be discussed with the help of figures.

2. Formulation of the problem

Consider the natural convection mass transfer flow of a viscous incompressible fluid past an oscillating porous plate with heat source in a porous medium in presence of a transverse magnetic field B_0. Let u and v be the velocity components in x- and y- directions respectively. All the physical variables are functions of y and t only. The Reynolds number is assumed to be very small and the induced magnetic field due to the flow is neglected with respect to the applied magnetic field and the pressure in the flow field is assumed to be constant. If v_0 denotes the suction/injection velocity at the plate, the equation of continuity is

$$\frac{\partial v}{\partial y} = 0 \tag{1}$$

Under the condition $y = 0$, $v = -v_0$ everywhere.

Now the governing boundary layer equations of the flow field in non-dimensional form are

$$\frac{\partial u}{\partial t} - v_0 \frac{\partial u}{\partial y} = \nu \frac{\partial^2 u}{\partial y^2} + g\beta(T - T_\infty) + g\beta^*(C - C_\infty) - \frac{\nu}{K_0}u - \frac{\sigma B_0^2}{\rho}u \tag{2}$$

$$\frac{\partial T}{\partial t} - v_0 \frac{\partial T}{\partial y} = k\frac{\partial^2 T}{\partial y^2} - S(T - T_\infty) \tag{3}$$

$$\frac{\partial C}{\partial t}-v_0\frac{\partial C}{\partial y}=D\frac{\partial^2 C}{\partial y^2}, \tag{4}$$

where g is the acceleration due to gravity , ν is the kinematic viscosity, k is the thermal diffusivity, K_0 is the permeability coefficient, β is the volumetric coefficient of expansion for heat transfer, β^* is the volumetric coefficient of expansion for mass transfer, ρ is the density , σ is the electrical conductivity of the fluid , T is the temperature, T_∞ is the temperature of the fluid far away from the plate, C is the concentration, C_∞ is the concentration of the fluid far away from the plate and D is the molecular diffusivity.

Now the first order velocity slip boundary conditions of the problem when the plate executes linear harmonic oscillations in its own plane are given by

$$u=U_0e^{i\omega t}+L_1\frac{\partial u}{\partial y},\quad T=T_w,\quad C=C_w \quad \text{at } y=0,$$

$$u\to 0,\ T\to T_\infty,\ C\to C_\infty \ \text{ as } y\to\infty \tag{5}$$

where $L_l=\frac{(2-m)}{m}L$ and $L=\mu\left(\frac{\pi}{2p\rho}\right)^{\frac{1}{2}}$ is the mean free path and m is the Maxwell's reflection coefficient.

We now introduce the following non-dimensional quantities

$$y^*=U_0\frac{y}{\nu},\ u^*=\frac{u}{U_0},\ T=\frac{T-T_\infty}{T_w-T_\infty},\ C=\frac{C-C_\infty}{C_w-C_\infty},\ t^*=U_0^2\frac{t}{\nu},\ v_0^*=\frac{V_0}{U_0},\ \omega^*=\frac{\nu\omega}{U_0^2},$$

$$S^*=\frac{\nu S}{U_0^2}\ \text{(Heat source parameter)},\ R=U_0\frac{L_1}{\nu}\ \text{(Rarefaction parameter)},$$

$$M=\frac{B_0}{U_0}\left(\frac{\nu\sigma}{\rho}\right)^{\frac{1}{2}}\ \text{(Hartmann number/ magnetic parameter)},\ P_r=\frac{\nu}{k}\ \text{(Prandtl number)},$$

$$K_p=\frac{K_0U_0^2}{\nu^2}\ \text{(Permeability parameter)},\ G_r=\nu g\beta\frac{(T_w-T_\infty)}{U_0^3}\ \text{(Grashof number for heat transfer)},$$

$$G_c=\nu g\beta^*\frac{(C_w-C_\infty)}{U_0^3}\ \text{(Grashof number for mass transfer)},$$

$$S_c=\frac{\nu}{D}\ \text{(Schmidt number)}. \tag{6}$$

Introducing the non-dimensional parameters mentioned above (6) in equations (2)-(4) and dropping the asterisks, the governing equations now reduce to the following non-dimensional forms:

$$\frac{\partial u}{\partial t}-v_0\frac{\partial u}{\partial y}=\nu\frac{\partial^2 u}{\partial y^2}+G_rT+G_cC-\left(M^2+\frac{1}{K_p}\right)u \tag{7}$$

$$\frac{\partial T}{\partial t}-v_0\frac{\partial T}{\partial y}=\frac{1}{P_r}\frac{\partial^2 T}{\partial y^2}-ST \tag{8}$$

$$\frac{\partial C}{\partial t} - v_0 \frac{\partial C}{\partial y} = \frac{1}{S_c} \frac{\partial^2 C}{\partial y^2} \tag{9}$$

The boundary conditions now reduce to

$$u = e^{i\omega t} + R\frac{\partial u}{\partial y}, \quad T = 1, \quad C = 1 \quad \text{at } y = 0,$$

$$u \to 0, \ T \to 0, \ C \to 0 \ \text{ as } y \to \infty. \tag{10}$$

3. Method of solution

For solving equations (7)-(9), we assume the following for the velocity, temperature and concentration distribution of the flow field.

$$u = u_0 + u_1 e^{i\omega t}, \tag{11}$$

$$T = T_0 + T_1 e^{i\omega t}, \tag{12}$$

$$C = C_0 + C_1 e^{i\omega t}. \tag{13}$$

Using equations (11)-(13) in equations (7)-(9) and separating the harmonic and non-harmonic terms, we get

$$u_0'' + v_0 u_0' - \left(M^2 + \frac{1}{K_p}\right) u_0 = -G_r T_0 - G_c C_0, \tag{14}$$

$$u_1'' + v_0 u_1' - \left(M^2 + \frac{1}{K_p} + i\omega\right) u_1 = -G_r T_1 - G_c C_1, \tag{15}$$

$$T_0'' + P_r v_0 T_0' + S P_r T_0 = 0, \tag{16}$$

$$T_1'' + P_r v_0 T_1' + (S - i\omega) P_r T_1 = 0, \tag{17}$$

$$C_0'' + S_c v_0 C_0' = 0, \tag{18}$$

$$C_1'' + S_c v_0 C_1' - i\omega C_1 = 0. \tag{19}$$

The corresponding boundary conditions are

$$u_0 = R\frac{\partial u_0}{\partial y}, \quad u_1 = 1 + R\frac{\partial u_1}{\partial y}, \quad T_0 = 1, \ T_1 = 0, \ C_0 = 1, \ C_1 = 0 \qquad \text{at } y = 0,$$

$$u_0 \to 0, \ u_1 \to 0, \ T_0 \to 0, \ T_1 \to 0, \ C_0 \to 0, \ C_1 \to 0 \ \text{ as } y \to \infty \tag{20}$$

Solving equations (14)-(19) under boundary conditions (20), we get the following solutions for velocity, temperature and the concentration distributions of the flow field.

$$T_0 = e^{\lambda_1 y}, \tag{21}$$

$$T_1 = 0, \tag{22}$$

$$C_0=e^{-Sc v_0 y}, \tag{23}$$

$$C_1=0, \tag{24}$$

$$u_0=A_1 e^{-\lambda_2 y}-A_2 e^{\lambda_1 y}-A_3 e^{-S_c v_0 y}, \tag{25}$$

$$u_1=A_4 e^{\lambda_4 y}, \tag{26}$$

where $\lambda_1=\frac{1}{2}\left[-P_r v_0-\sqrt{P_r^2 v_0^2-4SP_r}\right]$, $\lambda_2=\frac{1}{2}\left[P_r v_0+\sqrt{P_r^2 v_0^2-4P_r(S-i\omega)}\right]$,

$$\lambda_3=\frac{1}{2}\left[-P_r v_0+\sqrt{P_r^2 v_0^2-4P_r(S-i\omega)}\right],\quad \lambda_4=-\frac{1}{2}\left[-v_0+\sqrt{v_0^2+4\left(M^2+\frac{1}{K_p}+i\omega\right)}\right],$$

$$A_1=\frac{1}{(R\lambda_2+1)}\left[A_2(1-\lambda_1)-A_3(1-S_c v_0)\right],\quad A_2=\frac{G_r}{(\lambda_1+\lambda_2)(\lambda_1-\lambda_3)}$$

$$A_3=\frac{G_c}{(\lambda_2-S_c v_0)(\lambda_3+S_c v_0)},\quad A_4=\frac{1}{(1-R\lambda_4)}. \tag{27}$$

Using equations (21)-(26) in equations (11)-(13), the solutions for velocity, temperature and concentration distribution of the flow field are given by

$$u=A_1 e^{-\lambda_2 y}-A_2 e^{\lambda_1 y}-A_3 e^{-S_c v_0 y}+A_4 e^{\lambda_4 y+i\omega t}, \tag{28}$$

$$T=e^{\lambda_1 y}, \tag{29}$$

$$C=e^{-S_c v_0 y}. \tag{30}$$

Skin friction

The skin friction at the wall is given by

$$\tau=\left(\frac{\partial u}{\partial y}\right)_{y=0}=-\lambda_2 A_1-\lambda_1 A_2+S_c v_0 A_3+\lambda_4 A_4 e^{i\omega t}. \tag{31}$$

Heat flux

The rate of heat transfer or the heat flux at the wall in terms of Nusselts number is given by

$$N_u=\left(\frac{\partial T}{\partial y}\right)_{y=0}=\lambda_1. \tag{32}$$

4. Discussions and results

The effect of mass transfer on natural convection flow of a viscous incompressible electrically conducting fluid through a porous medium past an oscillating porous plate in with heat source in presence of a transverse magnetic field has been considered. The effects of the important flow parameters such as magnetic parameter *M*, heat source parameter *S*, Grashof number for mass transfer G_c, permeability parameter K_p, Schmidt number S_c on the velocity of the flow field have been discussed with the help of Figures 1-4.

4.1 Velocity field (u)

The velocity field suffers a change in magnitude with the variation of the flow parameters. The flow parameters responsible for this change in the velocity field are magnetic parameter *M*, heat source parameter *S*, Grashof number for mass transfer G_c, permeability parameter K_p, Schmidt number S_c. These variations in the velocity field are depicted in Figures 1-4.

4.2 Effect of magnetic parameter M

Figure 1 depicts the effect of magnetic parameter *M* on the velocity field for three different values of the magnetic parameter (M= 0, 3, 5). In the figure curve with M= 0 corresponds to the non-MHD flow. Comparing the curves of the figure, it is seen that the magnetic parameter decelerates the magnitude of the velocity of the flow field at all points due to the action of Lorentz force on the flow field.

4.3 Effect of heat source parameter S

The heat source parameter *S* plays a drastic role on the behaviour of the velocity field. The variations in the velocity field due to heat source parameter S is shown in Figure 2. In the figure curve with S=0 corresponds to the absence of heat source and the curves with S=0.3 and S=-0.3 correspond to the presence of heat source and heat sink in the flow field. A close observation on the curves of the Figure 2 shows that the heat source parameter increases the magnitude of the velocity at all points of the flow field.

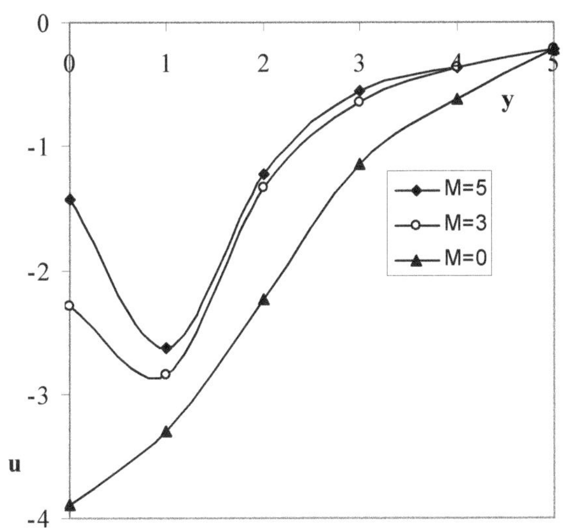

Figure 1. Velocity profiles against *y* for different values of *M* with R=0.3, G_r=3, G_c=3, S_c=0.22, K_p=2, P_r=0.71, S=0.5, v_0=2, ωt=π/2, ω=2

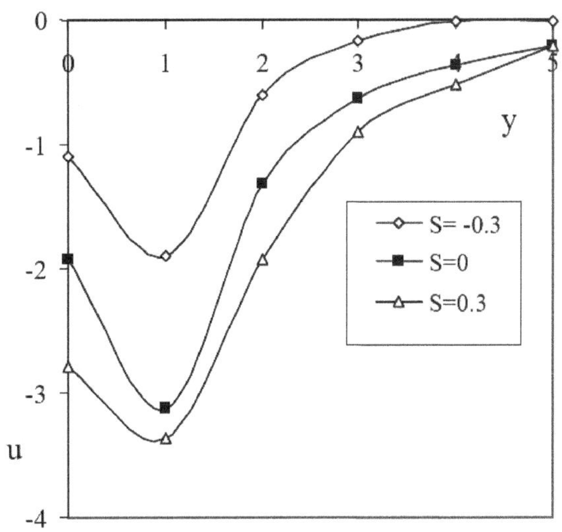

Figure 2. Velocity profiles against *y* for different values of S with M=2, R=0.3, G_r=3, G_c=3, S_c=0.22, K_p=2, P_r=0.71, v_0=2, ωt=π/2, ω=2

4.4 Effect of Grashof number G_c, permeability parameter K_p and rarefaction parameter R

The effects of rarefaction parameter *R*, Grashof number for mass transfer G_c and the permeability parameter K_p on the velocity of the flow field are depicted in Figure 3. A comparative study of the curves of Figure 3 shows that the effect of the above parameters is to enhance the magnitude of the velocity at all points of the flow field.

4.5 Effect of Schmidt number S_c

The presence of foreign mass in the flow field influences the velocity of the field to an appreciable extent. These effects have been shown in Figure 4. In the figure curve with S_c= 0 refers to the absence of foreign mass in the flow field. A growing S_c (heavier diffusing species) is seen to enhance the magnitude of the velocity of the flow field at all points.

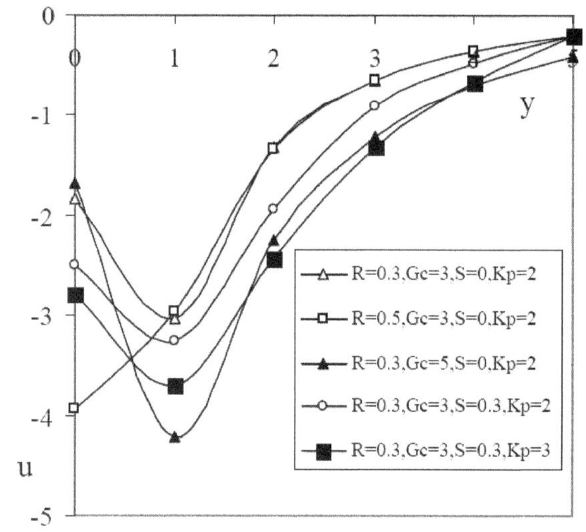

Figure 3. Velocity profiles against y for different values of G_c, K_p and R with M=2, G_r=3, P_r=0.71, S_c=0.22, v_0=2, ωt=π/2, ω=2

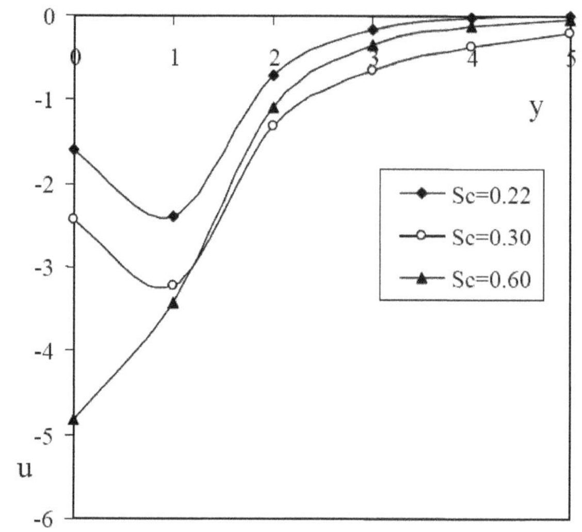

Figure 4. Velocity profiles against y for different values of S_c with M=2, R=0.3, G_r=3, G_c=3, K_p=2, P_r=0.71, S=0, v_0=2, ωt=π/2, ω=2

5. Conclusion

The above analysis points out the following interesting results of physical interest on the velocity of the flow field.

1. A growing magnetic parameter M retards the magnitude of the velocity of the flow field at all points due to the action of the Lorentz force acting on the flow field.
2. The heat source parameter S has an accelerating effect on the magnitude of the velocity of the flow field at all points.
3. The effect of growing Grashof number for mass transfer G_c and the permeability parameter K_p is to enhance the velocity (absolute value) of the flow field at all points.
4. An increase in Schmidt number S_c is to increase the magnitude of the velocity of the flow field at all points.
5. A growing rarefaction parameter R enhances the magnitude of the velocity of the flow field at all points.

References

[1] Gebhart B., Pera L. The nature of vertical natural convection flows resulting from the combined buoyancy effects thermal and mass diffusion. Int. J. Heat and Mass transfer, 1971; 14, 2025-2050,

[2] Georgantopoulos G. A., Koullias J, Goudas C. L., Courogenis C. Free convection and mass transfer effects on the hydro-magnetic oscillatory flow past an infinite vertical porous plate, J. Astrophysics Space Sciences. 1981; 74(2), 357-389.

[3] Hossain M. A, Begum R. A. Effect of mass transfer and free convection on the flow past a vertical plate. ASME J. heat transfer. 1984; 106, 664-668.

[4] Bejan A., Khair K.R. Heat and mass transfer by natural convection in a porous medium . Int. J. Heat Mass Transfer. 1985; 28, 909-918.

[5] Hossain M. A., Begum R.A. Effects of mass transfer on the unsteady flow past an accelerated vertical porous plate with variable suction. Astrophysics space Sci., 1985; 115, 145.

[6] Raptis A.A., Perdikis C.P. Oscillatory flow through a porous medium by the presence of free convective flow, Int. J. of Engg. Sci. 1985; 23(1), 51-55.

[7] Sattar M. A. Free and forced convection boundary layer flow through a porous medium with large suction. Int. J. Energy Research. 1993; 17, 1-7.

[8] Chamkha A. J. Hydromagnetic three-dimensional free convection on vertical stretching surface with heat generation or absorption. Int. J. Heat Fluid flow. 1999; 20, 84-92.

[9] Chamkha A. J., Khaled A. A. R., Hydromagnetic combined heat and mass transfer by natural convection from a permeable surface embedded in a fluid saturated porous medium. Int. J. Numerical Methods Heat and Fluid flow. 2000; 10(5), 455-476.

[10] Nagraju P., Chamkha A. J., Takhar H. S., Chandrasekhara B. C. Simultaneous radiative and convective heat transfer in a variable porosity medium. Heat Mass Transfer. 2001; 37, 243-250.

[11] Singh A. K., Singh A. K., Singh N. P. Heat and mass transfer in MHD flow of a viscous fluid past a vertical plate under oscillatory suction velocity. Ind. J. Pure Appl. Math., 2003;.34(3), 429-442.

[12] Hayat T., Mohyuddin M. R., Asghar S., Siddiqui A. M. The flow of a visco-elastic fluid on an oscillating plate. Z. Angew. Math. Mech., 2004; 84 (1), 65-70.

[13] Jain N. C., Gupta P. Unsteady hydromagnetic thermal boundary layer flow past an infinite porous surface in the slip flow regime. Ganita. 2005; 56(1), 15-25.

[14] Singh P., Gupta C.B. MHD free convective flow of viscous fluid through a porous medium bounded by an oscillating porous plate in slip flow regime with mass transfer. Ind. J. Theo.Phys., 2005;53(2), 111-120.

[15] Sharma P.R., Singh G. Unsteady MHD free convective flow and heat transfer along a vertical porous plate with variable suction and internal heat generation. Int. J. Appl. Math. and Mech. 2008; 4(5), 1-8.

[16] Das S. S., Satapathy A.,. Das J. K and. Panda J.P. Mass transfer effects on MHD flow and heat transfer past a vertical porous plate through a porous medium under oscillatory suction and heat source. Int. J. Heat and Mass Transfer, 2009, 52(25-26), 5962-5969.

[17] Das S. S., Tripathy U. K., Das J. K. Hydromagnetic convective flow past a vertical porous plate through a porous medium with suction and heat source, Int.J.Energy Environ. 2010; 1(3), 467-478..

[18] Das S. S., Parija S., Padhy R. K., Sahu M. Natural convection unsteady magnetohydrodynamic mass transfer flow past an infinite vertical porous plate in presence of suction and heat sink. Int. J.Energy and Environ. 2012; 3(2), 209-222.

[19] Das S. S., Saran M. R., Mohanty S., Padhy R. K., Finite difference analysis of hydromagnetic mixed convective mass diffusion boundary layer flow past an accelerated vertical porous plate through a porous medium with suction. Int J. Energy and Environ. 2014; 5(1), 127-138.

19

Experimental investigation and comparison of heat transfer coefficient of a two phase closed thermosyphon

I. Khazaee

Faculty of Mechanical and Energy Engineering, Shahid Beheshti University, A.C., Tehran, Iran.

Abstract
In this study, we investigated and reviewed the heat transfer equations of the evaporator and condenser of a two phase closed Thermosyphon as well as the differences between this equations for a working fluid and different conditions. The heat transfer limits of a two phase closed thermosyphon such as sonic, flooding (or entrainment) dry-out and boiling are investigated . A good agreement is observed between analytical results of this study and the analytical and experimental results of those available in the open literature. The heat transfer limits are very important in thermosyphons and have to be calculated accurately while the input heat must be lower than the heat transfer limits for the operations of thermosyphon without any problem. Also by using the experimental results a semi empirical equation for heat transfer coefficient of evaporator was proposed.

Keywords: Thermosyphon; Heat transfer limit; Flooding limit; Boiling limit; Dry-out limit.

1. Introduction
A two phase closed thermosyphon is a heat pipe needing no wicks to return the condensate working fluid from the condenser section to the evaporator one because of the gravity. However, the evaporator must be positioned at the lower part of the gravitation field, compared to the condenser. But, because of its simple structure, stable operating condition during steady state, the characteristic of a thermal diode and wide operating temperature range, it was applied to many industrial fields. The two phase closed thermosyphon is widely used because of its simple structure compared to the other types of heat pipes. Therefore, thermosyphons are being used in many applications such as : heat pipe, heat exchangers, cooling of electronic components, solar energy systems, deicing and snow melting. The analysis of two phase flow heat transfer in a two phase closed thermosyphon is very complicated. The condensation of heat transfer in the condenser was based on the Nusselt model and Faghri [1] compared it with some correlations. While in the evaporator, the nucleate boiling model made by Rohsenow in reference [2] was applied to predict the heat transfer coefficients and compared with the correlation suggested by Imura et al. [3].
Thermal performance of the thermosyphon is affected by many factors, such as the type of working fluid, filling ratio (ratio of volume of working fluid to volume of evaporator Section), aspect ratio (ratio of evaporator section length to inside diameter of pipe), inclination angle (from horizontal axis), operating pressure (or corresponding saturation temperature), and length of various sections of the pipe. Various heat transfer limits occur as a result of the dryout of liquid film in the evaporator. When the heat input to the evaporator is relatively high, intensive liquid film evaporation causes the vapor flow to move upwards, quickly, exceeding the flooding limit. The drying out of the liquid film may result from liquid

droplet entrainment from the liquid film by the interfacial shearing effect of high speed vapor flow, resulting in the entrainment limit.

2. Models of equations

The rate of heat transfer to the evaporator section of a thermosyphon can be calculated from the Eq. (1)

$$Q_{in} = V^2 / R = RI^2 \tag{1}$$

The rate of heat remove from the condenser section by the coolant water is obtained from Eq. (2)

$$Q_c = \dot{m} c_p (T_{o,w} - T_{i,w}) \tag{2}$$

The heat losses from the evaporator and the condenser are obtained by

$$Q_{loss} = \sigma A_t \varepsilon (T_{ins}^4 - T_\infty^4) + h A_t (T_{ins} - T_\infty) \tag{3a}$$

where ε = 0.085 and σ = $5.669 \times 10^{-8} W/m^2.K$ and free convection heat transfer coefficient on a vertical cylinder obtained from [2]:

$$Nu = \frac{hL_t}{k} = \left\langle 0.825 + \frac{0.387\, Ra^{1/6}}{\left[1 + \left(\frac{0.492}{\mathrm{Pr}}\right)^{9/16}\right]^{8/27}} \right\rangle^2 \tag{3b}$$

The experimental heat transfer coefficient of laminar film in condenser of a heat pipe is:

$$h_c = \frac{Q_c}{A_c (T_v - T_c)} \tag{4}$$

And the analytical correlation of the condenser section in reference [2] is given by:

$$\bar{h}_c = 0.943 \left\{ \frac{\rho_l g k_l^3 (\rho_l - \rho_v)\left[h_{fg} + 0.68 c_{pl} (T_v - T_c)\right]}{\mu_l L_c (T_v - T_c)} \right\}^{1/4} \tag{5}$$

This equation can be written in dimensionless form of average Nusselt and Reynolds number for liquid film:

$$Nu^{\bullet} = \frac{\bar{h}_c}{k_l} \left[\frac{\nu_l^2}{g} \left(\frac{\rho_l}{\rho_l - \rho_v} \right) \right]^{1/3} = 0.925\, \mathrm{Re}_{l,\max}^{-1/3} \qquad \mathrm{Re}_{l,\max} \prec 325 \tag{6}$$

$$\mathrm{Re}_{l,\max} = \frac{Q_c}{\pi D \mu_l h_{fg}} \tag{7}$$

Several correlations of heat transfer coefficient in evaporator section of a two phase closed thermosyphon were investigated. Rohsenow in reference [2] reported a model for nucleate boiling as:

$$q_s = \mu_l A h_{fg} \left[\frac{g(\rho_l - \rho_v)}{\sigma} \right]^{1/2} \left(\frac{c_{pl}(T_e - T_v)}{C_{sf} h_{fg} \Pr^n} \right)^3 \tag{8}$$

where $q_s = h_e(T_e - T_v)$

[For example, for water-copper $C_{s,f} = 0.0068$, $n = 1.0$ where $C_{s,f}$ is the solid-fluid constant]

And the correlation that Imura et al. [3] suggests for the evaporator section of a two phase closed thermosyphon is:

$$\bar{h}_e = 0.32 \left(\frac{\rho_l^{0.65} k_l^{0.3} c_{pl}^{0.7} g^{0.2} q_e^{0.4}}{\rho_v^{0.25} h_{fg}^{0.4} \mu_l^{0.1}} \right) \left(\frac{p_v}{p_{atm}} \right)^{0.3} \tag{9}$$

In which q_e is the applied heat flux in the evaporator section.

Although thermosyphons are very efficient heat transfer devices, they are subject to a number of heat transfer limitations. These limitations determine the maximum heat transfer rate that a thermosyphon can achieve under certain working conditions. These limits can arise from the cessation of capillary pumping and the attainment of sonic velocity in the heat pipe vapor.

The sonic limit is the highest possible heat transport rate that can be sustained in a heat pipe for a specific vapor temperature at the evaporator end of a heat pipe. The sonic limit is reached when the vapor leaving the evaporator attains sonic velocity. An expression of sonic limit is given by Dunn and Reay [4]:

$$Q_{c,90} = \rho_v h_{fg} A_v \sqrt{\frac{\gamma R_v T_v}{2(\gamma + 1)}} \tag{10}$$

where $Q_{c,90}$ is the heat transfer rate limits at a vertical position.

The correlation of flooding limit investigated by Faghri [1] is given as:

$$Q_{c,90} = K h_{fg} A \left[g\sigma(\rho_l - \rho_v) \right]^{1/4} \times \left[\rho_v^{-1/4} + \rho_l^{-1/4} \right]^{-2} \tag{11}$$

where

$$K = \left(\frac{\rho_l}{\rho_v} \right)^{0.14} \tanh^2 (Bo)^{1/4} \tag{12}$$

The dry out limit is reached in a two phase closed thermosyphon at the bottom of the evaporator section when the fill charge ratio is very small. For this, the entire amount of working fluid should be circulating throughout the thermosyphon.

Rosler et al. [5] used a thermosyphon with R113 as the working fluid employing filling ratios between 0.02 and 0.4 and experimentally and theoretically correlated an equation for dry out limit which is a function of filling ratio, the properties of fluid and heated length:

$$FR = 1 - \frac{1}{1 + C \left[\frac{(Q_{c,90} / A h_{fg})^2}{2\sigma\rho_v} \left(\frac{3 Q_{c,90} \mu_l L_e}{A \rho_l (\rho_l - \rho_v) g h_{fg}} \right)^{1/3} \right]^{3/4}} \tag{13}$$

where C is the exponential constant. This equation can be solved for $Q_{c,90}$ as a function of filling ratio and fluid properties and the evaporator length.

Busse [6] carried out a two dimensional analysis of viscous limit in a two phase closed thermosyphon and found that

$$Q = \frac{D_e^2 h_{fg} \rho_v P_v}{64 \mu_v L_v} \tag{14}$$

where P_v *and* ρ_v are the pressure and density of vapor and this expression changes with vapor characteristics and inside diameter of evaporator.

The transfer of heat into a heat pipe produces a temperature gradient across the wall of the tube. At the interface between the wall and the vapor space, the liquid temperature on the wall is equal to or greater than that of the adjacent saturated vapor. The liquid temperature rises steadily with distance from the interface, reaching a maximum value at the outer surface. The heat transfer rate corresponding to the incipient boiling condition is called the boiling limit.

Grobis et al [7] proposed empirical correlations for the maximum radial heat flux at the boiling limitation in a two phase closed thermosyphon.

$$\frac{Q_{c,90}}{Q_{c,\infty}} = C^2 \left[0.4 + 0.006 D_i \sqrt{\frac{g(\rho_l - \rho_v)}{\sigma}} \right]^2 \tag{15}$$

where $Q_{c,\infty}$ is the critical heat flux for pool boiling that is:

$$Q_{c,\infty} = 0.142 A \sqrt{\rho_v} \left[g \sigma (\rho_l - \rho_v) \right]^{1/4} \tag{16}$$

And

$$C = 0.538 \left(\frac{D_i}{L_c} \right)^{-0.44} \left(\frac{D_i}{L_e} \right)^{0.55} \psi^{0.13} \quad for \quad \psi \le 0.35$$
$$C = 3.54 \left(\frac{D_i}{L_c} \right)^{-0.44} \left(\frac{D_i}{L_e} \right)^{0.55} \psi^{-0.37} \quad for \quad \psi \succ 0.35 \tag{17}$$

3. Description of experiments procedures

For experimental investigation of heat transfer specifications in two phase closed thermosyphon a setup has been fabricated. Figure 1 shows the appearance of a two phase closed thermosyphon for testing and investigating the thermal performance.

In these experiments two copper tubes of 1000 mm length with inside diameter of 15 and 25 mm and 2 mm thickness were employed (Figure 1).

The working length of the thermosyphon consists of three parts; a lower part of 430 mm as the evaporator section, the middle part of 160 mm as the adiabatic section, and a upper part of 410 mm as the condenser section. The 410 mm long water jacket was surrounding the condenser section. Inlet and outlet connections located obliquely across each other to introduce swirl flow. Eight Ni-Cr thermocouples were installed mechanically to the surface of the evaporator and adiabatic and condenser to monitor the temperature distribution. A personal computer and a data logger were used to show the temperatures measured by thermocouples. The two phase closed thermosyphon was surrounded by 40 mm thickness of glass wool for insulating and stopping the heat transfer to the environment. Table 1 shows the specifications of two phase closed thermosyphon for this study.

An electrical resistance of 1000 W was employed to produce the heat input of evaporator and the accuracy of monitoring for voltage and electrical current was $\pm 2\%$. The temperature of inlet and outlet coolant water was measured by digital thermometer and the coolant mass flow rate of water was measured by a Rotameter. Also a vacuum pump was employed to produce vacuum condition inside the thermosyphon.

The measured parameters were: heat input of the evaporator section Q_{in}, heat output of the condenser section Q_c, temperature of inlet and outlet of the coolant water, the coolant mass flow rate of water, and the surface temperatures of evaporator, adiabatic and condenser sections. The experimental limits were as follows:

- Heat input (Q_{in}) between 100 and 200 W
- Aspect ratio (AR) 28.6 and 17.2
- Filling ratio (FR) between 15 and 100%
- Inclination angle (φ) between 15° and 90°
- Coolant mass flow rate ($\dot{m}$) between 0.00997 and 0.03 kg/s

At first, the experiments were performed taking AR as 28.6 for various filling ratio, inclination angle, heat input, and coolant mass flow rate and measure the pointed parameters and then the experiments were repeated for AR=17.2 .

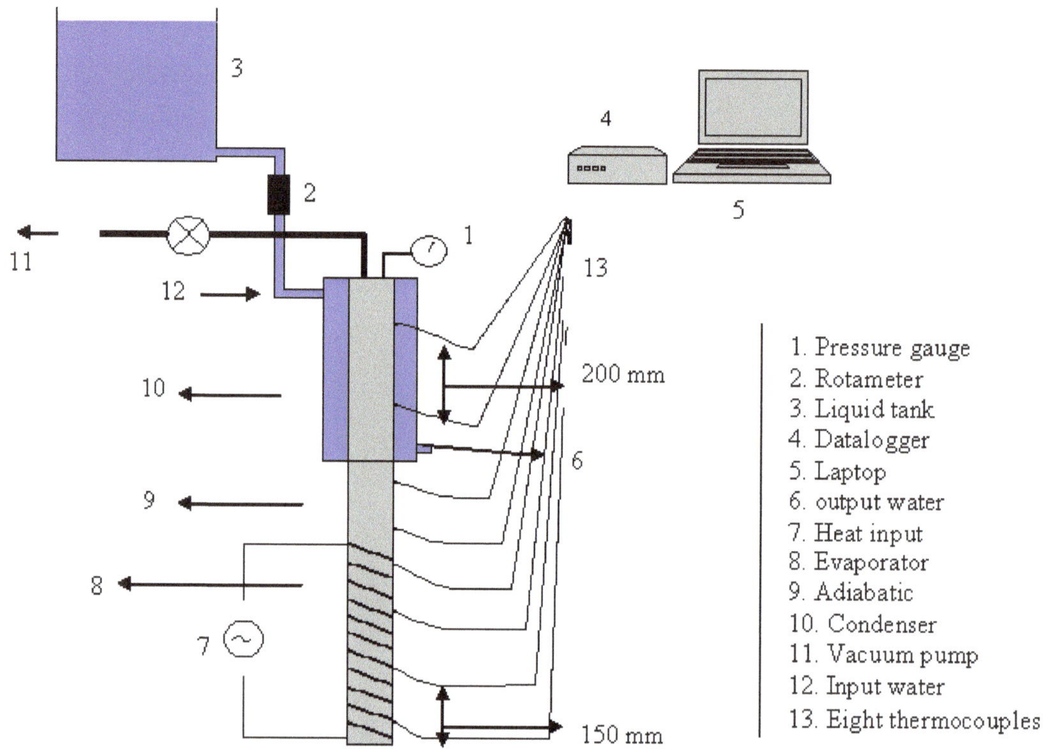

Figure 1. Experimental appearance of thermosyphon

Table 1. Specifications of the two phase closed thermosyphon

ID	OD	Evap L	Adia L	Cond L	Material	Working Fluid
15 mm	17 mm	430 mm	160 mm	410 mm	Copper	ethanol
25 mm	27 mm	430 mm	160 mm	410 mm	Copper	ethanol

4. Results and discussion

Figure 2 illustrates the experimental data of Park et al [8] for C_6F_{14} as the working fluid and three fill charge ratio. It can be seen that the Nusselt number increases by increasing the fill charge ratio. It can be explained that during the steady state condition, the heat transfers directly from the high temperature working fluid to the lower part of the condenser without evaporating process.

Noie et al. [9] investigated the heat transfer coefficient of condenser versus the inclination angle for different filling ratios. They used Eq(4) and found that the condensation heat transfer increases as the filling ratio increases and the maximum condensation heat transfer coefficient takes place at φ=30° for FR=22% and 30% and at φ=45° for FR=15%. Figure 3 shows their experimental data for heat transfer coefficient.

Figure 4 shows the heat transfer coefficient in evaporator versus heat flux for correlations of Rohsenow [2], Eq(8) for C_6F_{14}, $C_{s,f} = 0.004$ and Imura [3], Eq (9) and the experimental data of Park [8] for various fill charge ratios. This Figure shows that in about 1000-30000 (W/m^2) evaporator heat flux, the heat transfer coefficient increases from 300-3400 (W/m^2k) and the experimental data are at good agreements with Rohsenow [2] correlation.

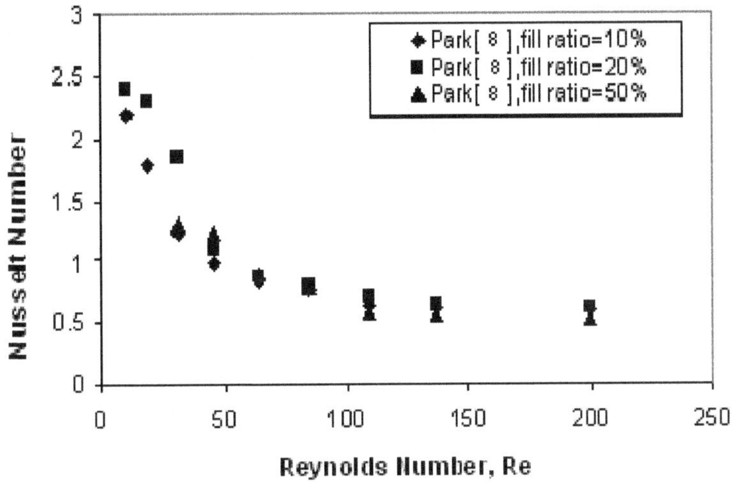

Figure 2. Nusselt number versus Reynolds number for three fill charge ratio for C_6F_{14} at $T_v = 50℃$

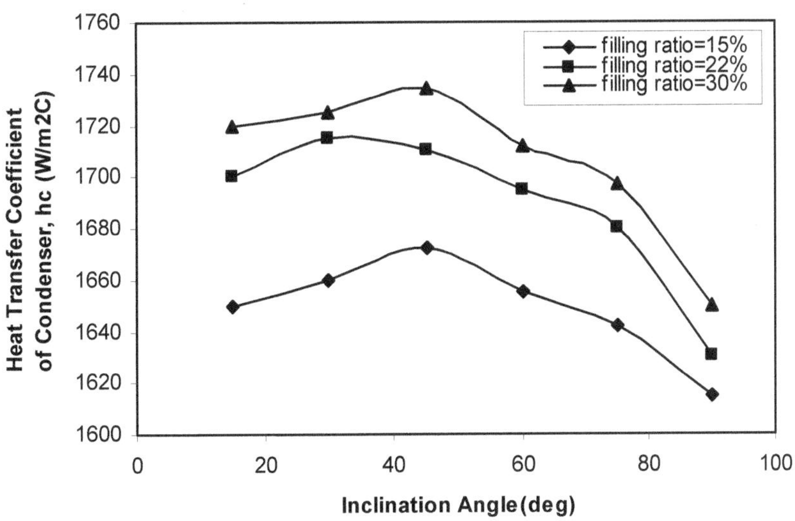

Figure 3. Heat transfer coefficient of condenser versus inclination angle for different filling ratios[9]

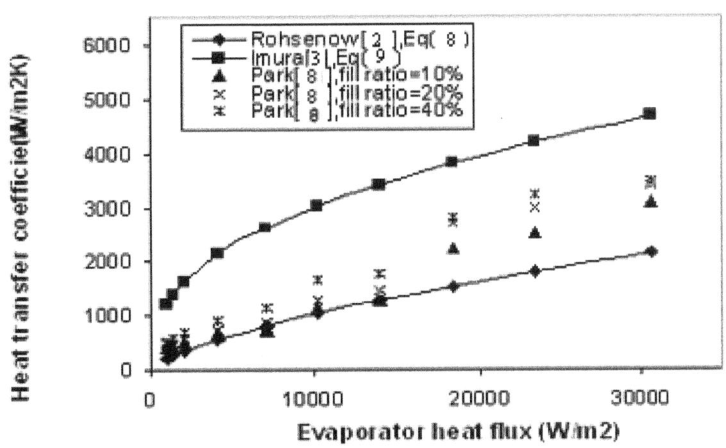

Figure 4. Heat transfer coefficient versus heat flux in evaporator for different fill charge ratios for C_6F_{14} *at* $T_v = 50℃$

Noie [10] investigated the heat transfer coefficient of evaporator of a two phase closed thermosyphon versus the average temperature of evaporator surface for different filling ratios and compared them with the correlations of Rohsenow [2] and Imura [3]. For his investigation, the experimental results were found to be in reasonable agreement with correlations of Rohsenow [2] and Imura [3]. Figure 5 shows the experimental data of Noie [10] for heat transfer coefficient of evaporator.

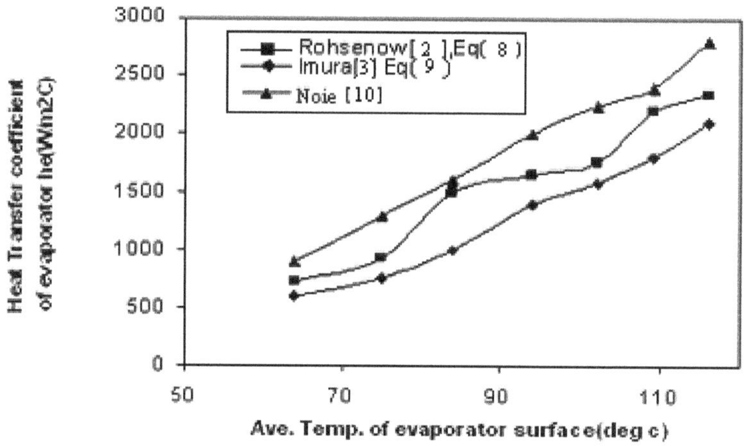

Figure 5. Heat transfer coefficient of evaporator versus average temperature of evaporator surface for different filling ratios

There is no scarcity of experimental data on limiting heat pipe heat transport rates in the literature. Such data are frequently presented along with analytical predictions of heat transport limits. In principle, if the calculated and experimental limits are in a reasonable agreement, the analytical methods may be considered to be valid.

Figure 6 presents the comparison of experimental results of Rosler et al [5] with the theoretical predictions of Eq (13) by the dryout regions and R113 as the working fluid. The continues lines at Figure 6 show the theoretical results for three different vapor temperature. As it can be seen, for small fill ratios ($FR \leq 80\%$), the experimental data [5] agree very well with the theoretical predictions [5].

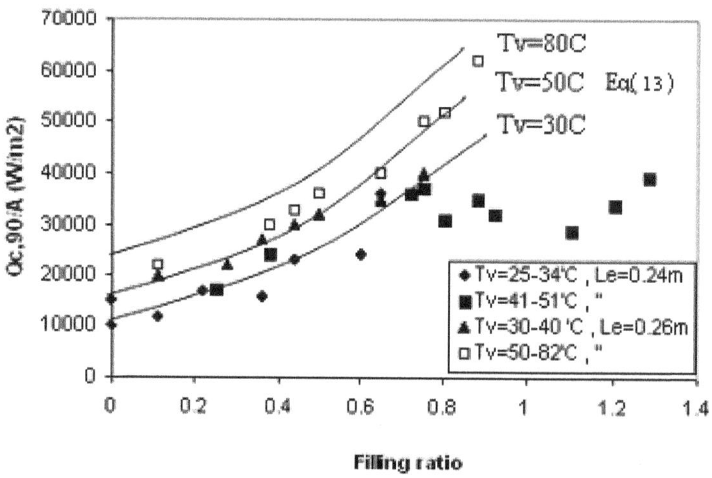

Figure 6. Dryout limit of experimental data[5] and the Eq(13) for R113 for different Tv

Figure 7 shows the comparison of boiling limit from Grobis [7], Eq (15) for R113 as the working fluid at $T_v = 30℃$ and the experimental data [7]. It can be seen that there are agreements with the analytical and the experimental data.

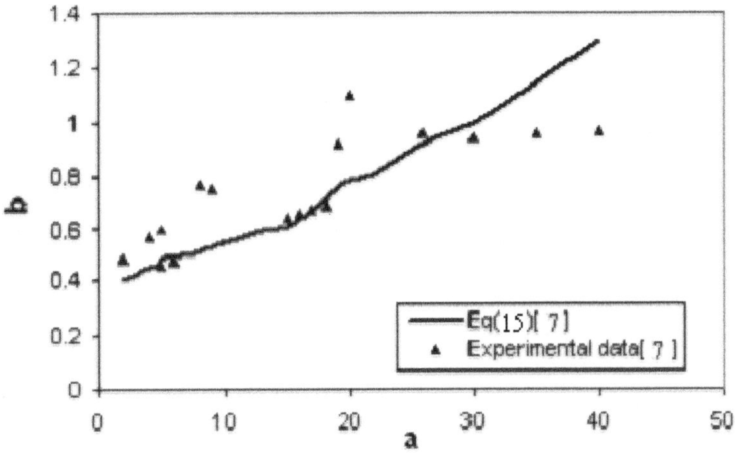

Figure 7. Boiling heat transfer for water in two phase closed thermosyphon where $a = R\sqrt{\frac{g(\rho_l - \rho_v)}{\sigma}}$ and $b = \frac{1}{C}\sqrt{\frac{Q_{c,90}}{Q_{c,\infty}}}$

5. A new correlation for heat transfer coefficient of evaporator

By doing the experiments, it is clear that changing Temperature and some thermodynamic properties affect the heat transfer coefficient of evaporator inside the thermosyphon.

The method of fitting used in this paper is the least square method. This method fits a set of data points (x_i, y_i) to a function that is a combination of any number of functions of the independent variable x. The goal of nonlinear regression is to determine the best-fit parameters for a model by minimizing a chosen merit function. Where nonlinear regression differs is that the model has a nonlinear dependence on the unknown parameters, and the process of merit function minimization is an iterative approach. The process is to start with some initial estimates and incorporates algorithms to improve the estimates iteratively. The new estimates then become a starting point for the next iteration. These iterations continue until the merit function effectively stops decreasing. The nonlinear model to be fitted can be represented by:

$$y = y(x;a) \tag{18}$$

The merit function minimized in performing nonlinear regression the following:

$$\chi^2(a) = \sum_{i=1}^{N}\left\{\frac{y_i - y(x_i;a)}{\sigma_i}\right\}^2 \tag{19}$$

where σ_i is the measurement error, or standard deviation of the *i*th data point. For understanding how the results calculated we have:

The *ith* predicted, or fitted value of the dependent variable Y, is denoted by $\hat{Y}_i$. This value is obtained by evaluating the regression model $\hat{Y} = f(X, \hat{\beta}_j)$, where $\hat{\beta}_j$ are the regression parameters, or variables. Then the residuals $(Y_i - \hat{Y}_i)$ and sum of the residuals $\sum_{i=1}^{n}(Y_i - \hat{Y}_i)$ calculated and then, the average of residuals and residual of sum of squares calculated.

SSE = Residual or Error Sum of Squares (Absolute) = $\sum_{i=1}^{n}(Y_i - \hat{Y}_i)^2$

SSE_R = Residual or Error Sum of Squares (Relative) $= \sum_{i=1}^{n}[(Y_i - \hat{Y}_i)^2 * W_i]$ where $W_i = 1/\sigma_i^2$ normalized so that $\sum_{i=1}^{n} W_i = n$.

σ_i = the standard deviation of the i_{th} data point Y_i and n is the number of data points, or observations. The principle behind nonlinear regression is to minimize the residual sum of squares by adjusting the parameters $\hat{\beta}_j$ in the regression model to bring the curve close to the data points. This parameter is also referred to as the error sum of squares, or *SSE*. If the residual sum of squares is equal to 0.0, the curve passes through every data point.

Thus, the correlation that proposed to indicate the effect of those parameters on the heat transfer coefficient of evaporator is:

$$h_e = a\left(\frac{q}{h_{fg}}\right)^b \left(\frac{\mu_l}{\sqrt{\frac{\sigma}{g(\rho_l - \rho_v)}}}\right)^c (C_p)(pr)^d (\rho_l)^e \tag{20}$$

In Eq (20) the constants a, b, c, d and e are undefined and by using software as Datafit which fits the results of experiment from one to more independent variables that in Table 2 the value, upper limit and lower limit of constants in Eq (20) was shown. Hence, by analyzing the results of experiments, Eq (20) converts into Eq (21) as

$$h_e = 8.09\frac{k_l^{1.59}\ \sigma^{1.09}\ \rho_l^{0.66}\ q^{0.94}}{h_{fg}^{0.94}\ (\rho_l - \rho_v)^{1.09}\ C_p^{0.59}\ \mu_l^{3.77}\ g^{1.09}} \tag{21}$$

Table 2. Value and limits of constants in Eq (20)

Variable	Value	Lower limit	Upper llimit
a	8/09E-20	-372/235	372/397
b	0/939901	0/380878	1/498924
c	-2/1827	-281/978	277/6128
d	-1/59859	-335/766	332/5689
e	0/661656	-715/295	716/6181

Figure 8 shows the comparison between the experimental and theoretical results of the heat transfer coefficient for AR= 17.2 FR=80% , φ=90° and $\dot{m} = 0.0199 kg/s$ for water as working fluid and it is clear that there are significant agreements with them.

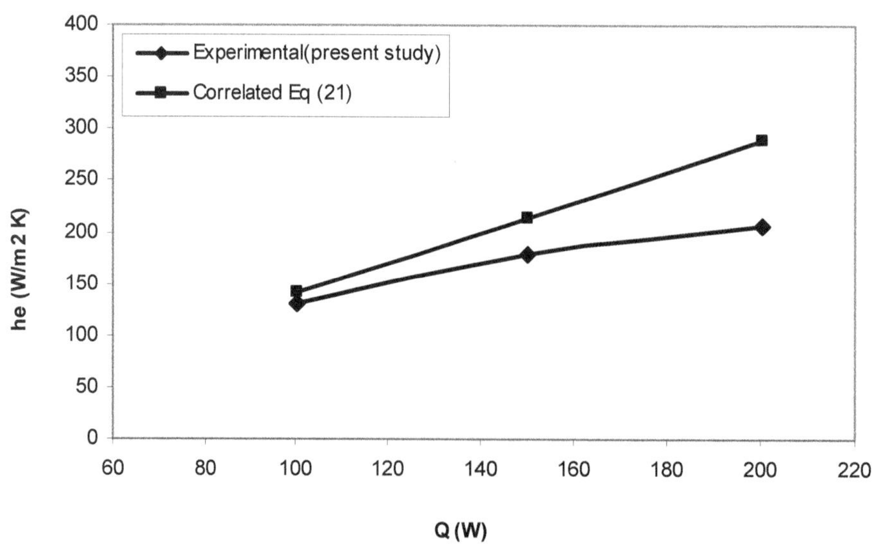

Figure 8. Comparison between experimental and correlated Eq

6. Conclusion

In this study the heat transfer coefficient of condenser and evaporator and the heat transfer limits were investigated. The analytical heat transfer coefficient and the heat transfer limits and the experimental data from several researches were compared with each other. Based on these findings, the following conclusions can be drawn:

1. There are good agreements between analytical and experimental results of previous works.
2. The condensation and evaporation heat transfer coefficient increases as filling ratio increases.
3. Maximum of the heat transfer coefficient of condenser is at about $40° < \varphi < 50°$.
4. The critical heat flux as the dryout limit increases as the fill charge ratio increases.
5. A new correlation for predicting the heat transfer coefficient of thermosyphon according to thermodynamic properties of working fluid was proposed and there is a good agreement between its results and the experimental results.

References

[1] Faghri,A., (1995) Heat pipe science and technology, Taylor& Francis, Washington, D.C.,USA.
[2] Incropera, F.P. and Dewitt, D.P., (2002) Introduction to heat transfer, 4th ed., Wiley.
[3] Imura, H. Sasaguchi ,K. and Kozai, H. (1983)Critical heat flux in a closed two phase thermosyphon, Int. J. Heat Mass Transfer, 26(8), pp. 1181-1188.
[4] Dunn, P.D.,and Reay, D.A.,Heat pipes, 3rd ed., Pergamon Press, Oxford,U.K.,1982.
[5] Rosler, S. Takuma, M. Groll, M.and Maezawa, S. (1987) Heat transfer limitation in a vertical annular closed two phase thermosyphon with small fill rates, Heat Recovery Systems,7(4), pp. 319-327.
[6] Busse, C.A., (1973) Theory of ultimate heat transfer limit of cylindrical heat pipes. Int. J. Heat and Mass Transfer,16, pp. 169-186.
[7] Grobis, Z.R. and Savchenkov, G.A. Low Temperature two phase closed thermosyphon Investigation, Proc. 2nd International Heat Pipe Conf., Bologna, Italy, pp. 37-45, 1976.
[8] Park, Y.J. Kang , H.K.and Kim, C.J. Heat transfer characteristics of a two-phase closed thermosyphon to the fill charge ratio, International Journal of Heat and Mass Transfer. 45, 4655-4661, 2002.
[9] Noie, S.H., Sarmastiemami, M.R., and Khoshnoodi, M., (2007) Effect of inclination angle and filling ratio on thermal performance of a two-phase closed thermosyphon under normal operating conditions, Heat Transfer Engineering, 28(4), 365-371.
[10] Noie, S.H., (2005) Heat transfer characteristics of a two phase closed thermosyphon, Applied Thermal Engineering,25, pp.495-506.

20

Optimization of operating parameters to maximize the current density without flooding at the cathode membrane interface of a PEM fuel cell using Taguchi method and genetic algorithm

S.S.L. Rao, A. Shaija, S. Jayaraj

Department of Mechanical Engineering, National Institute of Technology, Calicut-673601, India.

Abstract
A mathematical model was developed to investigate water accumulation at the cathode membrane interface by varying different operating parameters like fuel cell operating temperature and pressure, cathode and anode humidification temperatures and cathode stoichiometry. Taguchi optimization methodology is then combined with this model to determine the optimal combination of the operating parameters to maximize current density without flooding. Results of analysis of variance (ANOVA) show that fuel cell operating temperature and cathode humidification temperature are the two most significant parameters in the ratio of 56.07% and 27.89% respectively and also that higher fuel cell temperature and lower cathode humidification temperature are favourable to get the maximum current draw without flooding at the cathode membrane interface. The global optimum value of the operating parameters to maximize the current density without flooding was obtained by formulating as an optimization problem using genetic algorithm (GA). These results were compared with the results obtained using Taguchi method and it was found to be similar and slightly better.

Keywords: Cathode flooding; Genetic algorithm; PEM fuel cell; Taguchi analysis; Water management.

1. Introduction

Proton exchange membrane fuel cells (PEMFCs) are suitable for portable, mobile and residential applications, due to their inherent advantages, such as compact and lightweight, work at low temperatures with a high output power density, no pollutant emission and offer superior system start-up and shutdown performances. Despite the greater advantages of PEMFCs, their large scale commercialization is still hampered by their higher cost material and lower reliability and durability. Water management is one of the most critical key factors that hinder PEMFC to be competitive with portable and automotive applications and its optimization is very important to achieve maximum performance and durability. It is known that proton conductivity of Nafion membranes in PEMFC depends on its water content which is desirable. But excessive water can cause flooding in the catalyst layer and the Gas Diffusion Layer (GDL) of the cathode side, resulting in a higher mass transport resistance. Therefore, an accurate evaluation of the water content at the cathode membrane interface in the PEMFC is required to study the proper functioning of the PEMFC.

1.1 Water management

Water transport from and to the cathode catalyst layer (CCL) also called cathode membrane interface of a PEMFC involves several transport processes as shown in Figure 1. They include:

a) Water supplied through the humidification of cathode gas
b) Electro-osmotic drag of water from anode to cathode
c) Water generated through electrochemical reaction at the cathode catalyst layer
d) Water source from the membrane desorption occurs in the cathode side catalyst layer
e) Water source from the interfacial mass transfer between liquid and vapor phase during evaporation and condensation
f) Back diffusion of water from cathode to anode through the membrane
g) Hydraulic permeation of water through the membrane from cathode to anode due to the gas pressure gradient across the membrane
h) Water flux transported from cathode to anode through the PEM under capillary pressure
i) Water transported through the cathode gas diffusion layer under capillary pressure
j) Removal of water through the cathode outlet.

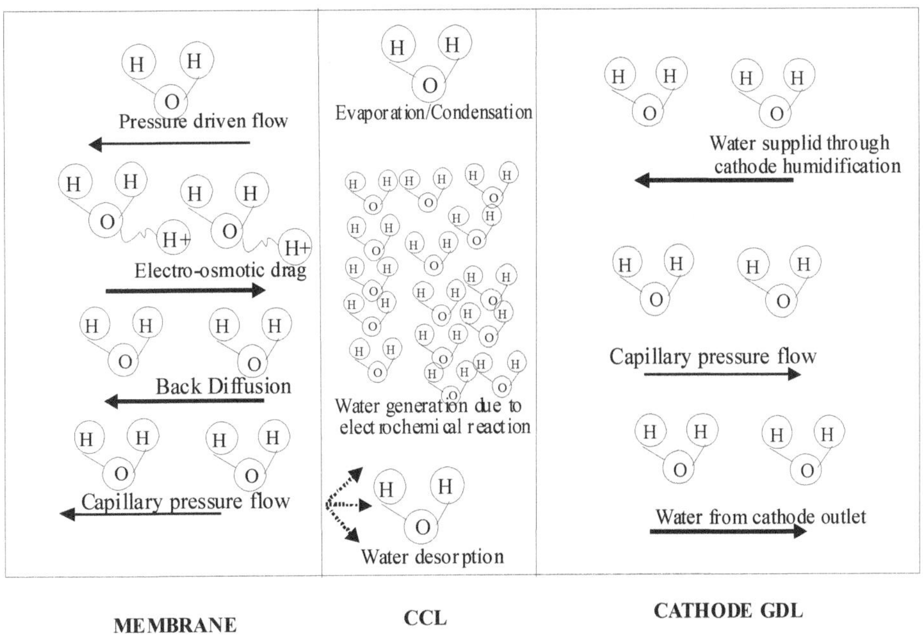

Figure 1. Water transport from and to the cathode catalyst layer

When a current is drawn from the fuel cell, the protons migrate through the membrane which is associated with a drag of water molecules from the anode to the cathode side depending on the water content in the membrane. This phenomenon is called electro-osmotic drag transport. Water produced at the CCL due to electrochemical reaction, water desorption from the cathode side of the membrane, water source from the liquid vapor interaction at the CCL and electro-osmotic drag transport together results in an accumulation of water at the cathode membrane interface. In turn, the water concentration in the cathode side increases significantly compared to the anode side during fuel cell operation. This concentration gradient causes back diffusion of water from CCL to anode side through the membrane which works against the drying of the membrane from the anode side. Hydraulic permeation of water through the membrane from cathode to anode can occur as a result of gas pressure gradient across the membrane or the capillary pressure gradient across the membrane.

Several researchers have worked to describe water transport in PEMFCs and the early contributions being of Bernardi [1] who presented a mathematical model of the PEMFC that identifies the operating conditions resulting in water balance. This model concludes that water balance is much more sensitive to changes in humidity of the inlet air than changes in humidity of fuel and flow rates of reactants. A water and heat management model for PEMFC developed by Nguyen and White [2] concludes that anode stream must be humidified to minimize the ohmic losses and when air is used instead of pure oxygen, the cathode stream must also be humidified. Wang *et al.* [3] studied experimentally the effect of operating parameters like fuel cell operating temperature and pressure, anode and cathode humidification temperatures on the performance of PEMFC. He *et al.* [4] proposed electronic flooding monitoring device to detect electrode flooding in PEMFCs and confirmed that increasing air flow rate or cell temperature increased the water removal rate from backing layers. Natarajan and Nguyen [5] considered the effects of flow rate, inlet humidity and temperature on the water flooding in the cathode. They concluded that higher fuel cell temperature and stoichiometric flow rates and lower inlet stream humidity facilitated better water removal resulting in higher net current.

The effect of operating parameters that affect the water generation and removal process, such as the operating temperature and relative humidity of anode and cathode streams was studied by Lin *et al.* [6] and found that water accumulation within the catalyst layer was more severe than the backing layer. Lim and Wang [7] studied the effect of hydrophobic content in GDL on the performance of PEMFC and showed that power performance varies strongly with the cathode humidification temperature and air flow rate. Ge *et al.* [8] presented a model for the absorption, desorption and water transport through the membrane and the simulation results shows that the mass transfer coefficient for absorption of water, which takes place at anode membrane interface is much lower than desorption of water, which takes place at cathode membrane interface.

Pasaogullari and Wang [9] developed theory of water transport and is applied to simulate flooding in PEMFC and its effects on cell performance. They concluded that the humidification level and flow rate of reactant streams are key parameters in controlling fuel cell performance and flooding depends not only on the material characteristics but also on the optimization of these operating parameters. A new scheme of water management based on redirection of water fluxes toward anode compartment has been presented by Kraytsberg and Ein-Eli [10] which discussed water transport through PEM under capillary pressure. Santarelli and Torchio [11] have done experimental analysis of the effects of operating variables on the performance of PEMFC and showed that at low fuel cell temperature, cathode humidification temperature should be less at high loads to avoid cathode flooding. Buaud *et al.* [12] conducted tests on a wide range of stoichiometry, for different values of current and stack temperatures using ambient air to detect the minimum value above which no flooding of the stack happens. They concluded that conditions to get flooding of the stack depend on the stack temperature, humidification of air and stoichiometry.

Dai *et al.* [13] presented a method to measure water transport through the membrane and GDL and identified the fuel cell operating conditions (current density, temperature, air stoichiometry and relative humidity) corresponding to membrane drying and flooding conditions for a particular GDL. Hussaini and Wang [14] visualized the process of cathode flooding in an operating PEMFC, where relative humidity of 26%, 42% and 66%, current densities of 0.2, 0.5 and 0.8 A/cm^2, and cathode stoichiometry from 2 to 4 are used. It was noticed that low current density and low stoichiometry are found to cause severe flooding under all considered humidity conditions. Accumulation of water in the cathode and anode channels of PEMFCs have been investigated by direct visualisation where water droplets and slugs were quantified over a range of operating conditions by Ous and Arcoumanis [15]. Increasing air stoichiometry and operating temperature were found to be very effective for the water extraction process. Falcao *et al.* [16] presented a simple steady state one dimensional water transport through a PEMFC with heat transfer effects. This model predicts that both anode and cathode streams must be humidified to avoid membrane dehydration, particularly at higher current densities. Misran *et al.* [17] investigated water transport characteristics in a single cell PEMFC at various operating pressures and temperatures. The results show that the electro-osmotic drag coefficient decreased and back-diffusion of water increased with temperature, hence less water presented at higher temperatures at the cathode side.

1.2 Taguchi method

Taguchi method is a powerful optimal design tool, which provides a simple, efficient and systematic approach to optimize designs for performance, quality and cost [18]. The main advantage of using Taguchi method for optimization is its use of orthogonal array (OA) for design simplifying the task of planning experiments greatly. An orthogonal array is a fractional factorial matrix, which assures a balanced comparison of levels of any factor or interaction of factors. Taguchi proposed signal to noise (SNR) ratios to find out the favourable levels of control factors. There are three different types of the performance characteristics in the analysis of the SNR, i.e. the lower-the-better, the higher-the-better and the nominal-the-best. For each type of the performance characteristics, a larger SNR corresponds to the better quality characteristics. The application of Taguchi method to PEMFC system has not received much attention in the quality analysis till recently. A very few researchers recently have employed

Taguchi method to obtain the optimal combination of parameters to improve the performance of PEM fuel cell.

Kaytakoglu and Akyalcin [19] employed Taguchi method to determine the optimum operating conditions of pressure and temperature of fuel cell, anode and cathode humidification temperatures and stoichiometry. They found that the operating pressure has the higher SNR to achieve maximum stack performance. However, higher operating pressures also require higher pumping power which reduces the net power. Yu *et al.* [20] employed the method of the design of experiments to obtain the optimal combination of six operating parameters (fuel cell temperature and pressure, anode and cathode humidification temperatures and anode and cathode stochiometric flow ratios) each having two levels. This study showed that operating pressure has the most significant effect on the fuel cell performance followed by the fuel cell temperature. Wu and Gu [21] conducted experiments with the Taguchi method to determine the optimal combination of six primary operating parameters including flow orientation of PEMFC. They concluded that the flow orientation is more significant than fuel and oxidant stoichiometries. Dalasm *et al.* [22] proposed a model to investigate the performance cathode catalyst layer (CCL) by statistical methods along with artificial neural network model. Thickness and membrane volume content in CCL were found to be significant parameters affecting its performance. Cheng *et al.* [23] studied the influence of flow channel design, operational temperature and relative humidity of cathode stream at three levels using Taguchi full factorial design on the performance of PEMFC based on computational fluid dynamic simulation. Results show that flow channel design has the most significant effect on the polarization curve followed by the cell temperature. Wu and Ku [24] used Taguchi methodology to achieve the maximum electrical power and minimum pressure drop of a PEMFC and the results show that out of the three levels, the optimal combination is cell temperature of 333 K (level 2), anode humidification temperature of 333 K (level 2), cathode humidification temperature of 313 K (level 1) and oxygen stoichiometric flow ratio of 1.2 (level 1) to achieve the maximum performance. Solehati *et al.* [25] investigated the significance of the key operating parameters with regard to the stack performance and efficiency using a computational study with Taguchi statistical method. The results show that high stack performance does not always associate with high stack efficiency due to higher parasitic load.

1.3 Genetic algorithm

Genetic algorithm (GA) [26] is a search technique used in computing the true or approximate solution to optimization and a search problem based on the biological principles proposed by Holland. A flow chart involving different steps of GA used to obtain the optimal parameters is shown in Figure 2. GA is based on the Darwinian law of survival of the fittest where the parents with stronger fitness ability will be evolved in the end. As in nature, the selection, reproduction, and mutation operations are performed probabilistically, based on the fitness of a chromosome, which measures the value of chromosome as a solution.

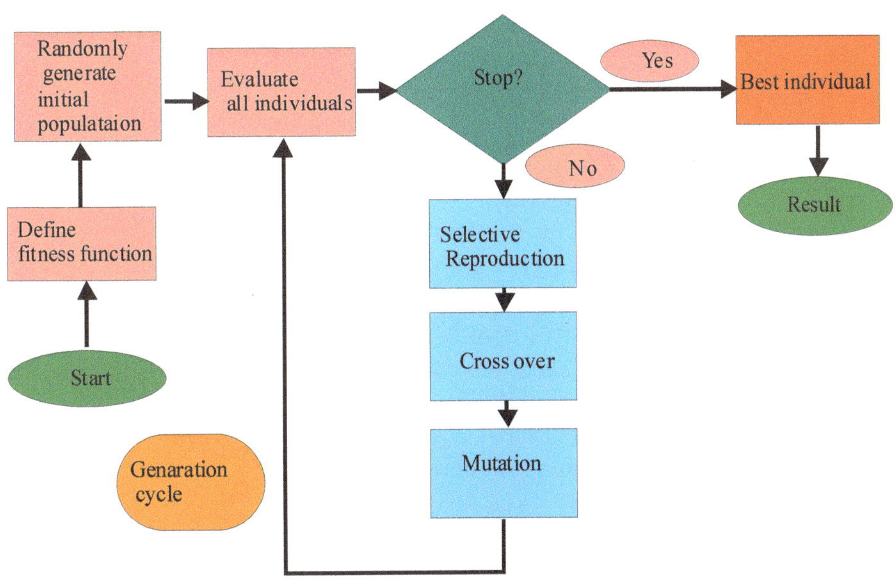

Figure 2. Flow chart of GA solution setup procedure

When two parents are selected to reproduce, the process where the resultant offspring end up having half the genes from one parent and half from the other is called crossover. The new offspring can be by random mutation in which one or more positions on the chromosomes are bit changed. After a number of generations, the best chromosome with the stronger fitness ability in the environment eventually will be evolved in the end. Therefore, GA is applied to avoid local point and reduce the time consumption to search the global optimum the performance of design factors. Recently, a very few work has been reported which deals with the optimization of design factors in PEMFC using GA.

Chang [27] proposed genetic algorithm neural network (GANN) model with and without Taguchi method to estimate the output voltage of PEMFC. The proposed GANN with Taguchi method was found to be better in the estimation of output voltage for PEMFC. Miao *et al.* [28] proposed methodology of meta-model design of optimization by combining Taguchi method with artificial neural network and GA to improve the performance of PEMFC. The results show that the improvement value of power density from Taguchi conditions to the meta-modeling approach was 5.44%. Cheng *et al.* [29] also employed meta-model optimal approach using Taguchi orthogonal array, radial basis function neural network and GA to improve power density in PEMFC. It was noticed that fuel cell and anode humidification temperatures were more significant as compared to the anode and cathode stoichiometry in improving the performance of the fuel cell.

Most of the models in the literature were found to focus on the parametric study of the effects of varying a single operating parameter at a time on water management and flooding of PEMFC, but evaluation of the influence of a single parameter, keeping other properties constant, is difficult. The major disadvantage of one factor method is that it fails to consider possible interaction between factors studied and results in a large cost when the number of experimental runs increases. Therefore, it is important to have an adequate optimization method to solve for the performance of the fuel cell. Although, there are numerous applications of Taguchi's experimental design method, only a few reported work deals with the optimum combination of the operating parameters analysis for maximizing the power density and efficiency of PEMFC. The main advantage of this method over other statistical methods is that, the parameters considered for the experimentation can be examined as controlling or non-controlling. However, no work has been reported in the literature for the optimum combination of the operating parameters for maximum current density without flooding at the cathode. Therefore, this work is an attempt to present the optimization of the operating parameters for maximum current density without cathode flooding using meta-model design of optimization by combining MATLAB simulation, Taguchi method and GA.

2. Methodology

Operating parameters such as temperature and pressure of the fuel cell, humidification temperature of cathode and anode gases and stoichiometry of the cathode gas are supposed to be the most influential parameters on the performance of the PEMFC [1, 5, 9, 13].

The temperature of the fuel cell has great influence on the cell performance and as the temperature increases from the room temperature, the electrochemical reaction rate also increases speeding up the diffusion of the reactant gases to the catalyst sites. Also, the condensation of water in the cell decreases and improves the water evaporation rate. However, when the cell temperature goes too high i.e., above 80°C, there would not be enough water in the membrane and becomes dry and fails to transmit protons through the membrane decreasing the performance of the fuel cell. Therefore, the operating temperature of the fuel cell has to be limited below 100°C and preferably not beyond 80°C. Many researchers have studied the performance of the fuel cell by varying the temperature from 40 to 80°C [20, 21, 24, 28, 29].

Water generated through electrochemical reaction at the CCL alone is insufficient to keep the PEM wet. Therefore, it is important to humidify the anode and cathode reactant gases. When the fuel cell operates at lower current densities, the effect of electro-osmotic drag is less than the back diffusion increasing the water content at the anode. But usually, the fuel cell operates at higher current densities where the effect of electro-osmotic drag becomes significant reducing the water content at the anode dehydrating the membrane at the anode side. The model of the Falcao *et al.* [16] predicted that both anode and cathode reactants must be humidified to avoid membrane dehydration, particularly at higher current densities. The problem of membrane dehydration at higher temperature could be overcome by external humidification. The present study considered the anode and cathode humidification temperatures from 40°C to 80°C [20, 21, 24, 28, 29], as the controlling factor in order to identify the most suitable humidification temperature.

Higher cathode stoichiometry removes the water at cathode membrane interface at elevated temperatures which improves the performance of the fuel cell. However, larger flow rates lead to enhance parasitic loses and reduce the overall performance. Moreover, higher stoichiometric coefficients adversely affect the cell performance by drying out membrane in low humidity operation. The cathode stoichiometric flow ratio from 1.4 to 3 [20, 24, 28, 29], has also been considered as a controlling factor in this study.

The performance of the fuel cell typically increases with operating pressure due to the increased rate of electro-chemical reaction which is proportional to the partial pressures of hydrogen and the oxygen. However, more water generates with increase in pressure at the cathode membrane interface which leads to reduce the limiting current density. The present study considered the fuel cell operating pressure from 1 to 3 bar [11, 19, 20], as the controlling factor.

The dynamic behaviour of the water is governed by the net accumulation rate of water at cathode membrane interface, which is the rate of generation minus rate of removal. The water is accumulated at the cathode membrane interface by electrochemical reaction, cathode reactant humidification, electro-osmotic drag phenomena and interfacial mass transfer between liquid and water vapor due to evaporation and condensation. On the other hand, water is removed from the cathode membrane interface by back diffusion of water from cathode to anode, hydraulic permeation of water through the membrane from the cathode to anode due to gas pressure gradient across the membrane, water transport through the gas diffusion layer under capillary pressure, water flux transported from cathode to anode through the PEM under capillary pressure and the maximum amount of water evaporated through the cathode outlet. In practice, anode and cathode pressures of the reactant gases are usually the same hence this effect has been neglected [30]. Since the capillary pressure through the GDL lies in very small range, this effect has also been neglected [31]. Temperature is assumed constant throughout the fuel cell; hence net water flux through this mode has been neglected.

Mathematical model has been developed for water balance at the cathode membrane interface using various transport equations. The five key parameters; temperature and pressure of the fuel cell, cathode and anode humidification temperatures and cathode stoichiometry with five levels for each parameter have been considered for this study. Taguchi method is used to determine the effect of each independent operating parameter on the dependant output variable and is also used to determine the optimum combination of all the design operating parameters for maximum current density without flooding at cathode membrane interface. L_{25} orthogonal array has been employed to analyze the five operating parameters with five levels. At each run in MATLAB, the current density at which the net water flux changes the sign i.e. flooding appears at the cathode membrane interface has been calculated. After the required data for the control factors are obtained, the larger-the-better SNR is employed to prioritize the operating parameters using MINITAB 16 statistical software. The sensitivity of each parameter is then analyzed by employing analysis of variance (ANOVA). Then the global optimization method GA is used to find the global optimum operating parameters to get the maximum current density.

3. Model development

As discussed earlier, the basic modes of water transport at the cathode membrane interface, viz., water supplied through the humidification of cathode gas, electro-osmotic drag of water from anode to cathode, water generated through electrochemical reaction at the CCL, water source from the membrane desorption which occurs in the cathode side catalyst layer, water source from the interfacial mass transfer between liquid and vapor phase during evaporation and condensation, back diffusion of water from cathode to anode through the membrane, hydraulic permeation of water through the membrane from cathode to anode due to the gas pressure gradient across the membrane, water flux transported from cathode to anode through the PEM under capillary pressure, removal of water through the cathode outlet. The amount of water produced by chemical reaction due to crossover of gasses could be neglected. As a general rule, the temperature driven flow will allow water to move through the membrane towards colder location. Temperature is assumed constant throughout the fuel cell; hence net water flux through this mode has been neglected. However, under start-up or shutdown conditions, where larger temperature gradients exist, the net water flux through this mode could be significant. A brief explanation of model equations used in the water balance at the cathode membrane interface has been discussed below.

3.1 Governing equations

The amount of water fed (W_{ch}) to the cathode stream through the humidification is given by [30]

$$W_{ch} = \frac{P_{sat} RH_{c,in}}{P\text{-}P_{sat} RH_{c,in}} \times \zeta_c \times (\frac{i \times A}{4 \times F}) \times (1 + \frac{x_{N2}}{x_{O2}}) \tag{1}$$

When current is drawn from the fuel cell, protons migrate from the anode to cathode carrying water molecules with them due to electro-osmotic drag and are more severe at higher current densities (W_o), which is given by [16]

$$W_o = \frac{\eta_d \times i \times A}{F} \tag{2}$$

where η_d is the electro-osmotic drag coefficient representing the average number of water molecules transported per proton and is given by [30]

$$\eta_d = \begin{cases} 1 & \text{for } \lambda \le 14 \\ \frac{1.5}{8} \times (\lambda - 14) + 1 & \text{for } \lambda > 14 \end{cases} \tag{3}$$

where λ is the water content in the membrane defined as the ratio of the number of water molecules to the number of charged sites. The relationship between water content in the Nafion membrane, λ and water vapor activity, a was experimentally measured at 30°C by Springer *et al.* [32]

$$\lambda_{303} = 0.043 + 17.81a - 39.85a^2 + 36a^3 \tag{4}$$

Later, Kulkovisky [33] developed an equation for water content of the membrane which is based on the water uptake data at 80°C

$$\lambda_{353} = 0.3 + 0.6a[1 - \tanh(a - 0.5)] + 3.9\sqrt{a}\,[1 + \tanh(\frac{a - 0.89}{0.23})] \tag{5}$$

In this model, since the water content of the membrane (λ) is assumed as a function of temperature, it is interpolated between these values as:

$$\lambda = \lambda_{303} + \frac{\lambda_{353} - \lambda_{303}}{50}(T - 303) \tag{6}$$

where a is water vapor activity in the membrane [34]

$$a = \frac{C_{wm} R_u T}{P_{sat}} \tag{7}$$

where P_{sat} is saturated vapor pressure of water in bar can be expressed as a function of temperature in °C [32]

$$\log P_{sat} = -2.1794 + 0.02953 \times t - 9.1837 \times 10^{-5} \times t^2 + 1.4454 \times 10^{-7} \times t^3 \tag{8}$$

In a fuel cell, the water at the CCL comes from the three sources, water generated from the electrochemical reaction, membrane desorption in the catalyst layer and the mass transfer between liquid and vapor phases during evaporation and condensation.

Water generated (W_g) from the electrochemical reaction is given by [1, 2]

$$W_g = \frac{i \times A}{2F} \tag{9}$$

Water desorption from the cathode side membrane face (W_d) is determined by equation [34, 35]

$$W_d = 10^{-4} \times A \times \delta_{cl} \times \gamma_d \frac{\rho_m}{M_m}(\lambda_e - \lambda) \tag{10}$$

The mass transfer coefficient for the desorption of water from the cathode side membrane face is determined by the equation [8, 35]

$$\gamma_d = a_d \times f_v \times \exp\left(2416\left(\frac{1}{303} - \frac{1}{T}\right)\right) \tag{11}$$

where, f_v is the volume fraction of the water in the electrolyte membrane and is calculated as [8, 34]

$$f_v = \frac{\lambda V_w}{V_m + \lambda V_w} \tag{12}$$

λ_e is the equilibrium membrane water content which is evaluated by [35]

$$\lambda_e = \begin{cases} 0.43 + 17.18a - 39.85a^2 + 36a^3 & 0 < a < 1 \\ 14 + 1.4(a-1) & 1 \le a \le 3 \end{cases} \tag{13}$$

The third source term is the interfacial mass transfer between liquid and water vapor due to evaporation and condensation is determined by the following equation [34]

$$W_l = \frac{A_{pore} A \delta_{cl} \Gamma_s \Gamma_m}{10^6 M_w} \sqrt{\frac{R_u T}{2\pi M_w}}(\rho_w - \rho_{sat}) \tag{14}$$

where A_{pore} is the pore surface area per unit volume, Γ_m is an uptake coefficient that accounts for the combined effects of heat and mass transport limitations in the vicinity of the liquid/vapor interface and Γ_s is accommodation coefficient similar to Γ_m. P_{sat} is the saturated water vapor density and is related to the saturation pressure P_{sat}, which is given by the equation

$$\rho_{sat} = \frac{P_{sat}}{R_u T} \tag{15}$$

The water moves from the anode to cathode due to electro-osmotic drag, water generated due to electrochemical reaction, desorption from the cathode side of the membrane and interfacial mass transfer between liquid and water vapor at the cathode catalyst layer builds up water at the cathode. Back diffusion phenomena occur from cathode to anode since the water concentration at the cathode is generally far higher than that at the anode. The back diffusion flux depends on diffusion coefficient, water concentration at the both sides of the membrane and thickness of the membrane and is determined by the equation [2, 17]

$$W_{bd} = 10^{-4} A \times D_w \left(\frac{C_{wc} - C_{wa}}{\delta_m}\right) \tag{16}$$

The water diffusion coefficient (D_w) is expressed by the membrane water content and temperature of the fuel cell [17]

$$D_w = D_\lambda \exp(2416(\frac{1}{303} - \frac{1}{T})) \tag{17}$$

where the coefficient D_λ is not constant, it is a function of water content of the membrane (λ) as given below [17]

$$D_\lambda = \begin{cases} 10^{-10} & \text{for } \lambda < 2 \\ 10^{-10}(1+2(\lambda-2)) & \text{for } 2 \le \lambda \le 3 \\ 10^{-10}(3-1.67(\lambda-3)) & \text{for } 3 < \lambda < 4.5 \\ 1.25\times 10^{-10} & \text{for } \lambda \ge 4.5 \end{cases} \tag{18}$$

The membrane water content at cathode and anode side (C_{wc} and C_{wa} respectively) depends on channel water activities through the water sorption equilibrium values and upon current and are determined by the following equations [36]

$$C_{wc} = \frac{1}{2}(\xi_+ + \xi_-) \tag{19}$$

$$C_{wa} = \frac{1}{2}(\xi_+ - \xi_-) \tag{20}$$

$$\xi_+ = C^*_{wc} + C^*_{wa} + \frac{i}{2\bar{\gamma} C_f F} \tag{21}$$

$$\xi_- = \frac{C^*_{wc} - C^*_{wa} + \dfrac{5i}{2\bar{\gamma} C_f F}}{1 + \dfrac{D_w e^{-2436/T} \xi_+}{\delta_m \bar{\gamma}}} \tag{22}$$

where C^*_{wc}, C^*_{wa} are water sorption equilibrium values at cathode and anode based on the water activities of the corresponding channel [2, 17]

$$C^*_{wk} = \begin{cases} \dfrac{\rho_m}{M_m}(0.043 + 17.81a_k - 39.85a_k^2 + 36a_k^3) & \text{for } a_k \le 1 \\ \dfrac{\rho_m}{M_m}(14 + 1.4(a_k - 1)) & \text{for } a_k > 1 \end{cases} \tag{23}$$

where

$$a_k = \frac{P_{sat,k}}{P_{sat}} \tag{24}$$

Water flux (W_c) transported from cathode to anode through the membrane under capillary pressure may be worth to be considered and is determined by the equation [10]

$$W_c = 10^{-4} \frac{A \times \rho_w \times k_p \times \sigma_m^w}{\delta_m \times \mu \times M_w \times r_m^w} \tag{25}$$

where σ_m^w and r_m^w are the surface tensions at the water membrane interface and radius of curvature at the water membrane surface respectively. M is the pore water viscosity and is a function of temperature [30]

$$\mu = 1.002 \times 10^{\left[\frac{(1.3272(293-T) - 0.001053(293-T)^2)}{(T-168)}\right]^{-3}} \tag{26}$$

The maximum amount of water vapor carried from the cathode outlet (W_{co}) is calculated as [14]

$$W_{co} = \frac{i \times A}{2F}\left[(2.38\zeta_c - 0.5)\frac{P_{sat}}{P - P_{sat}} - 2.38\zeta_c \frac{P_{sat} RH_{c,in}}{P - P_{sat} RH_{c,in}}\right] \tag{27}$$

The water flow towards the cathode membrane interface is taken as positive and from the cathode membrane interface is taken as negative.

Hence, the net water balance equation at the cathode membrane interface is given by:

$$W = W_{ch} + W_o + W_g + W_l - W_d - W_{bd} - W_c - W_{co} \tag{28}$$

4. Results and discussions

The dynamic behaviour of the water is governed by the net accumulation rate of water at cathode membrane interface, which is the rate of generation minus rate of removal. The water is accumulated at the cathode membrane interface by electro-chemical reaction, anode and cathode humidification, electro-osmotic drag phenomena and interfacial mass transfer between liquid and water vapor due to evaporation and condensation. On the other hand, water is removed from the cathode membrane interface by back diffusion of water, water flux through the PEM under capillary pressure from cathode to anode, water desorption from the cathode side membrane face and the amount of water evaporated through the cathode outlet.

The five key parameters; temperature and pressure of the fuel cell, cathode and anode humidification temperature and cathode stoichiometry with five levels for each parameter which are shown in Table 1 have been considered for this study. Taguchi method is used to find the sensitivity of each parameter and to find the optimum combination of the design factors for maximum current draw without flooding at cathode membrane interface. L25 orthogonal array has been employed to analyze the five operating parameters with five levels. A1 represents factor A (fuel cell temperature) at level 1, A2 represents factor A at level 2 and so on. The parameters used for the MATLAB simulation are shown in Table 2. At each run in MATLAB, the current density at which the net water flux changes the sign i.e. flooding appears at the cathode membrane interface has been calculated which are shown in Table 3.

Table 1. Design factors and levels for Taguchi OA L25 design

Factor	Parameters	L-1	L-2	L-3	L-4	L-5
A	Fuel cell operating temperature (°C)	40	50	60	70	80
B	Cathode humidification temperature (°C)	40	50	60	70	80
C	Anode humidification temperature (°C)	40	50	60	70	80
D	Cathode stoichiometry	1.4	1.8	2.2	2.6	3
E	Fuel cell operating pressure (bar)	1	1.5	2	2.5	3

Table 2. Parameters and their values used for the analysis

Parameter	Value	Reference
Desorption parameter of water, a_d	$4.59x10^{-5}$	[8, 34]
Molar volume of water, V_w	$1.8x10^{-5}$ m^3/mol	[8]
Molar volume of dry membrane, V_m	$5.5x10^{-4}$ m^3/mol	[8]
Equivalent weight of dry membrane, M_m	1.1 kg/mol	[2]
Dry density of the membrane, ρ_m	$1.98x10^3$ kg/m^3	[2, 34]
Uptake coefficient, Γ_s	0.001	[34, 35]
Accommodation coefficient, Γ_m	0.01	[34, 35]
Pore surface area per unit volume, A_{pore}	20 m^2/cm^3	[34, 35]
Thickness of catalyst layer, δ_{cl}	0.001 cm	[34]
Thickness of membrane, δ_m	0.0220 cm	[8]
Water transfer coefficient, $\bar{\gamma}$	5.7x10-^{6} m/s	[36]
Fixed charge concentration, C_f	1200 mol/m^3	[6, 36]
Membrane hydraulic permeability, k_p	$1.5x10^{-20}$ m^2	[10]
Surface tension at the water membrane interface, σ_m^w	$54x10^{-3}$ N/m	[10]
Radius of curvature at the water membrane surface, r_m^w	$3x10^{-9}$ m	[10]

Table 3. Numerical results obtained from the MATLAB analysis for the Taguchi OA L25 design

Exp No.	A	B	C	D	E	Current density (A/cm^2)
1	1	1	1	1	1	0.608
2	1	2	2	2	2	0.5635
3	1	3	3	3	3	0.507
4	1	4	4	4	4	0.44
5	1	5	5	5	5	0.3675
6	2	1	2	3	4	0.7445
7	2	2	3	4	5	0.7285
8	2	3	4	5	1	0.429
9	2	4	5	1	2	0.5345
10	2	5	1	2	3	0.4705
11	3	1	3	5	2	0.9905
12	3	2	4	1	3	0.8715
13	3	3	5	2	4	0.822
14	3	4	1	3	5	0.7555
15	3	5	2	4	1	0.2335
16	4	1	4	2	5	1.1335
17	4	2	5	3	1	1.3425
18	4	3	1	4	2	0.9655
19	4	4	2	5	3	0.759
20	4	5	3	1	4	0.8025
21	5	1	5	4	3	2.0825
22	5	2	1	5	4	1.683
23	5	3	2	1	5	1.2595
24	5	4	3	2	1	1.068
25	5	5	4	3	2	0.671

4.1 Taguchi parameter design

In this study, the quality characteristic is the maximum current density without flooding, which belongs to a larger-the better characteristic and is calculated as follows [24]:

$$\frac{S}{N} = -10\log\frac{1}{n}\left(\sum_{i=1}^{n}\frac{1}{y_i^2}\right) \tag{29}$$

Where n is the number of current density sets and y_i is the maximum current density value without flooding for the ith data set. The analysis was done using MINITAB -16, which is statistical software to determine the main effect of each independent operating parameter on the dependant output variable and it is also used to determine the optimum combination of the operating parameters. Figure 3 shows the SNR response graph derived from the factors at different levels from Table 3, which shows the relationship between the level of each operating parameter and its corresponding impact on the output, is discussed in detail below.

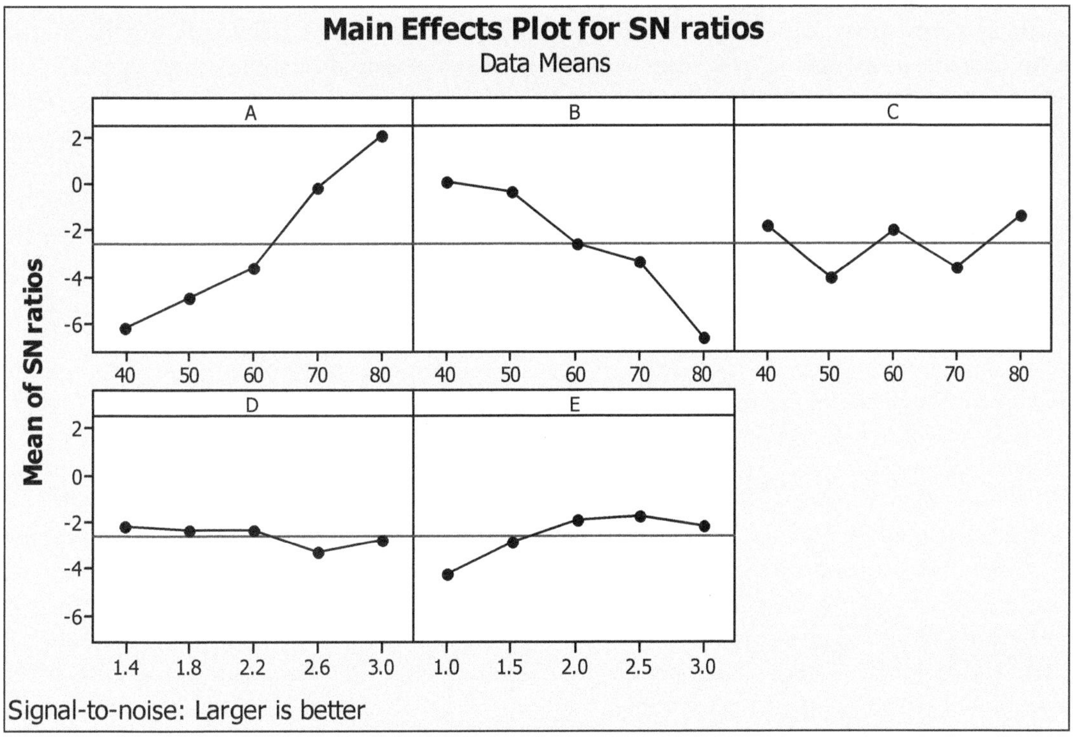

Figure 3. Response graph of main effect plots of SNR value

The maximum current draw without flooding at the cathode membrane interface is found to increase with increase in the temperature of the fuel cell at all levels as shown in the Figure 3(A). Higher temperature increases vapor pressure and volume flow rates leading to higher rate of evaporation of water. Also, the surface tension and viscosity of water decreases with increase in the cell temperature, allowing water to be flushed out more easily. Wang *et al.* [3] observed that the limiting current density of the fuel cell increases with the cell temperature from 50 to 80°C, which indicates enhancement in water transport due to the increasing diffusivity with temperature. Also, the water diffusion coefficient increases with the fuel cell temperature, which improves the water transport by back diffusion.

As the cathode humidification temperature increases, more water is carried through the inlet air to the cathode membrane interface and also the capacity of the gas stream to remove water by evaporation decreases. Due to this, more water remain in the cathode membrane interface, which causes flooding even at low current draw. Therefore, it has been observed that the maximum current density without flooding decreases with increase in cathode humidification temperature at all levels as shown in Figure 3(B). Wang *et al.* [3] observed that the limiting current density of the fuel cell decrease with increase in cathode humidification temperature which indicates that the flooding appears at lower current densities. Natarajan and Nguyen [5] also observed that the performance of cathode was found to decrease with the increasing inlet stream humidity. With increase in the humidity of the cathode gas, more water vapor enters the cell with air increasing the partial pressure and hence decreasing the water carrying capability of the cathode gas, and also, more water condenses increasing the accumulation of water at the cathode membrane interface. Low humidity operation can benefit the cell performance by reducing mass transport limitation at higher current densities. Lim and Wang [7] have studied power density versus current density under different cathode humidification temperatures experimentally. The peak power density increases from 0.48 to 0.6 W/cm^2 with decreasing humidification temperature of cathode gas, indicating the cathode flooding can be alleviated by using air feed of lower humidity.

The maximum current draw without cathode flooding was found to fluctuate when the anode humidification temperature increases from 40 to 80°C as observed in Figure 3(C). Flooding at the cathode advances when the anode humidification temperature increases from 40 to 50°C and 60 to 70°C. This is because of net water content at the anode side of the membrane increases with increase in anode humidification and hence decreases the back diffusion of water and more water accumulates at cathode which advances the flooding at the cathode. But the flooding is delayed when the anode humidification increases from 50 to 60°C and 70 to 80°C. This might be back diffusion of water from cathode prevails the electro-osmotic drag transport due to increase in water diffusion coefficient with the water content of the membrane as reported by Ge *et al.* [12]. The possibility of membrane dehydration particularly at higher current densities is overcome by the high amount of external humidification.

The maximum current draw without flooding was found to slightly decrease with the cathode stoichiometry from level 1 (1.4) to level 4 (2.6) and then increase slightly from level 4 to 5 (3), as shown in Figure 3(D). But in totality, the effect of cathode stoichiometry was almost negligible compared to the effects of other parameters indicating that its influence is insignificant in maximizing the current density without flooding. It was common to use higher stoichiometric air flow rates to remove excess water from the cathode membrane interface, but at the same time, a very high stoichiometric air flow rates increases the parasitic losses and adversely affect the cell performance by drying the membrane in low humidity operation. Dia *et al.* [13] observed that less water could be removed from the cell at high flow rates at a humidification temperature of 70°C. This indicates that increasing the stoichiometric air flow rate does not always increase the performance of the PEMFC.

The current density without flooding was found to increase with the cell pressure from level 1 (1 bar) to level 4 (2.5 bar) and then decrease from level 4 to 5 (3 bar), as shown in Figure 3(E). This is because high pressures force the oxygen and hydrogen to come in contact with the membrane which leads to the reducing of the mass transport losses. This result agrees well with Santarelli and Torchio [11] who have reported that when the cell pressure increases, the maximum value of power curve shifts to higher current densities up to 2.5 bar due to the corresponding shift of limiting current density. From 2.5 to 3 bar, decrease in limiting current density was observed due to the accumulation of water at the cathode side.

The optimum levels of the operating parameters in obtaining the maximum current density without flooding is obtained with the highest SNR value, implying larger-the-better performance characteristic, calculated using equation (29). The main effects plot for SNR graph (Figure 3) shows that the factor levels should be maintained at A5 (80°C), B1 (40°C), C1 (80°C), D1 (1.4) and E4 (2.5 bar).

4.2 Analysis of variance (ANOVA)

A statistical analysis of variance (ANOVA) was performed to understand the significance of the operating parameters in maximizing the current density without flooding, the results of which are listed in Table 4. The F-value for each parameter is the ratio of the mean square deviation of that parameter to the mean square error and is used to determine the significance of each input operating parameter on the output. In general, F-value less than 1 implies that the effect of input parameter is less than the error associated and therefore could be ignored and F-value above 4, the effect of variable is quite significant [22]. As could be observed from Table 4, it could be inferred that fuel cell temperature (A) and cathode humidification temperature (B) have the largest influence on the flooding at the cathode membrane interface followed by the anode humidification temperature (C). The effect of pressure of the fuel cell (E) has little significance on the flooding whereas cathode stoichiometry (D) has no significance.

The effects of the operating parameters on maximization of current density without flooding are also explained using the Pareto chart, as shown in Figure 4. Pareto analysis [29] is a creative way of looking at causes of problems because it helps stimulate thinking and organize thoughts.

Pareto chart is a useful tool to prioritise the vital parameters and highlight those which are most important for further action. Pareto chart is a bar chart in which contribution of operating parameters with bars representing percentage of contribution by the operating parameters are arranged in descending order of their importance. A line graph of cumulative percentage of contribution is also drawn in the same chart. It is obvious from the Figure 4 that, highest contribution to maximize the current draw without flooding at cathode membrane interface is the temperature of the fuel cell (56.07%) followed by cathode humidification temperature (27.89%) and anode humidification temperature (8.45%). The total contribution of these three operating parameters is 92.41%. Therefore, these three vital factors have to be focussed and analysed to maximize the current density without flooding.

Miao *et al.* [28] have stated that operating temperature of the fuel cell was the most significant parameter on the performance of the PEMFC followed by cathode humidification temperature and anode humidification temperature. Kraytsberg and Eli [10] have reported that, the current density of fuel cell which operates without flooding at room temperature is 0.48 A/cm^2 which can be increased to 1.2 A/cm^2 by increasing the temperature to 80°C. High operating temperature of the fuel cell was found to be very effective for water removal process reducing the flooding and improving the performance of the fuel cell. Hence, temperatures of the fuel cell, cathode and anode humidification temperatures are the most significant factors which affect the performance of the PEMFC [24, 29].

Table 4. Results of the ANOVA for the L_{25} design for the maximum current density without flooding

Factor	DOF	SS	MS	F	P	Percentage contribution
A	4	2.42014	0.60504	17.46	0.008	56.07
B	4	1.20380	0.30095	8.69	0.030	27.89
C	4	0.36483	0.09121	2.63	0.186	8.45
D	4	0.02513	0.00628	0.18	0.937	0.01
E	4	0.16345	0.04086	1.18	0.438	3.79
Error	4	0.13853	0.03465			3.79
Total	24	4.31594				

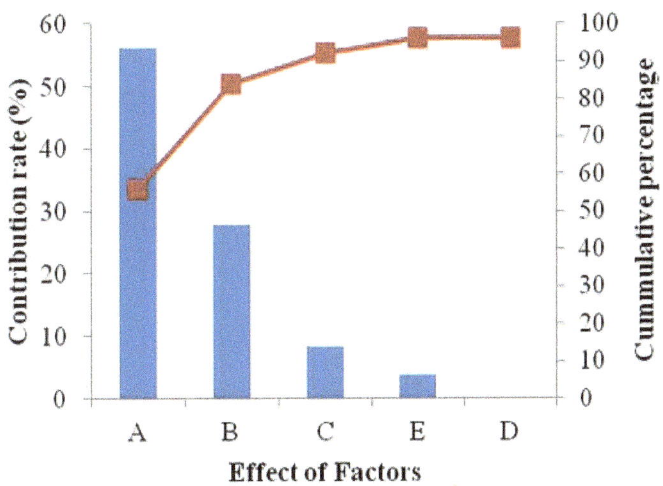

Figure 4. Pareto chart for contribution of operating parameters to maximize the current density

4.3 Optimization by GA

Genetic algorithm is a global optimization method to converge on a global optimum value among several possible local optimums. The global optimum value of the operating factors such as fuel cell temperature (A), cathode humidification temperature (B), anode humidification temperature (C) cathode stoichiometry (D) and fuel cell operating pressure (E) to maximize the current density without flooding can be efficiently obtained by formulating it as an optimization problem in the standard mathematical format as shown below and solving it using MATLAB.

A, B, C, D and E being different parameters, the objective function was to maximize the current density without flooding with limits of A, B, C, D and E, as follows:

$40\,^{o}C \le A \le 80\,^{o}C$

$40\,^{o}C \le B \le 80\,^{o}C$

$40\,^{o}C \le C \le 80\,^{o}C$

$1.4 \le D \le 3$

$1\text{ bar} \le E \le 3\text{ bar}$

The range of operating parameters considered in this study are same as that used in the Taguchi method to compare the results. To solve a global optimization problem with GA approach, several critical parameters such as size of the population, crossover rate, number of generations and mutation function should be carefully decided [29]. This study uses a population size of 20, cross over rate of 0.8, 100 generations and mutation function as adaptive feasible. Screen shot depicting the optimized convergence result obtained using GA for maximum current density is shown in Figure 5.

The final optimization results of the operating factors for maximum current density without flooding from the Taguchi OA method and GA optimization method is compared in Table 5. This table shows that the results obtained by GA provide global optimization solution with small changes in the operating pressure and cathode humidification temperature as compared to values obtained using Taguchi method.

The results show that the optimum conditions to get the maximum current draw without flooding are; fuel cell operating temperature of 80°C, cathode humidification temperature of 40.017°C, anode humidification temperature of 80°C, cathode stoichiometry is 1.4 and fuel cell operating pressure is 2.55 bar. This indicates that higher fuel cell temperature and anode humidification temperature, lower cathode humidification temperature and cathode stoichiometry and pressure around 2.5 bar are favourable to get the maximum current draw without flooding at the cathode membrane interface.

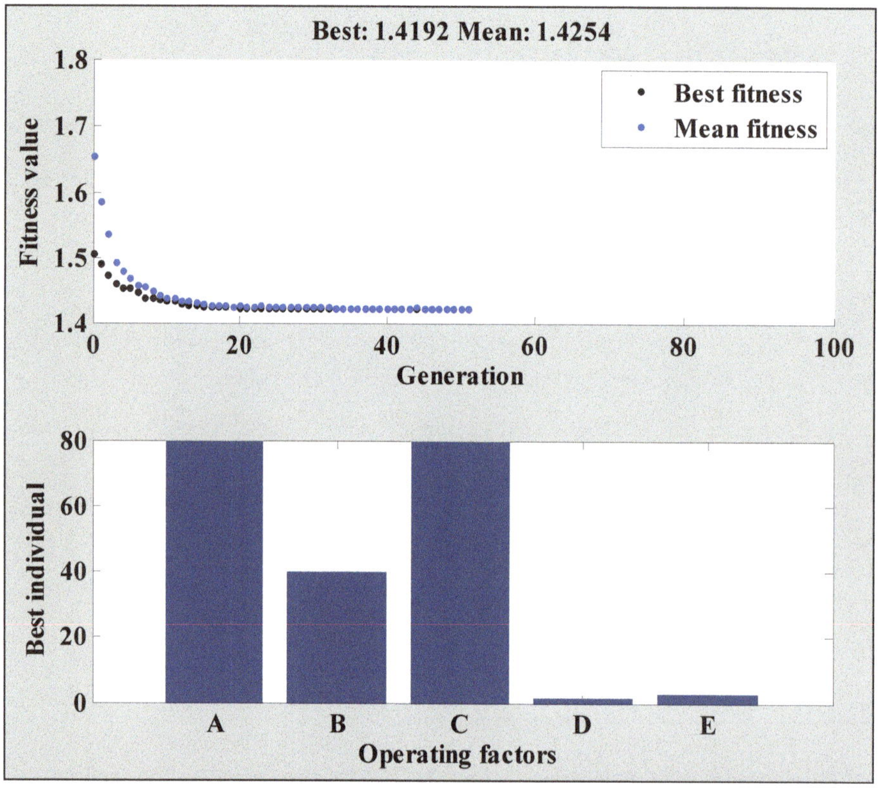

Figure 5. Convergence result obtained using GA iterations

Table 5. Optimization results of Taguchi OA method and GA

Experiment combination	A (°C)	B (°C)	C (°C)	D	E (bar)	Maximum current density without flooding (A/cm^2)
Taguchi OA method	80	40	80	1.4	2.5	1.4185
GA	80	40.017	80	1.4	2.55	1.4192

5. Conclusion

In this work, a novel methodology where mathematical modelling together with statistical methods and GA has been applied to optimize the operating parameters to maximize the current density without flooding at the cathode membrane interface. The five key parameters; temperature and pressure of the fuel cell, cathode humidification temperature, anode humidification temperature and cathode stoichiometry with five levels for each parameter have been considered for this study. Taguchi method is used to determine the effect of each independent operating parameter on the dependant output variable and is also used to determine the optimum combination of all the design operating parameters for maximum current density without flooding at cathode membrane interface. The sensitivity of each parameter is then analyzed by employing analysis of variance (ANOVA) method followed by the global optimization method GA to find the global optimum operating parameters to get the maximum current density.

The following conclusions could be drawn based on the results:

- The main effects plot for SNR graph shows the optimum combination of fuel cell temperature and pressure of 80°C and 2.5 bar respectively, cathode and anode humidification temperature of 40°C and 80°C respectively and cathode stoichiometry of 1.4 to get the maximum current density without flooding.
- The results of ANOVA show that, fuel cell temperature (56.07%) and cathode humidification temperature (27.89%) have the largest influence followed by the anode humidification temperature (8.45%) and fuel cell pressure (3.79%). Cathode stoichiometry has no significance on the maximum current density without flooding at the cathode membrane interface.
- GA provides global optimization solution with small changes in the operating pressure and cathode humidification temperature as compared to values obtained using Taguchi method. The maximum current density without flooding obtained by GA (1.4192 A/cm^2) is slightly higher than obtained using Taguchi OA method (1.4185 A/cm^2).
- The results show that the higher fuel cell temperature and lower cathode humidification temperature are favourable to get the maximum current draw without flooding at the cathode membrane interface.

This analysis is more useful where fuel cell operates under high current density which is nearer to the limiting current density particularly in automotive applications. High current density operation is promising solution to realize the cost reduction and compactness of the stack used in PEM fuel cell for wide applicability to automobiles.

Nomenclature

a	activity of water vapor in the membrane
a_d	desorption parameter
a_k	activity of water vapor in stream k
A	active area, cm^2
C_f	fixed charge concentration, mol/m^3
C_{wa}	membrane water content at the anode side, mol/m^3
C_{wc}	membrane water content at the cathode side, mol/m^3
C_{wm}	molar concentration of water in the membrane, mol/m^3
D_w	diffusion coefficient of water, m^2/s
F	Faraday constant, 96487 C per eq.
i	current density, A/cm^2
k_p	hydraulic permeability of the membrane, m^2
M_w	molar mass of water, kg/mol
P	fuel cell operating pressure, kPa
P_{sat}	saturation pressure of water at fuel cell temperature, kPa
$P_{sat,K}$	saturation pressure of water at anode or cathode, kPa
R_u	universal gas constant, 8.314 J/mol-K
$RH_{c,in}$	relative humidity of cathode inlet gas
t	fuel cell operating temperature, °C
T	fuel cell operating temperature, K
W	mass flow rate of water or accumulation of water, mol/s
x	mole fraction of the gas

Greek symbols

δ_{cl}	thickness of the catalyst layer, m
δ_m	thickness of the membrane, m
$\bar{\gamma}$	water mass transfer coefficient, m/s
λ	water content or local ratio H_2O/SO_3^- in the membrane
ρ_w	density of water, kg/m^3
ζ_c	cathode stoichiometry

References

[1] Bernardi D.M. Water- Balance Calculations for Solid- Polymer- Electrolyte Fuel Cells. Journal of the Electrochemical Society. 1990, 137(11), 3344-3350.

[2] Nguyen T.V., White R. E. A Water and Heat Management Model for Proton Exchange Membrane Fuel Cells. Journal of the Electrochemical Society. 1993, 140(8), 2178-2186.

[3] Wang L., Husar A., Zhou T., Liu H. A parametric study of PEM fuel cell performances. International Journal of Hydrogen Energy. 2003, 28, 1263-1270.

[4] He W., Lin G., Nguyen T.V. Diagnostic Tool to Detect Electrode Flooding in Proton- Exchange-Membrane Fuel Cells. AIChE Journal. 2003, 49(12), 3221-3228.

[5] Natarajan D., Nguyen T. V. Three- dimensional effects of liquid water flooding in the cathode of a PEM fuel cell. Journal of Power Sources. 2003, 115, 66-80.

[6] Lin G., He W., Nguyen T.V. Modeling Liquid Water Effects in the Gas Diffusion and Catalyst Layers of the Cathode of a PEM Fuel Cells. Journal of the Electrochemical Society. 2004, 151(12), 1999-2006.

[7] Lim C., Wang C.Y. Effects of hydrophobic polymer content in GDL on power performance of a PEM fuel cell. Electrochemica Acta. 2004, 49, 4149-4156.

[8] Ge S., Li X., Yi B., Hsing I.M. Absorption, Desorption, and Transport of Water in Polymer Electrolyte Membranes for Fuel Cells. Journal of the Electrochemical Society. 2005, 152(12), 1149-1157.

[9] Pasaogullari U., Wang C.Y. Two-Phase Modeling and Flooding Prediction of Polymer Electrolyte Fuel Cells. Journal of the Electrochemical Society. 2005, 152(2), 380-390.

[10] Kraytsberg A., Eli Y.E. PEM FC with improved water management. Journal of Power Sources. 2006, 160, 194-201.

[11] Santarelli M.G., Torchio M.F. Experimental analysis of the effects of the operating variables on the performance of a single PEMFC. Energy Conversion and Management. 2007, 48, 40-51.

[12] Buaud F., Lelandais D., Auvity B. Evidence of a non-dimensional parameter controlling the flooding of PEMFC stack. International Journal of Hydrogen Energy. 2008, 33, 2765-2773.

[13] Dai W., Wang H., Yuan X.Z., Martin J.J., Luo Z., Pan M. Measurement of the water transport rate in a proton exchange membrane fuel cell and influence of the gas diffusion layer. Journal of Power Sources. 2008, 185, 1267-1271.

[14] Hussaini I.S., Wang C.Y. Visualization and quantification of cathode channel flooding in PEM fuel cells. Journal of Power Sources. 2009, 187, 444-451.

[15] Ous T., Arcoumanis C. Visualisation of water accumulation in the flow channels of PEMFC under carious operating conditions. Journal of Power Sources. 2009, 187, 182-189.

[16] Falcao D.S., Oliveira V.B., Rangel C.M., Pinho C., Pinto A.M.F.R. Water transport through a PEM fuel cell: A one-dimensional model with heat transfer effects, Chemical Engineering Science. 2009, 64, 2216-2225.

[17] Misran E., Hassan N.S.M., Daud W.R.W., Majlan E.H., Rosli M.I. Water transport characteristics of a PEM fuel cell at various operating pressures and temperatures. International Journal of Hydrogen Energy. 2013, 38, 9401-9408.

[18] Ross, P.J. Taguchi Techniques for Quality Engineering. McGraw-Hill, 1996.

[19] Kaytakoglu S., Akyalcm L. Optimization of parametric performance of a PEMFC. International Journal of Hydrogen Energy. 2007, 32, 4418-4423.

[20] Yu W.L., Wu S.J., Shiah S.W. Parametric analysis of the proton exchange membrane fuel cell performance using design of experiments. International Journal of Hydrogen Energy. 2008, 33, 2311-2322.

[21] Wu H.W., Gu H.W. Analysis of operating parameters considering flow orientation for the performance of a proton exchange membrane fuel cell using the Taguchi method. Journal of power sources. 2010, 195, 3621-3630.

[22] Dalasm N.K.H., Ahadian S., Fushinobu K., Okazaki K., Kawazoe Y. Prediction and analysis of the cathode catalyst layer performance of proton exchange membrane fuel cells using artificial neural network and statistical methods. Journal of power sources. 2011, 196, 3750-3756.

[23] Cheng S.J., Miao J.M., Wu S.J. Investigating the effects of operational factors on PEMFC performance based on CFD simulations using a three-level full-factorial design. Renewable Energy. 2012, 39, 250-260.

[24] Wu H.W., Ku H.W. Effects of modified flow field on optimal parameters estimation and cell performance of a PEM fuel cell with the Taguchi method. International Journal of Hydrogen Energy. 2012, 37, 1613-1627.

[25] Solehati N., Bae J., Sasmito A.P. Optimization of operating parameters for liquid-cooled PEM fuel cell stacks using Taguchi method. Journal of industrial and Engineering Chemistry. 2012, 18, 1039-1050.

[26] Goldberg D.E. Genetic Algorithms in search, Optimization, and Machine learning. Pearson Education, 2003.

[27] Chang K.Y. The optimal design for PEMFC modelling based on Taguchi method and genetic algorithm neural networks. International Journal of Hydrogen Energy. 2011, 36, 13683-13694.

[28] Miao J.M., Cheng S.J., Wu S.J. Metamodel based design optimization approach in promoting the performance of proton exchange membrane fuel cells. International Journal of Hydrogen Energy. 2011, 36, 15283-15294.

[29] Cheng S.J., Miao J.M., Wu S.J. Use of metamodeling optimal approach promotes the performance of proton exchange membrane fuel cell (PEMFC). Applied Energy. 2013, 105, 161-169.

[30] Mench M.M. Fuel cell engines. John Wiley & Sons, 2008.

[31] Zamel N., Li X. Effective transport properties for polymer electrolyte membrane fuel cells- with a focus on the gas diffusion layer. Progress in energy and combustion science. 2013, 39, 111-146.

[32] Springer T.E., Zowodzinski T.A., Gottesfeld S. Polymer Electrolyte Fuel Cell Model. Journal of the Electrochemical Society. 1991, 138(8), 2334-2342.

[33] Kulikovsky A.A. Quasi- 3D Modeling of Water Transport in Polymer Electrolyte Fuel Cells. Journal of the Electrochemical Society. 2003, 150(11), 1432-1439.

[34] Wu H. Mathematical Modelling of Transient Transport Phenomena in PEM fuel cells. PhD diss., University of Waterloo, 2009.

[35] Cao T.F., Lin H., Chen L., He Y. L., Tao W.Q. Numerical investigation of the coupled water and thermal management in PEM fuel cell. Applied Energy. 2013, 112, 1115-1125.

[36] Berg P., Promislow K., Poerre J.S., Stumper J., Wetton B. Water management in PEM Fuel cells. Journal of the Electrochemical Society. 2004, 151(3), 341-353.

21

Using downscaled NCEP/NCAR reanalysis data for wind resource mapping

Arnau Bayon-Barrachina[1], Jeroen van Beeck[2], Fernando Martinez-Alzamora[1], Petra Amparo Lopez-Jimenez[1]

[1] Universitat Politècnica de València, Camí de Vera, s/n, 46022 - València, Spain.
[2] Von Karman Institute for Fluid Dynamics, Ch. de Waterloo, 74, 1640, Rhode-Saint-Genese, Belgium.

Abstract
Two problems arise when estimating numerically wind energy resources, namely: The low resolution of global-scale atmospheric models and the scarcity of meteorological observations to be used as boundary conditions for smaller-scale models. Downscaling techniques were developed to overcome these issues. These methods consist of using global-scale model output as boundary conditions of smaller-scale models. In this research a downscaling tool to refine NCEP/NCAR Reanalysis data using WAsP is presented. The downscaling technique proposed consists of extrapolating the wind climate at a given point using meteorological observations from another point by means of the WAsP model. Then, using time-marching NCEP/NCAR Reanalysis wind velocity values, on-line refined profiles of wind velocity and direction can be obtained. In order to assess the accuracy of the described tool, data of two episodes of 48 hours are downscaled and compared to meteorological observations at two different climates. A sensitivity analysis is performed in order to assess the effect that atmospheric stability and terrain roughness, among others, exert on the results. Results are not as accurate as expected, probably due to atmospheric instability or other factors neglected by the model. However, the main trends are followed when validating the model output using field measurements.

Keywords: Wind energy; Numerical modeling; Downscaling; NCEP/NCAR reanalysis data; WAsP.

1. Introduction

Atmospheric flow is one of the most complex and challenging to describe in the field of Fluid Mechanics. Indeed, its study led Edward Lorenz to postulate the widely known Chaos Theory. The author described a deterministic chaos as a system in which, despite the governing equations are well known, the number of degrees of freedom involved is so large that it is extremely sensitive to initial conditions. For this reason, exact long-term forecasting is not feasible [1].

Nevertheless, different approaches exist in order to provide sufficiently reliable weather and climate predictions. Most of them are based on numerical modeling, although some authors also conducted wind tunnel experiments [2]. The main difference among numerical atmospheric models is the scale at which they work, as a very wide range of scales coexist in the atmospheric flow. Such a large difference between coexisting-phenomenon scales makes that, when it comes to assessing wind energy resources over a given area, no mathematical model provides a general definitive solution.

On the one hand, micro-scale models of atmospheric flow are the most widely used in the field of wind resource estimation. Nevertheless, as stated in [3], they require data from a dense coverage of meteorological stations as input, which is often not available. On the other hand, larger-scale models, according to Badger *et al.* [4], provide currently a worldwide set of data on overall weather conditions. However, they cannot be used to estimate wind farm sites for power production purposes. This is due to the fact that micro-scale orography and roughness features are not explicitly resolved by those models: results are grid-averaged at too large resolutions. Therefore, despite meso-scale models can increase their horizontal resolution down to 2km, the mismatch between numerical weather predictions and actual field observations can only be reduced to a certain extent [5].

In order to overcome this problem, the downscaling techniques arise as an attempt to combine the advantages of both global and smaller-scale models (i.e. high accuracy and data availability, respectively). Such techniques represent a case of numerical nesting, as they consist in taking output data from large-scale models and adding information at scales smaller than the original grid resolution. Their final purpose is to refine the results from large-scale models, in which local features are not explicitly resolved, using a smaller-scale model.

The downscaling methods fall into three general types or approaches, namely: statistical, physical and hybrid. The statistical downscaling methods in their pure form are an attempt to find the relationship among a series of variables in a non-deterministic way, i.e. disregarding the physical laws that govern the phenomenon under study. The so-called physical, dynamic or deterministic methods use the laws of fluid mechanics and physics of the atmosphere to obtain a good estimate of wind conditions. The third family of downscaling methods consists of the so-called physical-statistical or hybrid methods which, according to Giebel *et al.* [6], turn out to be the most successful ones. In this kind of methods, despite they are mostly statistically-based, certain knowledge on the physical behavior of the atmosphere is used to improve the results. Many authors, such as Roy and Kok [5], claim that downscaling wind speed techniques should always combine both physical and statistical approaches in order to achieve a good accuracy.

The aim of this research is to develop a downscaling tool in order to refine global-scale model results by means of a micro-scale model. The proposed approach belongs to the aforementioned hybrid downscaling methods. In this case, the global-scale model used is the NCEP/NCAR Reanalysis. This dataset is provided by the U.S. National Center for Environmental Prediction (henceforth, NCEP) and the National Center for Atmospheric Research (henceforth, NCAR). The micro-scale model used is the commercial code WAsP (Wind Atlas Analysis and Application Program).

Many authors have successfully developed and validated wind forecasting methods based on the downscaling concept [7]. The most similar is that of Frank *et al.* [8], in which also NCEP/NCAR Reanalysis data are combined with WAsP. However, unlike the method proposed in this article, Frank *et al.* [8] use Karlsruhe Atmospheric Meso-scale Model (KAMM) to process the NCEP/NCAR Reanalysis data. Other authors, such as Stephen *et al.* [9] and Essa and Embaby [10], obtained good results but their approaches are more probability-based. Also using black box models, Bechrakis and Sparis [11] and Barbounis *et al.* [12] conducted accurate wind climate estimations using artificial neural networks (ANN). For a more detailed discussion on wind resource assessment, several extensive reviews are available in the literature [13-15].

2. Materials and methods

2.1 NCEP/NCAR Reanalysis data

The NCEP/NCAR Reanalysis [16] is the dataset used as global-scale model data in this model. It consists of a continuously-updated gridded dataset, which provides a wealth of atmospheric variables (air temperature, humidity, pressure, wind velocity, etc.) consistent in time and space at respective resolutions down to 6h in time and 2.5° in horizontal coordinates. In the Y-axis (N-S), this is approximately 280km, whereas in the X-axis (W-E), it represents approximately 200km at mid-European latitudes. As regards vertical resolution, data are defined at 17 different pressure levels[1]. The NCEP/NCAR Reanalysis time coverage ranges from 1948 to the present.

Once data have been harvested and assimilated, a reanalysis process is carried out. This process consists of blending global circulation models (GCM) with observations, in order to detect inconsistencies in data spanning long periods of time so that a good description of the state of the atmosphere is achieved.

[1] Vertical coordinates are defined by pressure levels instead of geometric heights.

The most relevant variables provided by the model are the wind velocity horizontal components (U and V, respectively). These variables are taken at a single grid point at the 17 pressure levels available spanning from 1000mb to 10mb every 6h. Obviously, not every wind velocity value is relevant in order to determine the near-surface wind resources (e.g. pressure level 10mB is generally dozens of kilometers above the ground level). For this reason, which pressure levels are taken into account is a relevant decision (this topic is discussed below).

2.2 WAsP

WAsP is a commercial code developed by the Danish Risø National Laboratory and it is widely used to predict wind climate and wind energy resources. This program is an implementation of the Wind Atlas methodology, whose details are explained in depth in [17]. The main advantage of WAsP is that, thanks to the use of semi-empirical linearized equations instead of the Navier-Stokes Equations, the model can be run in a regular computer using reasonable computation times. It is worth remarking that, despite its severe simplifications, WAsP yields good results.

The WAsP most basic assumption is, according to Landberg *et al.* [3], that for a specific micro-scale area, the overall wind conditions (i.e. the geostrophic wind climate) change so slowly that the wind climate can be extrapolated from a meteorological station (e.g. Point A) to any point within the same the region (e.g. Point B) just taking into account the local effects of both points.

The so-called local or site effects are the effects exerted on wind climate by surrounding obstacles and terrain roughness and orography in the vicinity of the points under study. They are taken into account by WAsP as follows: the observed wind climate (henceforth, OWC) at Point A is affected by the particularities of the surrounding topography, the terrain roughness and the shelter produced by close obstacles. In order to extrapolate this wind climate to Point B, the wind conditions in Point A have to be "cleaned" of the site effects, resulting the so-called regional wind climate (henceforth, RWC). This generalized wind climate, "the wind climate for standard conditions given by flat terrain of uniform roughness" [4], is considered extrapolable to all points within the same region. Therefore, to estimate the wind resource at Point B, it only has to be affected by Point B site effects (see Figure 1).

2.3 Model limitations

As models are just simplified representations of the reality, it is crucial to be aware of their limitations since, to a certain extent, they explain the mismatch between model output and observed values. In the case of WAsP, as in most models, one of the main limitations is the scale: its computational domain must extend horizontally at least 10km away from all the points under study [8]; otherwise internal boundary layers (henceforth, IBLs) may develop out of the modeled region. On the other hand, larger scales are not desirable as global-scale effects, which are not taken into account by WAsP (e.g. thermally driven winds, etc.), may play an important role. Eq. 1 [17] determines the IBL height (h') as a function of a roughness change horizontal distance (x).

$$\frac{h'}{z_0'}\left(\ln\frac{h'}{z_0'}-1\right)=0.9\frac{x}{z_0'} \tag{1}$$

Where z_0' is the larger roughness length of the roughness change that creates the IBL. The height above which IBL effects are no longer perceived (h_{max}) can be estimated using Eq. 1. In this case a roughness length of $z_0' = 1.00m$ and a fetch of $x = 10km$ are considered. This height threshold turns out to be $h_{max} = 1440m$.

Another important source of bias in models like WAsP, which are based on the assumption of potential flow, is terrain ruggedness. As Landberg *et al.* [3] report, flow over rugged or complex terrain, where separation is likely to occur, will in many cases not be modeled accurately. Several authors, such as Wood [18], consider complex terrain all surfaces steeper than 0.3. Typical annual energy prediction errors for wind turbines are about 10% in normal conditions, whereas in rugged terrain larger errors can be expected [8]. To overcome this violation of the potential flow assumption, correction algorithms are often implemented [3]. Indeed, this method was successfully applied and compared to CFD results by Berge *et al.* [19], obtaining no significant differences in accuracy.

Another known issue of WAsP is the flow over forested areas, which represent an unfortunate type of terrain for wind turbines due to the high turbulence intensity and shear of the flow over the tree canopy. The difficulty to predict the characteristics of such flows, as Manning *et al.* [20] report, has led to a lack of consensus on how to model forestry in WAsP among the wind-energy scientific community and industry. Another relevant issue of the WAsP model is atmospheric instability: despite its effects are corrected in the model implementation, this phenomenon is still a significant source of error [21].

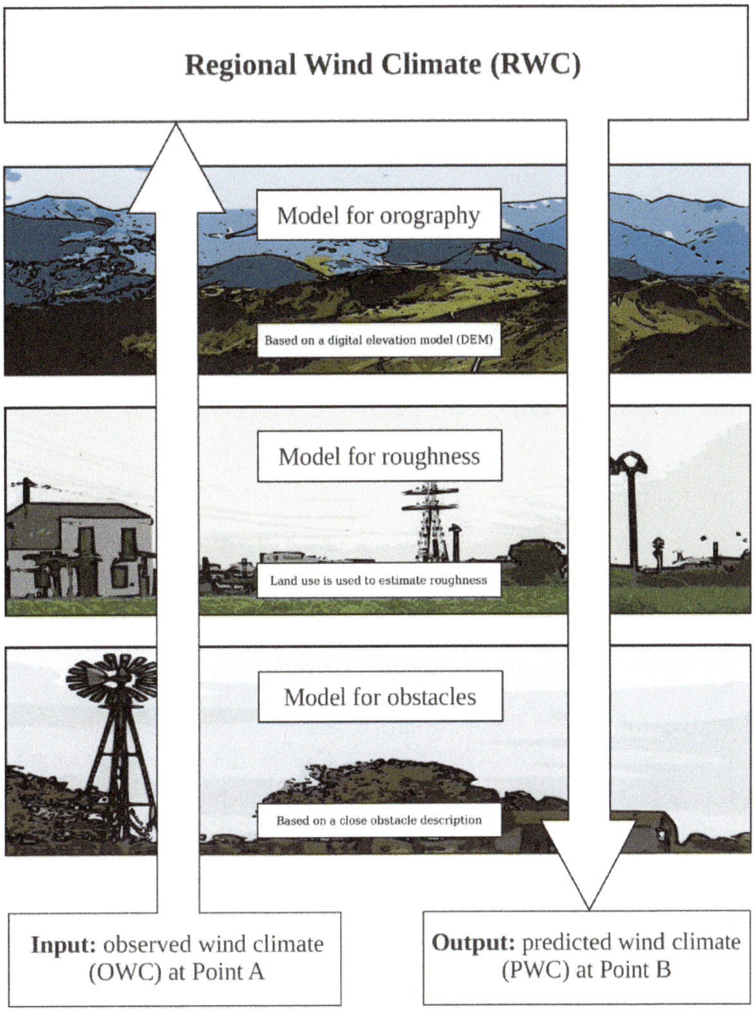

Figure 1. Scheme of the Wind Atlas methodology; self-elaborated from [17]

2.4 Downscaling process

The final goal of the described downscaling tool is to obtain a set of time-marching direction-wise wind velocity profiles out of NCEP/NCAR Reanalysis data (global-scale model) using WAsP (micro-scale model) to refine them. Thus, WAsP provides high-resolution direction-wise time-averaged (static) wind velocity profiles whereas the NCEP/NCAR Reanalysis introduce the time variability of wind velocity and direction. Figure 2 outlines the whole downscaling process and how NCEP/NCAR Reanalysis data and WAsP model output are combined.

The link between the wind velocity profiles provided by WAsP and the NCEP/NCAR Reanalysis data is made assuming that at a given point of the terrain, at a given reference height (z_{ref}) and at a given time, the wind condition is the same in both the refined and non-refined datasets (i.e. WAsP and NCEP/NCAR Reanalysis) and so they are reciprocally convertible, as site-surrounding features do not affect the wind condition. That reference height has to be in the vertical meso-scale in order to be a reliable liaison between both models. How it is estimated is discussed in detail below. Eq. 2 summarizes in an elegant way how both models, WAsP and NCEP/NCAR Reanalysis, are combined. Indeed, this simple equation is the core of the entire downscaling process:

$$u_k(t_j, z_i) = \frac{u_{W,k}(z_i)}{u_{W,k}(z_{ref})} u_{N,k}(t_j, z_{ref}) \tag{2}$$

where:

- $u_k(t_j, z_i)$ is the resulting refined wind velocity profile for direction k at time step t_j,
- $u_{W,k}(z_i)$ is the average wind velocity profile provided by WAsP for direction k,
- $u_{W,k}(z_{ref})$ is the average wind velocity provided by WAsP for direction k at z_{ref},
- $u_{N,k}(t_j, z_{ref})$ is the wind velocity provided by NCEP/NCAR Reanalysis at direction k, at z_{ref} and t_j.

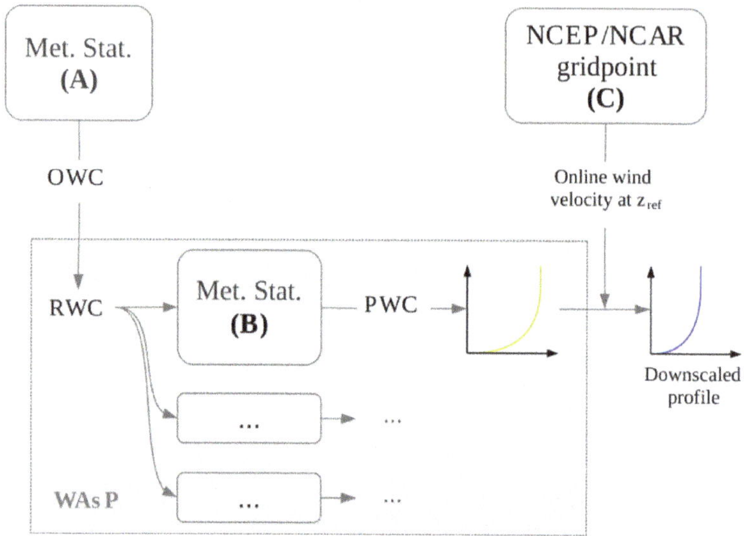

Figure 2. Downscaling scheme

Another way to see it is the following: the first term of Eq. 2 provides a long-term averaged wind velocity profile, which is adimensionalized using the velocity at the reference height (u_{ref}). Thus, the second term can be seen as a factor representing the wind magnitude and direction with respect to time, which multiplied by the dimension-less profile yields the downscaled wind velocity profile at a given timestep. Figure 3 shows an example of a time-averaged WAsP profile $u_{W,k}(z_i)$ and a resulting profile $u_k(t_j, z_i)$. As mentioned above, it is worth pointing out that, when dealing with NCEP/NCAR Reanalysis data, a problem arises: wind profiles are referred to pressure levels instead of geometric height. To overcome this problem, a conversion has to be carried out. This operation and the wealth of secondary operations which have to be performed are explained in the following lines.

There are several ways to transform the vertical coordinate of the profiles provided by the NCEP/NCAR Reanalysis (e.g. hydrostatic equation, barometric formula, etc.) but they are normally based on the assumption of constant temperature or constant vertical temperature gradient. Such assumption is only valid in certain stable regimes, but cannot be applied to all cases. Hence, another approach is necessary. Although NCEP/NCAR Reanalysis data provides temperature values (which makes possible an accurate hydrostatic approach), there is a simpler and more elegant way to convert wind velocity data with respect to pressure levels to geometric-height profiles: via the geopotential height.

The geopotential height is a "gravity-adjusted height" which is widely used in numerical weather prediction (NWP) as it allows neglecting centrifugal forces and air density effects, which are complex to model. Indeed, the geopotential height difference between two consecutive pressure levels is proportional to the mean temperature in that range and such relation is used to perform the aforementioned conversion. The geopotential height is defined as follows:

$$z_g = \frac{\Phi}{g_0} \tag{3}$$

Where g_0 is the gravity acceleration at the sea level in function of latitude (ϕ), which can be estimated in a highly accurate way by means of the World Geodetic System (WGS84) expression:

$$g_0 = 9.780327 \frac{1+0.00193185 \sin^2\phi}{\sqrt{1+0.00669438 \sin^2\phi}} \tag{4}$$

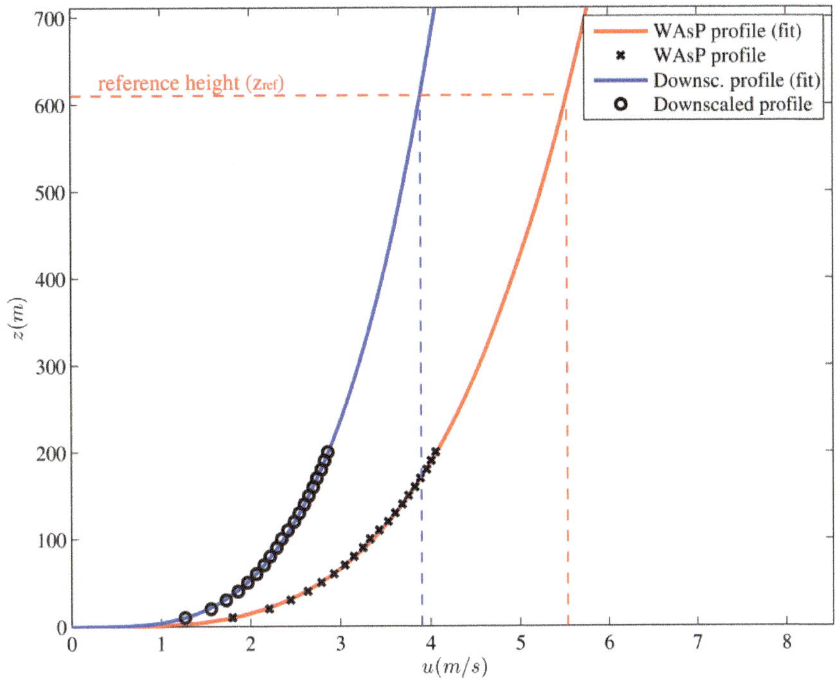

Figure 3. Example of wind velocity profiles (time-averaged and downscaled)

And where Φ is the geopotential, defined as follows:

$$\Phi = \int_0^H g(\phi, z)\, dz \qquad (5)$$

Being $g(\phi, z)$ the gravity acceleration in function of latitude (ϕ) and height (z), which can be computed using the following expression:

$$g(\phi, z) = g_0 \left(\frac{r}{r+H}\right)^2 \qquad (6)$$

Where r is the mean Earth radius, defined under a spheroidal approach in function of the latitude (ϕ) and the Earth Equatorial and Polar Radii (according to the WGS84: $r_e = 6{,}378{,}137m$ and $r_p = 6{,}356{,}752\text{m}$, respectively), as follows:

$$r = \sqrt{\frac{(r_e^2 \cdot \cos\phi)^2 + (r_p^2 \cdot \sin\phi)^2}{(r_e \cdot \cos\phi)^2 + (r_p \cdot \sin\phi)^2}} \qquad (7)$$

And H is the the total geometric height above the sea level, i.e. the sum of the terrain elevation above the sea level and the height above the ground level. Bearing this in mind, Vedel [22] demonstrates that the relation between geopotential and geometric height is the following:

$$H = \frac{z_g Q}{r + z_g} \qquad (8)$$

Being $Q = r \cdot g_0/(9.80665\,\text{m/s}^2)$ the gravity ratio. Since NCEP/NCAR Reanalysis data provides both wind velocity and geopotential height at the same given point, pressure level and time, using Eq. 8, wind velocities can be easily expressed as a function of the geometric height. Once geometric heights and horizontal velocities are known, wind velocity profiles can be represented.

As regards WAsP data, only the predicted wind climate (PWC) at heights from 10m to 200m (at 10m-intervals) have to be introduced into the downscaling tool. The PWC must be computed at the point under study using the RWC obtained at any point within the same region.

To do so, long-term averaged climatic data (i.e. wind atlas) are obtained from the European Wind Atlas [17]. Terrain orography is obtained by means of GIS tools out of the GTOPO30 dataset, a global digital elevation model (DEM) in geographic coordinates (datum WGS84) with a horizontal grid spacing of 30" and a vertical resolution of 1m, which is provided by the U.S. Geological Survey [23]. Terrain roughness length was obtained out of the CORINE Land Cover 2000 maps provided by the European Environment Agency [24], with an approximated horizontal resolution of 3". Again, in order to adapt the roughness maps (raster) to WAsP format (vector-polygon), GIS software has to be used. As regards surrounding obstacles, they are not modeled sensu stricto. Instead, their effect is implemented as equivalent roughness lengths according to the land use. Once PWCs are obtained, the points of the profiles are respectively fitted to potential curves in order to inter- and extrapolate other profile points.

At this point, it is necessary to determine the reference height (z_{ref}) at which both data, WAsP output and the NCEP/NCAR Reanalysis data, can be correlated. Recalling Eq. 1, in the most unfavorable case considered in Sec. 2.2, the average maximum height at which site effects are perceived in WAsP is approximately $h_{max} = 1440m$. According to Troen and Petersen [17], in coastal regions, wind conditions at a pressure level of 850mb (approximately, a height of 1500m) can be considered representative of the geostrophic wind conditions.

One may think that this should be the reference height at which information can be exchanged between both models. Nevertheless, bearing in mind that WAsP profile highest point is by default 200m, at 1500m such profiles may be no longer representative of near-surface conditions. In addition, the geostrophic wind condition is the "site-effect free" wind condition and therefore it cannot be compared to WAsP predicted wind climates (site-effect affected). Indeed, some preliminary studies performed by the authors have demonstrated that larger errors occur when downscaling using $z_{ref} = 1500m$).

For this reason, as lower points (1000mb and 925mb) are often available in the NCEP/NCAR Reanalysis data, the criterion to establish a reference height (at which wind velocities are assumed to be reciprocally convertible, i.e. z_{ref}) is the NCEP/NCAR Reanalysis data lowest available point since, although such point may be out of the WAsP vertical scale, it is generally relatively close.

It is important pointing out that, as such profiles are provided with respect to pressure levels instead of geometric heights, vertical coordinates are not constant from one time step to another. Indeed, the pressure level 1000mB may happen to be below the terrain surface in many cases (especially, at mountainous regions, where surface pressure is always below 1000mB). Therefore, the reference height used is the lowest NCEP/NCAR Reanalysis data point available.

Once WAsP data and the NCEP/NCAR Reanalysis data have been properly processed, the downscaling process itself can take place and Eq. 2 is resolved. The points of the resulting profile $u_k(t_j, z_i)$ are the final results of the downscaling tool. The resulting profiles, as it is done to WAsP-profile points, are fitted to a potential curve in order to compute velocities at points different from those at which profiles are defined.

2.5 Validation

In order to assess the accuracy of the results yielded by the developed downscaling tool, two episodes were tested in two different locations, namely: the period between 17th and 19th of March 2011 in Wideûmont (Belgium) and the period between 22 th and 24 th of July 2011 in Carcaixent (Spain). The necessary field observations are provided by the Institute Royal Météorologique (IRM)[2] and the Agencia Estatal de Meteorología (AEMET)[3], respectively. Recalling the nomeclature of the general scheme of the downscaling tool used in Figure 2, in each case study the selected points are the following:

Test case 1:

- Point A: Saint-Hubert (Belgium).
- Point B: Wideûmont (Belgium).
- Point C: Lat.: 50N - Long.: 5E.

Test case 2:

- Point A: Carcaixent (Spain).
- Point B: Carcaixent (Spain).
- Point C: Lat.: 40N - Long.: 0.

---

[2] Data available at: http://www.meteo.be/meteo/view/fr/123386-Observations+-+meteo.html

[3] Data available at: \url{ftp://ftpdatos.aemet.es/datos_observacion/

The mast height of the weather stations is 10m and their surroundings are free from obstacles that cannot be modeled as roughness. The distance between points in case number 1 ranges from 15km to 30km, which is within the WAsP computational domain. However in case number 2 points are separated by distances up to more than 100km. The main advantage of the latter is that the station from which the regional wind climate (RWC) is obtained (Point A) and that where the validation data are collected (Point B) match, unlike in case number 1.

3. Analysis of results

3.1 Wind velocity

In order to validate the proposed downscaling tool, wind velocity and direction were estimated for two two-day episodes at two different sites in Europe. Figure 4 shows the wind velocity evolution computed using the downscaling tool and that observed at the meteorological stations of Carcaixent and Wideûmont, respectively.

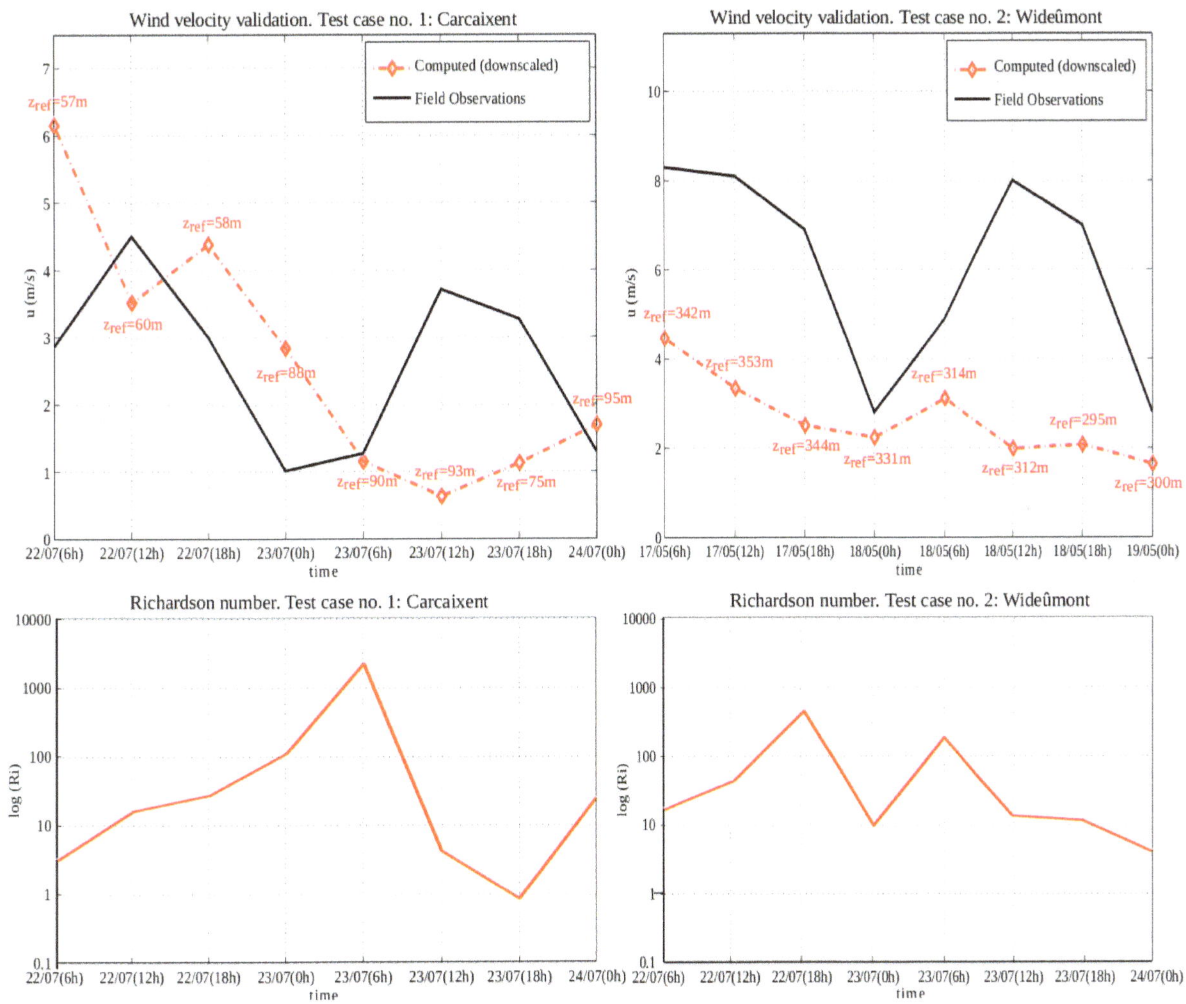

Figure 4. Validation of wind velocities computed by the downscaling tool with 6h-averaged meteorological observations and comparison to Richardson number values

At a first sight, it can be seen that there is a large mismatch between the values obtained by downscaling meteorological data and the field measurements. Nevertheless, in some cases, some trends are followed. E.g.: in both cases, a clear daily pattern of recurrence arises. Therefore, the following step is to find out how the offset and scale factor between computed and observed values can be reduced to the lowest extent (i.e. calibrating the model). In the Wideûmont case, a systematic underestimate of velocities occurs, which may be caused by several reasons; one of them is an overestimate of the terrain roughness length, but also a poor description of the surrounding obstacles of the meteorological station. Nevertheless, other reasons are likely involved in the still large mismatch between computed and observed values. In this regard, the large scale difference between WAsP and the NCEP/NCAR Reanalysis data may play a paramount role in the lack of accuracy of computations.

The root mean squared error (RMSE) in the Carcaixent case is 2.0m/s, which is approximately 76% of the observed mean value, whereas in the case of Wideûmont, the RMSE is 3.9m/s (37% of the mean observed value).

3.2 Wind direction

As regards the estimate of wind directions, the results are slightly more encouraging. Figure 5 shows respectively the values computed by the downscaling tool and those observed at meteorological stations.

The mismatch is in general very low, especially, if one takes into account that data are treated in a discrete way (in 30°-sectors). That means that two points which, apparently, are separated by 30° may be either almost overlapping or separated by almost 60°. Besides, the height difference may in some cases introduce bias due to the Coriolis forces.

The RMSE comparing downscaled and observed wind directions is in the Carcaixent case 45°, which represents 12% of the complete wind rose (360°). In the Wideûmont case, the RMSE is 34° (9%). Errors up to 30° (one sector) are generally accepted in wind direction predictions. In this case, the results obtained are slightly above that threshold.

As in the case of the wind velocities, a daily pattern of recurrence arises. It is important pointing out that the wind direction is determined by the NCEP/NCAR Reanalysis data (provided at an approximate height of up to 500 or 1000m and different location) and compared to meteorological station observations (measured at a height of 50m or less). This explains the aforementioned pattern: wind directions at different heights tend in general to the same values during periods of atmospheric stability, whereas they diverge when the atmosphere is instable. Indeed, back to the validation of the downscaling tool results, one can see that the highest differences between observed and computed values (provided at different heights) arise by night.

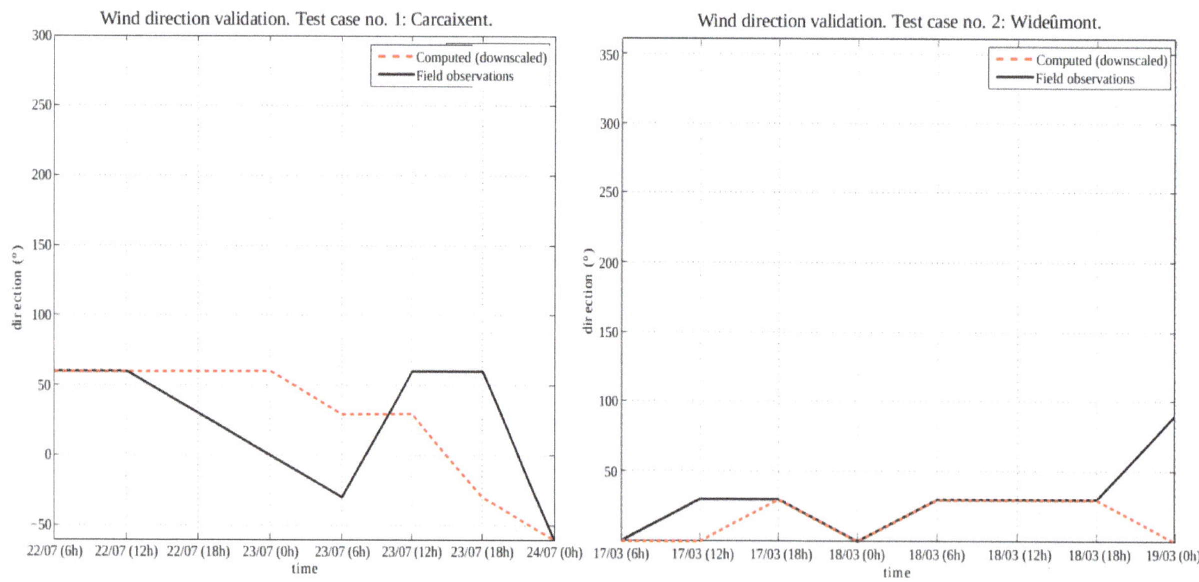

Figure 5. Validation of wind directions computed by the downscaling tool with 6h-averaged meteorological observations

A last attempt to find the causes of the large mismatch between computed and observed values is done by estimating the Richardson number, i.e. the ratio between the potential to the kinetic energy of a flow. It is computed at a given height z as follows:

$$Ri = \frac{g}{\theta}\frac{\frac{d\theta}{dz}}{\left(\frac{du}{dz}\right)^2} \quad (9)$$

Where g is the gravity, $\theta = T \cdot p/p_0$ is the potential temperature (being p_0 the pressure at on the ground level[4] and p and T the pressure and the temperature at z) and u is the velocity at z.
In atmospheric flow studies, the Richardson number is widely used in NWP as it is also a flow stability indicator. The correspondence between Richardson number values and atmospheric stability is the following: values below zero indicate stable conditions, whereas positive values indicate unstable conditions. When the Richardson number tends to zero the condition is neutral. No significant correlation is found between the Richardson number at a given timestep and the corresponding model-observation mismatch.

4. Conclusions

A method to downscale NCEP/NCAR Reanalysis data using WAsP is developed. This allows to estimate wind energy resources where meteorological data are scarce or not available at all. The downscaling process is validated using observed meteorological data of two meteorological stations in Europe, namely: Carcaixent (Spain) and Wideûmont (Belgium). A large mismatch between the results concerning wind velocities is found, although the main trends are eventually followed by the model. As regards wind direction, better results are achieved.
The development of new downscaling methods, according to several authors, such as Badger *et al.* [4], opens up the possibility to apply global-scale model data in a more sophisticated way and using higher resolutions. In a near future, when computational resources are more powerful, more accurate numerical models and more reliable downscaling techniques can provide good-quality estimates. This is an asset in places where meteorological data are scarce or not available at all. E.g. they can be useful in wind-energy programs in underdeveloped countries. Even in developed countries, the development of this kind of tools is interesting, as meteorological observation time-series have to be generally purchased to institutions.
At last, a future work proposal can be outlined. As mentioned above, a mismatch between computed and observed wind velocity evolution is found, although the main trends are followed in some cases. For this reason, it is important to find out the reason of this mismatch in order to figure out how to reduce it. As regards wind velocity, computed values tend in most cases to underestimate the observed values. Likely, improving the simulations with WAsP would increase the accuracy of the final results. Therefore, better elevation and roughness maps could be used and an in-situ description of the obstacles surrounding the points under study should be performed.
Another action that would likely reduce errors would be refining NCEP/NCAR Reanalysis data by means of WRF or a similar model. In this case, the latter model would be used as an intermediate step in the whole downscaling chain. Under a scheme NCEP/NCAR Reanalysis data → WRF → WAsP a more gradual scale reduction throughout all the scales, both in time and space, would likely improve results.
Also the whole downscaling method may be extended to the use of NCEP/NCAR Reanalysis data from more than one point simultaneously, as other downscaling tools do, e.g. KAMM/WAsP Method [8]. It would be interesting finding out if some statistically-weighted system would yield better results.
In addition, the same approach could be applied to other global datasets. According to Lileo and Petrik [25], the global dataset used in this project (NCEP/NCAR Reanalysis data) show significant time inconsistencies at some grid points of Earth. This can affect drastically the energy production estimates, leading to errors up to 14% with respect to more consistent datasets, such as MERRA-grid data.
As NCEP/NCAR Reanalysis data, the ECMWF Reanalysis Dataset (published by European Center for Medium-Range Weather Forecasts) is also freely available for research purposes, but significantly more consistent. Hence, developing a similar downscaling protocol to refine ERA data using WAsP for wind-energy assessment purposes would be a useful and challenging task.

Acknowledgements
This research was conducted thanks to the funding provided by the Von Karman Institute for Fluid Dynamics (Belgium) and the "VALi+D PhD Program" of the Generalitat Valenciana (Spain).

References

[1] Lorenz, E. N. Deterministic nonperiodic flow. J. Atmospheric Sciences. 1963, 20(2), 130-141.

---

[4] The values were computed using the values at the two lowest NCEP/NCAR Reanalysis points, as points at the ground level are not available.

[2] Hunter, R. & Elliot, G. Wind-diesel systems: a guide to the technology and its implementation. Cambridge University Press, 1994.

[3] Landberg, L., Mortensen, N., Dellwik, E., Badger, J., Corbett, J.-F., Rathmann, O. & Myllerup, L. Long-term (1-20 years) prediction of wind resources (WAsP). Introduction to wind technology. 2006, 1-21.

[4] Badger, J., Larsen, X. G., Mortensen, N. G., Hahmann, A. N., Hansen, J. C. & Ejsing Jorgensen, H. A universal meso-scale to micro-scale modelling interface tool. 2010 European Wind Energy Conference and Exhibition, 2010.

[5] Rooy, W. d. & Kok, K. On the use of physical and statistical downscaling techniques for NWP model output. De Bilt: KNMI, 2002.

[6] Giebel, G., Landberg, L., Madsen, H., Nielsen, T. & Nielsen, H. Short-term (30 min-72 hrs) prediction for power output from wind farms. Introduction to Wind Technology. 2006, 1-19.

[7] Verkaik, J., Jacobs, A., Tijm, A. & Onvlee, J. Local Wind Speed Estimation by Physical Downscaling of Weather model forecasts. Submitted to J. Wind Engineering and Industrial Aerodynamics. 2006.

[8] Frank, H. P., Rathmann, O., Mortensen, N. G. & Landberg, L. The numerical wind atlas-the KAMM/WAsP method. 2001.

[9] Stephen, B., Galloway, S., McMillan, D., Anderson, L. & Ault, G. Statistical profiling of site wind resource speed and directional characteristics. IET Renewable Power Gen. 2013, 7(6), 583-592.

[10] Essa, K. S. & Embaby, M. Statistical evaluation of wind energy at Inshas, Egypt. Wind Engineering. Multi-Science, 2005, 29(1), 83-88.

[11] Bechrakis, D. A. & Sparis, P. D. Correlation of wind speed between neighboring measuring stations. Energy Conversion, IEEE Transactions on. IEEE, 2004, 19(2), 400-406.

[12] Barbounis, T. G., Theocharis, J. B., Alexiadis, M. C. & Dokopoulos, P. S. Long-term wind speed and power forecasting using local recurrent neural network models. Energy Conversion, IEEE Transactions on. IEEE, 2006, 21(1), 273-284.

[13] Prasad, R., Bansal, R. & Sauturaga, M. Some of the design and methodology considerations in wind resource assessment. IET Renewable Power Generation. IET, 2009, 3(1), 53-64.

[14] Thogersen, M. L., Motta, M., Sorensen, T. & Nielsen, P. Measure-correlate-predict methods: case studies and software implementation. European Wind Energy Conference & Exhibition, 2007.

[15] Beaucage, P., Brower, M. C. & Tensen, J. Evaluation of four numerical wind flow models for wind resource mapping. Wind Energy. Wiley Online Library, 2012.

[16] Kalnay, E., Kanamitsu, M., Kistler, R., Collins, W., Deaven, D., Gandin, L., Iredell, M., Saha, S., White, G., Woollen, J. & others The NCEP/NCAR 40-year reanalysis project. Bulletin of the American meteorological Society. 1996, 77(3), 437-471.

[17] Troen, I. & Petersen, E. L. European wind atlas. Riso National Laboratory, 1989.

[18] Wood, N. The onset of separation in neutral, turbulent flow over hills. Boundary-Layer Meteorology. Springer, 1995, 76(1-2), 137-164.

[19] Berge, E., Gravdahl, A. R., Schelling, J., Tallhaug, L. & Undheim, O. Wind in complex terrain. A comparison of WAsP and two CFD-models. Proceedings from EWEC, 2006, 27.

[20] Manning, J., Whiting, R., LeBlanc, M. & Hancock, P. Optimising WAsP Wind Flow Modelling for Wind Resource Assessment around Forested Areas. Conference for the Engineering Doctorate in Environmental Technology, Brunel University, Brunel (United Kingdom), 2005.

[21] Bowen, A. & Mortensen, N. Exploring the limits of WAsP, the Wind Atlas Analysis and Application Program. European Wind Energy Conference and Exhibition (EWEC), 1996.

[22] Vedel, H. Conversion of WGS84 geometric heights to NWP model HIRLAM geopotential heights. Danish Meteorological Institute, 2000.

[23] EROS Data Center, E. GTOPO30: Global 30 Arc-Second Elevation Data Set. Available at: http://edcdaac.usgs.gov/gtopo30/gtopo30.html (accessed January 15 th 2011), 1996.

[24] European Environment Agency, E. Corine Land Cover Maps. Available at http://www.dataforwind.com/services/roughness_service.html (accessed July 15 th 2011), 2000.

[25] Lileo, S. & Petrik, O. Investigation on the use of NCEP/NCAR, MERRA and NCEP/CFSR reanalysis data in wind resource analysis. Sigma. EWEC, 2011, 1(2).

Permissions

All chapters in this book were first published in IJEE, by International Energy and Environment Foundation (IEEF); hereby published with permission under the Creative Commons Attribution License or equivalent. Every chapter published in this book has been scrutinized by our experts. Their significance has been extensively debated. The topics covered herein carry significant findings which will fuel the growth of the discipline. They may even be implemented as practical applications or may be referred to as a beginning point for another development.

The contributors of this book come from diverse backgrounds, making this book a truly international effort. This book will bring forth new frontiers with its revolutionizing research information and detailed analysis of the nascent developments around the world.

We would like to thank all the contributing authors for lending their expertise to make the book truly unique. They have played a crucial role in the development of this book. Without their invaluable contributions this book wouldn't have been possible. They have made vital efforts to compile up to date information on the varied aspects of this subject to make this book a valuable addition to the collection of many professionals and students.

This book was conceptualized with the vision of imparting up-to-date information and advanced data in this field. To ensure the same, a matchless editorial board was set up. Every individual on the board went through rigorous rounds of assessment to prove their worth. After which they invested a large part of their time researching and compiling the most relevant data for our readers.

The editorial board has been involved in producing this book since its inception. They have spent rigorous hours researching and exploring the diverse topics which have resulted in the successful publishing of this book. They have passed on their knowledge of decades through this book. To expedite this challenging task, the publisher supported the team at every step. A small team of assistant editors was also appointed to further simplify the editing procedure and attain best results for the readers.

Apart from the editorial board, the designing team has also invested a significant amount of their time in understanding the subject and creating the most relevant covers. They scrutinized every image to scout for the most suitable representation of the subject and create an appropriate cover for the book.

The publishing team has been an ardent support to the editorial, designing and production team. Their endless efforts to recruit the best for this project, has resulted in the accomplishment of this book. They are a veteran in the field of academics and their pool of knowledge is as vast as their experience in printing. Their expertise and guidance has proved useful at every step. Their uncompromising quality standards have made this book an exceptional effort. Their encouragement from time to time has been an inspiration for everyone.

The publisher and the editorial board hope that this book will prove to be a valuable piece of knowledge for researchers, students, practitioners and scholars across the globe.

List of Contributors

Chiyembekezo S. Kaunda
Waterpower Laboratory, Department of Energy and Process Engineering, Norwegian University of Science and Technology, Trondheim 7491, Norway

Cuthbert Z. Kimambo
Department of Mechanical and Industrial Engineering, University of Dar es Salaam, P.O. Box 35091, Dar es Salaam, Tanzania

Torbjorn K. Nielsen
Waterpower Laboratory, Department of Energy and Process Engineering, Norwegian University of Science and Technology, Trondheim 7491, Norway

T. P. Upadhyay
University of Tromsø - Arctic University of Norway, School of Business and Economics, Campus Alta, Follumsvei 31, N-9509 Alta, Norway

J. H. Greibrokk
University of Tromsø - Arctic University of Norway, School of Business and Economics, Campus Alta, Follumsvei 31, N-9509 Alta, Norway

Zemin Ding
Institute of Thermal Science and Power Engineering, Naval University of Engineering, Wuhan 430033, China
Military Key Laboratory for Naval Ship Power Engineering, Naval University of Engineering, Wuhan 430033, China
College of Power Engineering, Naval University of Engineering, Wuhan 430033, China

Lingen Chen
Institute of Thermal Science and Power Engineering, Naval University of Engineering, Wuhan 430033, China
Military Key Laboratory for Naval Ship Power Engineering, Naval University of Engineering, Wuhan 430033, China
College of Power Engineering, Naval University of Engineering, Wuhan 430033, China

Yanlin Ge
Institute of Thermal Science and Power Engineering, Naval University of Engineering, Wuhan 430033, China
Military Key Laboratory for Naval Ship Power Engineering, Naval University of Engineering, Wuhan 430033, China
College of Power Engineering, Naval University of Engineering, Wuhan 430033, China

Fengrui Sun
Institute of Thermal Science and Power Engineering, Naval University of Engineering, Wuhan 430033, China
Military Key Laboratory for Naval Ship Power Engineering, Naval University of Engineering, Wuhan 430033, China
College of Power Engineering, Naval University of Engineering, Wuhan 430033, China

Wathsala Perera
Telemark University College, Porsgrunn, Norway

Deshai Botheju
Agility Group AS, Sandefjord, Norway

Rune Bakke
Telemark University College, Porsgrunn, Norway
Tel-Tek, Porsgrunn, Norway

I.G. Yashchenko
Institute of Petroleum Chemistry, Siberian Branch of Russian Academy of Sciences, 4, Akademichesky Ave., 634021, Tomsk, Russia

Y. M. Polishchuk
Institute of Petroleum Chemistry, Siberian Branch of Russian Academy of Sciences, 4, Akademichesky Ave., 634021, Tomsk, Russia

Mustapha Faraji
Physics Department, LPMMAT Laboratory, Faculty of Sciences Ain Chock, Hassan II University, PO 5366-Maarif, Casablanca- Morocco

Arturo Del Valle-Carrasco
New Mexico State University, Department of Industrial Engineering, MSC 4230/ECIII, PO Box 30001, Las Cruces NM, 88003-8001, USA

Delia J. Valles-Rosales
New Mexico State University, Department of Industrial Engineering, MSC 4230/ECIII, PO Box 30001, Las Cruces NM, 88003-8001, USA

Luis C. Mendez
Autonomous University of Ciudad Juarez, Department of Industrial Engineering and Manufacturing Av. Del Charro, 450 Nte, Ciudad Juarez, Chihuahua, 32315, Mexico

Alejandro Alvarado-Iniesta
Autonomous University of Ciudad Juarez, Department of Industrial Engineering and Manufacturing Av. Del Charro, 450 Nte, Ciudad Juarez, Chihuahua, 32315, Mexico

Faisal M. AL- Ghathian
Faculty of Engineering Technology, Al-Balqa' Applied University, P.O. Box 179, Tafila, 66110, Jordan

Pakamas Chetpattananondh
Department of Chemical Engineering, Faculty of Engineering, Prince of Songkla University, Hat Yai, Songkhla, 90112, Thailand

Sumate Chaiprapat
Department of Civil Engineering, Faculty of Engineering, Prince of Songkla University, Hat Yai, Songkhla, 90112, Thailand

Chaisri Suksaroj
Department of Civil Engineering, Faculty of Engineering, Prince of Songkla University, Hat Yai, Songkhla, 90112, Thailand

Chao Wang
Institute of Thermal Science and Power Engineering, Naval University of Engineering, Wuhan 430033, China
Military Key Laboratory for Naval Ship Power Engineering, Naval University of Engineering, Wuhan 430033, China
College of Power Engineering, Naval University of Engineering, Wuhan 430033, China

Lingen Chen
Institute of Thermal Science and Power Engineering, Naval University of Engineering, Wuhan 430033, China
Military Key Laboratory for Naval Ship Power Engineering, Naval University of Engineering, Wuhan 430033, China
College of Power Engineering, Naval University of Engineering, Wuhan 430033, China

Yanlin Ge
Institute of Thermal Science and Power Engineering, Naval University of Engineering, Wuhan 430033, China
Military Key Laboratory for Naval Ship Power Engineering, Naval University of Engineering, Wuhan 430033, China
College of Power Engineering, Naval University of Engineering, Wuhan 430033, China

Fengrui Sun
Institute of Thermal Science and Power Engineering, Naval University of Engineering, Wuhan 430033, China
Military Key Laboratory for Naval Ship Power Engineering, Naval University of Engineering, Wuhan 430033, China
College of Power Engineering, Naval University of Engineering, Wuhan 430033, China

E. Eftimie
Department of Product Design, Mechatronics and Environment, Transilvania University of Braşov, Eroilor 29, 500036, Romania

N. Eftimie
Department of Manufacturing Engineering, Transilvania University of Braşov, Eroilor 29, 500036, Romania

M. R. Abouseba
Sepehr Institute of Isfahan, Isfahan, Iran

J. Khodakarami
Engineering Department, Ilam University, Ilam, Iran

Jones S. Silva
Inst Pesquisas Hidráulicas, Univ Fed Rio Grande do Sul, Porto Alegre, Brazil

Alexandre Beluco
Inst Pesquisas Hidráulicas, Univ Fed Rio Grande do Sul, Porto Alegre, Brazil

Luiz Emílio B. de Almeida
Inst Pesquisas Hidráulicas, Univ Fed Rio Grande do Sul, Porto Alegre, Brazil

Rajat Gupta
Department of Mechanical Engineering, National Institute of Technology, Silchar, Assam, PIN-788010, India

Rituraj Gautam
Department of Mechanical Engineering, National Institute of Technology, Silchar, Assam, PIN-788010, India

Siddhartha Sankar Deka
Department of Mechanical Engineering, National Institute of Technology, Silchar, Assam, PIN-788010, India

M. S. EL-Wazery
Department of Production Engineering and Mechanical Design, Faculty of Engineering, Menoufiya University, Shebin El-Kom, Egypt

Oktay Arikan
Department of Electrical Engineering, Yildiz Technical University, Esenler, Turkey

Evren Isen
Department of Electrical & Electronics Engineering, Kirklareli University, Kirklareli, Turkey

Cagri Kocaman
School of Civil Aviation, Ondokuz Mayis University, Samsun, Turkey

Bedri Kekezoglu
Department of Electrical Engineering, Yildiz Technical University, Esenler, Turkey

Albert A. Koenig
ARB Geowell, 100 Four Falls Corporate Center Suite 215, West Conshohocken, PA 19428, USA

Martin F. Helmke
West Chester University, Department of Geology and Astronomy, 207 Merion Science Center, West Chester, PA, 19383, USA

S. S. Das
Department of Physics, KBDAV College, Nirakarpur, Khurda-752 019(Odisha), India

S. Mishra
Department of Physics, Christ College, Mission Road, Cuttack-753 001(Odisha), India

P. Tripathy
Department of Physics, Centurion University, Paralakhemundi, Gajapati-761 211(Odisha), India

I. Khazaee
Faculty of Mechanical and Energy Engineering, Shahid Beheshti University, A.C., Tehran, Iran

S. S. L. Rao
Department of Mechanical Engineering, National Institute of Technology, Calicut-673601, India

A. Shaija
Department of Mechanical Engineering, National Institute of Technology, Calicut-673601, India

S. Jayaraj
Department of Mechanical Engineering, National Institute of Technology, Calicut-673601, India

Arnau Bayon-Barrachina
Universitat Politècnica de València, Camí de Vera, s/n, 46022 - València, Spain

Jeroen van Beeck
Von Karman Institute for Fluid Dynamics, Ch. de Waterloo, 74, 1640, Rhode-Saint-Genese, Belgium

Fernando Martinez-Alzamora
Universitat Politècnica de València, Camí de Vera, s/n, 46022 - València, Spain

Petra Amparo Lopez-Jimenez
Universitat Politècnica de València, Camí de Vera, s/n, 46022 - València, Spain